France and the Mediterranean

Modern French Identities

Edited by Peter Collier

Volume 86

PETER LANG
Oxford· Bern· Berlin· Bruxelles· Frankfurt am Main· New York· Wien

Emmanuel Godin and
Natalya Vince (eds)

France and the Mediterranean

International Relations,
Culture and Politics

PETER LANG
Oxford· Bern· Berlin· Bruxelles· Frankfurt am Main· New York· Wien

Bibliographic information published by Die Deutsche Nationalbibliothek
Die Deutsche Nationalbibliothek lists this publication in the Deutsche
Nationalbibliografie; detailed bibliographic data is available on the
Internet at http://dnb.d-nb.de.

A catalogue record for this book is available from the British Library.

Library of Congress Cataloging-in-Publication Data:

France and the Mediterranean : international relations, culture and
politics / Emmanuel Godin and Natalya Vince (eds).
  p. cm. -- (Modern French identities ; 86)
 Includes bibliographical references and index.
 ISBN 978-3-0343-0228-9 (alk. paper)
 1. Mediterranean Region--Relations--France. 2.
France--Relations--Mediterranean Region. I. Godin, Emmanuel. II. Vince,
Natalya, 1981-
 DE85.5.F8F72 2012
 303.48'24401822--dc23
                                     2012000527

ISSN 1422-9005
ISBN 978-3-0343-0228-9

© Peter Lang AG, International Academic Publishers, Bern 2012
Hochfeldstrasse 32, CH-3012 Bern, Switzerland
info@peterlang.com, www.peterlang.com, www.peterlang.net

All rights reserved.
All parts of this publication are protected by copyright.
Any utilisation outside the strict limits of the copyright law, without
the permission of the publisher, is forbidden and liable to prosecution.
This applies in particular to reproductions, translations, microfilming,
and storage and processing in electronic retrieval systems.

Printed in Germany

For George and Patricia Vince

Contents

Acknowledgements — xi

List of Abbreviations — xiii

List of Tables and Figures — xvii

EMMANUEL GODIN AND NATALYA VINCE
Introduction: France and the Mediterranean in 2011 — 1

PART I International Relations: France and the Mediterranean — 17

JEAN-YVES MOISSERON AND MANAR EZZAT BAYOUMI
The Mediterranean: A Contested Concept — 19

JEAN-ROBERT HENRY
France's Mediterranean Policy:
Between National Ambitions and European Uncertainties — 37

IVANO BRUNO
French Foreign Policy in the Mediterranean:
Europeanization between Collective and National Approaches — 69

DAVID STYAN
The Revival of Franco-Libyan Relations — 105

PART 2   Cultural Dialogue and Cross-Fertilization   131

GÉRALD ARBOIT

France and Cultural Diplomacy in the Mediterranean   133

PATRICIA CAILLÉ

On the Confusion between Films about Women and
Films by Women in the Construction of a Regional Cinema:
The Maghreb vs. the Mediterranean   163

MARY BEN BONHAM

Hybrid Cultural Significance for the Bibliothèque Municipale à
Vocation Régionale de Marseille   193

PART 3   Politics of Ethnicity and Postcolonial Memory   217

BRUNO LEVASSEUR

French *Artistes Engagés* Transcending the Politics of 'Race' and
'Ethnicity' in the Republic (1980s–2000s)   219

ED NAYLOR

'A system that resembles both colonialism and the invasion of
France': Gaston Defferre and the Politics of Immigration in 1973   249

ANDREA BRAZZODURO

Postcolonial Memories of the Algerian War of Independence,
1955–2010: French Veterans and Contemporary France   275

NATALYA VINCE

Questioning the Colonial Fracture: The Algerian War as a
'Useful Past' in Contemporary France and Algeria   305

	ix
Notes on Contributors	345
Index	349

Acknowledgements

We first submitted this book to Peter Lang on 13 January 2011. The next day, President Ben Ali was forced to flee Tunisia with part of his family and found refuge in Saudi Arabia. Throughout the Arab world, protests (as in Algeria and Morocco), 'revolutions' (as in Tunisia and Egypt), civil war (Libya) and ferocious repression (Syria) have extensively altered the political contexts of countries on the southern shores of the Mediterranean. On its northern shores the impact of what is now referred to as the Arab Spring have been no less consequential: France, as other European countries, has had to find ways to ensure that her past support for Arab dictators and their regime did not damage her influence and interests in the Mediterranean.

The Arab spring has thus compelled us to revise the volume we first submitted in January 2011 and if it is too early to draw unambiguous conclusions about how Franco-Mediterranean relations have been altered since January 2011, we hope that it will provide some insights into the complex relations that France and the French entertain with the Mediterranean at many different levels.

The Association for the Study of Modern and Contemporary France (ASMCF), the French Embassy, London, and the Centre for European and International Studies Research (CEISR) at the University of Portsmouth offered their generous financial and logistic support to organise the 2009 ASMCF conference on 'France and the Mediterranean'. This conference provided the intellectual impetus for the completion of this volume. As such, we would like to thank Professor Máire Cross (ASMCF president), Professor Philippe Lane (attaché for higher education at the French embassy in London), and Professor Tony Chafer (director of CEISR) for their encouragement and support.

We are also grateful to people who have helped, informed and suggested revisions and amendments during the editing on this volume: in particular, we would like to thank Walid Benkhaled, Tony Chafer, Martin Evans, Hélène Gill, Jim House, Barbara Lebrun, Margaret Majumdar and Joseph McGonagle. Helen Dixon translated the chapters by Jean-Robert Henry and Gérald Arboit and made insightful stylistic comments on other chapters, as did Eleanor Waterhouse. Finally, we are also indebted to our publishers for their trust and patience – Graham Speake, Hannah Godfrey and Isabel James as well as the anonymous reviewers for the initial book proposal and its subsequent versions. Of course, we take full responsibility for any inaccuracies and mistakes that may be found in the text.

Abbreviations

AA	Association Agreement
ACP	African, Caribbean and Pacific Countries
ACSE	Agence nationale pour la cohésion sociale et l'égalité des chances
AIF	Agence intergouvernementale de la Francophonie
ALECSO	Arab League Education, Cultural and Scientific Organisation
ALF	Anna Lindh Foundation
ASAF	Association soutien à l'armée française
BMVR	Bibliotheque Municipale à Vocation Régionale
BNF	Bibliotheque Nationale de France
CCFL	Chambre de Commerce Franco-Lybienne
CDCAFN	Cercle pour la défense des combattants d'Afrique française du Nord
CDM	Comité de Défense des Marseillais
CEA	Commissariat à l'Énergie Atomique
CFDT	Confédération Française Démocratique du Travail
CGT	Confédération Générale du Travail
CIEEMG	Commission Interministérielle pour l'Étude des Exportations de Matériel de Guerre
CIMADE	Comité Inter-Mouvements Auprès Des Evacués
CIRTEF	Conseil International des Radios et Télévisions francophones
CNC	Centre national de la cinématographie
CRI	Collectif des Rapatriés Internautes
CSCM	Conference for Security and Co-operation in the Mediterranean
DGA	Délégation Générale de l'Armement

DGCID	Direction générale de la cooperation internationale du développement
DOM-TOM	Départements d'Outre-Mer – Territoires d'Outre-Mer
EADS	European Aeronautic Defence and Space Company
EC	European Community
EEC	European Economic Community
EGCM	Etats généraux culturels méditerranéens (Mediterranean Cultural Forum)
EMP	Euro-Mediterranean Partnership
ENI	Ente Nazionale Idrocarburi *(but not used as an acronym any more)*
ENP	European Neighbourhood Policy
EU	European Union
EUFOR	European Operational Rapid Force
EUFOR/RCA	European Union Force in Central African Republic
FFPAVS	Fonds Francophone de Production Audiovisuelle du Sud
FIS	Front Islamique du Salut
FLN	Front de Libération Nationale
FN	Front National
FNACA	Fédération Nationale des Anciens Combattants d'Algérie
FTA	Free Trade Area
GIA	Groupe Islamique Armé
GMP	Global Mediterranean Policy
IMF	International Monetary Fund
IRCICA	Research Centre for Islamic History, Art and Culture
IREMAM	Institut de Recherches et d'Études du Monde Arabe et Musulman
ISESCO	Islamic Educational, Scientific and Cultural Organization
MCF	Mediterranean Cultural Forum (*Etats généraux culturels méditerranéens*)
MEDA	Morocco, Egypt, Jordan and Tunisia
MENA	Mediterranean and North African [partners/ region]

Abbreviations

MEP	Member of the European Parliament
MEPP	Middle Eastern Peace Process
MIR	Mouvement des indigènes de la République
MODEM	Mouvement Démocrate
MTA	Mouvement des Travailleurs Arabes
NATO	North Atlantic Treaty Organization
NGO	Non-Governmental Organisation
OAS	Organisation Armée Secrète
OECD	Organisation for Economic Co-operation and Development
OIF	Organisation internationale de la Francophonie
ONEC	Organisation des enfants de shuhada (martyrs)
ONM	Organisation nationale des mujahidin (combatants)
OPEC	Organization of the Petroleum Exporting Countries
ORTF	Office de radiodiffusion-télévision française
PACA	Provence, Alpes, Côte d'Azur
PS	Parti Socialiste
SODEPAU	Solidaritat per al Desenvolupament i la Pau
TEU	Treaty of the European Union
UDR	Union pour la Défense de la République
UfM	Union for the Mediterranean
UM	Union of the Mediterranean
UMP	Union pour un Mouvement Populaire
UN	United Nations
UNC	Union Nationale de Combattants
UNESCO	United Nations Educational, Scientific and Cultural Organization
WMD	Weapons of Mass Destruction

Tables and Figures

Table 1	Plan of Action at Alexandria, 21 January 2008	145
Table 2	The Ten Mediterranean Cultural Projects, 8 December 2008	150

Figure 1	BMVR, Marseille: view of glass-marble façade including Alcazar name on historic entrance. Photo: Mary Bonham, 2008.	194
Figure 2	Excerpt from 'A Description of Marseilles' from *Marseille. Chambre du Conseil de l'Hôtel de Ville* (London, 1722; Gale: Eighteenth Century Collections Online).	197
Figure 3	BMVR, Marseille: view of atrium 'inner street'. Photo: Mary Bonham, 2008.	203
Figure 4	Museum of Science and Industry: view of greenhouses. Photo: Mary Bonham, 2008.	209
Figure 5	BMVR, Marseille: view of interior of double façade. Photo: Mary Bonham, 2008.	209

EMMANUEL GODIN AND NATALYA VINCE

Introduction: France and the Mediterranean in 2011

On 27 February 2011 President Nicolas Sarkozy reshuffled the French cabinet. Justifying his ministerial choices in a seven-minute televised address, Sarkozy made the link between domestic politics and the 'immense upheaval' on the 'other side of the Mediterranean' explicit from the outset:

> Arab peoples are taking their destiny in hand, overthrowing regimes which, having been agents of emancipation during the era of decolonization, had become instruments of servitude. Since the end of colonization, all Western states and successive French governments have maintained economic, diplomatic and political relations with these regimes, despite their authoritarian character, because in the eyes of all they seemed to be ramparts against religious extremism, fundamentalism and terrorism.[1]

This *mea culpa* dispensed with, paralleling Sarkozy's policy errors with those of his presidential predecessors and conveniently submerging French strategic myopia in shared Western responsibility, Sarkozy hailed a new era in France's relations with North Africa:

> This change is historic. We must not be afraid. It carries extraordinary hope because it has been accomplished in the name of our most precious values, human rights and democracy. For the first time in history, these can triumph on both sides of the Mediterranean.

---

1   'Allocution radiotélévisée du Président sur la situation internationale', 27 February 2011, http://www.elysee.fr/president/les-actualites/discours/2011/allocution-radiotelevisee-du-president-sur-la.10756.html?search=allocution&xtmc=allocution&xcr=7 (accessed 1 October 2011).

If the mistakes of the past were collective, Sarkozy argued that France's historical and geographical ties with the region and unique civilizational message positions the *Hexagone* – and by inference Sarkozy – as a major player in a more radiant Mediterranean future. This special role is nevertheless to be realized not through bilateral relations but through multilateral instruments. In his reshuffle address, Sarkozy referred to a 'recasting' of the Union for the Mediterranean (UfM), the much maligned French – and now European Union – initiative, which began life with the recently deposed Egyptian president Hosni Mubarak as joint president and the now beleaguered Syrian president Bashar al-Assad as special guest at both the July 2008 launch meeting and Paris's Bastille Day celebrations. Sarkozy also suggested in his speech recourse to the European Council for a collective response to stem the inevitable migratory flows resulting from the Arab Spring.

Despite Sarkozy's claims of historic change and a new era, there is little that lends itself better to analyses of continuity than French politicians – of all parties – claiming a new start. Nowhere is this perhaps more true than in the overlapping spheres of Franco-African, Franco-Mediterranean and Franco-Arab relations. Sarkozy's 2011 reshuffle speech encapsulates entrenched themes in French policymakers' – admittedly often hazy – conceptualizations of 'the Mediterranean': it is both a proximate space in which ambitious foreign policy can flourish and the worryingly close location of an invading 'other' which needs to be contained and controlled. Both these fantasies are of domestic importance, with the potential to garner electoral support for 'statesman-like' politicians and generate interminable media and political debate on immigration and national identity. The ministerial changes of 27 February and subsequent events – notably the Franco-British-led military involvement in Libya and Franco-Italian challenges to the Schengen agreement – have fuelled this French Mediterranean imaginary.

In the reshuffle, the foreign, defence and interior ministries changed hands – clearly demonstrating what Sarkozy believes to be the strategically important areas for both France in the world and Sarkozy in the 2012 elections. Notably, out went foreign minister Michèle Alliot-Marie, embroiled in scandal over her use of a private jet belonging to an ally of the deposed

Tunisian president Zine El Abidine Ben Ali whilst on holiday, and under fire for her offer to the Tunisian and Algerian regimes of French 'savoir faire' in maintaining law and order.[2] In came Alain Juppé, judged a 'safe pair of hands' and notably a panacea for increasingly vocal criticisms, not least amongst diplomats and Quai d'Orsay civil servants, of the Elysée's ad-hoc and haphazard foreign policy making under Sarkozy.[3] The controversial Ministry for Immigration, Integration, National Identity and Codevelopment, created by Sarkozy in 2007, was disbanded and immigration reattached to the Ministry for the Interior, under the president's close adviser Claude Guéant, on the right of the right-wing Union pour un Mouvement Populaire (UMP). Despite the death of the ministry, Sarkozy's flirtation with extreme-right discourse has thus continued, with Guéant declaring that 'uncontrolled immigration' has led to 'the French' sometimes not feeling 'at home' and referring to the French intervention in Libya as a 'crusade'.[4]

On the eve of the extraordinary European Council meeting on the Libyan situation on 11 March 2011, and with the help of public intellectual and media celebrity Bernard-Henri Lévy, Sarkozy scrambled to make France the first state to officially recognize the National Transitional Council of the anti-Gaddafi opposition. The move – symptomatic of the highly mediatized, voluntarist style of Sarkozy's 'hyper-presidency' – was apparently without the approval or knowledge of Juppé, at the time in Brussels.[5] Both France and Britain initially attracted domestic plaudits for their military

2   E. Martin, 'Faut-il s'étonner du silence français?', *Un monde libre* (14 January 2011).
3   Marly group, 'La voix de la France a disparu dans le monde', *Le Monde* (22 February 2011). In 2010, two former foreign ministers, including Juppé, wrote an opinion piece for *Le Monde* in which they criticized the impact of budget cuts on the functioning of French diplomacy. A. Juppé and H. Védrine 'Cessez d'affaiblir le Quai d'Orsay!', *Le Monde* (7 July 2010).
4   'Immigration: Guéant suscite la polémique', *Europe 1*, 17 March 2011, http://www.europe1.fr/Politique/Immigration-Gueant-suscite-la-polemique-458409/ (accessed 1 October 2011); 'Le Talk: Carl Meeus reçoit Claude Guéant', 21 March 2011 http://www.lefigaro.fr/le-talk/2011/03/21/01021-20110321ARTFIG00452-claude-gueant-invite-du-talk.php (accessed 1 October 2011).
5   A. Cabana, 'Quand Juppé menace Sarkozy', *Le Point* (17 March 2011).

intervention in Libya on 19 March 2011 against the forces of Muammar Gaddafi, the long-time international pariah whom both countries had so recently rehabilitated. Ostensibly installing a no-fly zone and protecting civilians under UN resolution 1973, the novelty of this 'humanitarian' mission with no clear exit strategy quickly began to wear off.[6] Germany was sceptical from the outset. Russia was highly critical of French admissions in late June of arming anti-Gaddafi forces, describing it as a 'serious violation' of UN resolution 1970.[7] Sarkozy also – unsuccessfully – sought to resist NATO taking over the military command of the no-fly zone, despite the president's instrumental role in orchestrating France's reintegration into NATO's military command in 2009.[8] On the international scene, Sarkozy is clearly seeking to regain the initiative as a credible and decisive player, legitimizing and rendering financially viable unilateral policy through multilateral instruments whilst simultaneously foregrounding France's precursory contribution in defining international action. This task is rendered all the more difficult by a plethora of other actors also seeking to give the appearance of redefining their foreign policy, including the United States, as demonstrated by Barack Obama in his 19 May speech on the Middle East and the EU, which on 29 June created the post of EU Special Representative for the Southern Mediterranean Region. Spanish diplomat Bernardino

---

6   The France Inter radio station press review on 18 March 2011 refers to plaudits in the *Daily Mail* (UK), *La Stampa* (Italy), the *Times of Malta* and *Le Figaro* (France) amongst others, http://www.franceinter.fr/chro/larevuedepresse/102614 (accessed 1 October 2011). A survey of 1,006 French people aged over eighteen between 24 and 26 May 2011 showed that 55 per cent were in favour of military intervention in Libya, down from 66 per cent at the start of the intervention three months previously. Ifop for *Dimanche Ouest France*, 'L'approbation de l'intervention militaire en Libye', 26 May 2011, http://www.ifop.com/media/poll/1513-1-study_file.pdf (accessed 1 October 2011).

7   'Livraisons d'armes aux rebelles libyens: accords et désaccords', *L'Express* (1 July 2011).

8   K. Willsher, 'Sarkozy Opposes NATO Taking Control of Libya Operation', *The Guardian* (22 March 2011).

Introduction: France and the Mediterranean in 2011

Leon was promptly selected to fill the post.[9] The choice of a Spaniard for this role merits reflection. The freshly appointed secretary general of the UfM, Moroccan diplomat Youssef Amrani, officially welcomed the development; Algerian newspaper *Liberté* interpreted it as another obstacle to French attempts to shape policy in the region.[10]

As France rushed to bid adieu to selected North African *anciens régimes*, the border controls which Ben Ali and Gaddafi enforced for the EU collapsed. The Mediterranean Sea started to fill with ever-increasing numbers of migrants, with thousands arriving on the small Italian island of Lampedusa. A year away from 2012, the toxic tandem of immigration and crime, which dominated the 2002, and to a lesser extent 2007, presidential elections, was once again pushed to the fore. Seeking to capitalize on favourable poll showings and having already argued that France should pass bilateral agreements with Spain and Italy in order to 'push back into international waters migrants who want to enter Europe', Front National leader and MEP Marine Le Pen visited Lampedusa on 14 March. Accompanied by Mario Borghezio, an Italian MP and member of the right-wing anti-immigrant Lega Nord, Le Pen attacked what she described as the European failure to deal with the problem.[11] As the Italian authorities on the mainland issued migrants with

---

9 Obama's Middle East speech, 19 May 2011. Available at: http://news.bbc.co.uk/1/hi/8083250.stm (accessed 1 October 2011); 'Catherine Ashton Proposes Bernardino Leon as New EU Special Representative for the Southern Mediterranean Region' (29 June 2011), http://www.consilium.europa.eu/uedocs/cms_data/docs/pressdata/EN/foraff/123285.pdf (accessed 1 October 2011).

10 Secretariat of the Union for the Mediterranean, 'Secretary General Amb. Amrani: "The Secretariat stands ready to work in close co-operation ..."' (8 July 2011), http://www.ufmsecretariat.org/en/secretary-general-amb-amrani-the-secretariat-stands-ready-to-work-in-close-cooperation-with-both-the-eu-special-representative-and-the-task-force-to-promote-joint-regional-projects/ (accessed 1 October 2011); D. Bouatta, 'Union pour la Mediterranée: UE saborde les efforts du président français', *Liberté* (11 July 2011).

11 'Marine Le Pen, président du Front National, "On peut repousser humainement des bateaux dans les eaux internationales"', *RTL* (1 March 2011), http://www.rtl.fr/emission/l-invite-de-rtl/ecouter/marine-le-pen-presidente-du-front-national-on-peut-repousser-humainement-des-bateaux-dans-les-eaux-internationales-7664513013

temporary visas, many began to make their way to France, where they had family ties and/or linguistic knowledge. Franco-Italian tensions reached a critical point, and France began to make noises about the temporary suspension of the Schengen agreement. On 26 April 2011, Sarkozy and the Italian prime minister Silvio Berlusconi sent a joint letter to the presidents of the European Council and European Commission demanding a revision of Schengen. On 4 May the European Commission published a document suggesting the Schengen agreements could be temporarily suspended under 'very exceptional circumstances' although the mechanisms for enabling this continue to be debated.[12] The salience of populist discourse and policies on immigration on both sides of the Alps indicates that this is not solely a French phenomenon and that 'le Sarkozysme' may not be a French specificity.[13] In a broader context, the EU and its founding principles of free exchange and free movement are currently being pushed to their limits by crises in the Mediterranean region, not just in North Africa and the Middle East, but also on the EU's own Mediterranean shores with the spectre of economic collapse in Greece and its potential impact on the Eurozone.

The first versions of the chapters in this volume were all written before the momentous events of 2011 and where appropriate they have been revised accordingly. It is impossible at this stage to judge the full impact of the Arab Spring, but developments so far have highlighted the pertinence of the key themes which this volume engages with: ruptures and continuities in French discourse and policy in 'the Mediterranean', the exceptionalism or otherwise of French action and relations in the region and the ability or desire of nation-states on either side of the Mediterranean to function outside of the global economic system and/or international organizations such as NATO, EU, UN and IMF. The Arab Spring has also highlighted

(accessed 1 October 2011); 'Opération communication pour Marine Le Pen à Lampedusa', *Le Monde* (14 March 2011).
12 Y. Pascouau, 'Internal Border Controls in the Schengen Area: Much Ado About Nothing?', *European Policy Centre* (28 June 2011), http://www.epc.eu/documents/uploads/pub_1309_internal_border_controls_in_the_schengen_area_-_much_ado_about_nothing.pdf (accessed 1 October 2011).
13 N. Hewlett, *The Sarkozy Phenomenom* (Exeter: Imprint Academic, 2011).

the importance of using frameworks of analysis which go beyond state-level relations. Alliot-Marie's misreading of the Tunisian situation was symptomatic of a wider European and North American obsession, accentuated in the past decade, with securing commercial and security interests, paying little attention to the growing gulf between rulers and the ruled across North Africa and the Middle East. Authoritarian rulers had also demonstrated their ability to speak the same language of commercial exchange, immigration control and anti-Islamism. After a decade of popular, political and sometimes academic discourse in which North African youth had two dominant and often intersecting manifestations – the illegal migrant and/or Islamist radical – it is perhaps of little surprise that US and European governments were taken off guard by an apparently leaderless movement whose frames of reference were not couched in terms of their attitudes towards 'the West', either as Eldorado or infidel. This volume thus also places an emphasis on 'bottom up' approaches to understanding 'France in the Mediterranean', including the perspectives of non-elite south Mediterranean actors and cultural and social networks operating at local as well as European and transnational levels.

What is the Mediterranean?

One of the key conceptual themes which this volume seeks to engage with is the idea of the Mediterranean, its multiple meanings in political and cultural discourses as well as its usefulness as a framework for analysis. What does the Mediterranean mean? For whom? In which contexts? Taking a long historical perspective, Jean-Yves Moisseron and Manar Ezzat Bayoumi argue in their chapter that the Mediterranean is a concept produced by the north which has little resonance on southern shores, apart from as an instrumental tool to acquire EU funding. For the populations of the southern shore – as opposed to its elites – the somewhat lyrical invocations of a common Mediterranean culture remain largely abstract, if not irrelevant. Moisseron and Ezzat Bayoumi suggest that the 'Arab World' or the 'Muslim World' may provide more appropriate frameworks of analysis and alternative geographical schematizations to 'the Mediterranean', which appeals first and foremost to a Westernized elite.

Jean-Robert Henry argues that whilst in the Arab world a French *politique arabe* is a meaningful and useful term, equated with pro-Arab positions, the concept of a 'Mediterranean' policy 'sells' better to France's European partners. One might also argue that *la politique arabe* is perceived as a compromise between 'disreputable regimes' whilst 'Mediterranean policy' could be conceived of as a form of increased rapprochement with Israel, presented as 'the only democracy in the Middle East'. In the following chapter, Ivano Bruno demonstrates that it is now impossible to think of France's position in the Mediterranean without taking into account the EU's Mediterranean policy framework. Without denying that bilateral relations remain crucial to articulate and support France's objectives in the region, Bruno analyses how France has used the EU as a complementary, if not more efficient track, to realize her Mediterranean ambition and to extend her influence well beyond a 'Maghrebi Mediterranean' when seeking, for instance, closer relations with Egypt, or engaging with regional actors such as the African Union. The objectives of such a policy may be understood as a desire to readjust the balance of power within the EU, with France contesting Spain's claims to leadership in the field, and seeking in the south the sort of influence that Germany has carved for herself in the east since 1989. Analysing the renewal of Franco-Libyan relations between 2003 and 2010 in the final chapter of Part 1, David Styan emphasizes that the Mediterranean is a fairly poor conceptual tool. During this period, Gaddafi's pan-African geopolitical objectives partly converged with French defence and foreign policy objectives in Africa, such as in Chad and Sudan, and French relations with Libya, notably on issues of security and immigration, were also part of a wider European agenda. As such, European and African policy frameworks may be more useful and indeed, more important, than French and Mediterranean ones. The increasing involvement of actors external to North Africa, the Middle East and their immediate neighbours – including China, Russia and the United States – also needs to be taken into account. 'The Mediterranean' is thus a contested concept, but rather than dismissing

Introduction: France and the Mediterranean in 2011   9

it as a category of analysis, this volume seeks to examine why and how it is contested and elucidate what insights this gives us into the intersections of different types of actors in a loosely defined 'region'.

Ruptures and Continuities in French Foreign Policy

Sarkozy's 2007 presidential election campaign was based on the theme of 'rupture' in domestic and foreign policy. Henry, Styan and Bruno offer contrasting diagnostics, but agree that this rupture has failed to materialize; instead, they insist on the continuity between Sarkozy's objectives and those of previous presidents. In this respect, Styan maintains, long-term perspectives bring to the fore the essential tenets of France's objectives in the Mediterranean and beyond, that too strong a focus on initiatives well publicized by the media fail to capture. Sarkozy – and his special adviser Henri Guaino – hastily enunciated a Mediterranean policy without much thought or coherence, and France and other European states' reluctance to dissuade Israel from its January 2009 Gaza offensive threw the project into disarray almost immediately. The fact that France, its fellow European countries and the EU as an institution all seemed to have been caught on the back foot by the revolts in spring 2011 has done little to strengthen their credibility as proactive – rather than reactive – players in the region. In the end, Sarkozy's Mediterranean initiatives from the UfM to the present are interesting because they provide insights into the president's policymaking and discursive style. Certainly, Sarkozy's relationship with the region is less personalized than that of his predecessor Jacques Chirac, whose network of contacts with heads of state of often unsavoury regimes has been characterized as openly paternalistic rather than part of any foreign policy grand strategy.[14] In his chapter Gérald Arboit emphasizes that we

---

14   See E. Aeschimann and C. Boltanski, *Chirac d'Arabie* (Paris: Grasset, 2006), and A. Youssef, *L'Orient de Jacques Chirac: la politique arabe de la France* (Paris: Editions

should carefully examine who discourses are aimed at and the domestic contexts in which they develop. He explains that France's tendency to develop enthusiastic discourses about her 'special mission in the service of humanity' is particularly acute when a sense of national insecurity and decline prevail. He argues that France's Mediterranean cultural policy is also a means to project at home a sense of universal grandeur, a discursive performance which bears little relation to what is actually happening on the ground, as foreign policy objectives and a lack of funding put severe limits to France's Mediterranean *rayonnement*.

Human Ties: Migration and Memory

A Braudelian approach of the Mediterranean questions the usual dichotomy between centre and periphery, north and south, national and local and prefers to stress the density of networks which mediate exchanges across the Mediterranean.[15] In this volume, Jane-Robert Henry contrasts the relative mediocrity of exchanges between Mediterranean states with the richness of relations linking individuals and communities on each side of the sea. What gives substance and warmth to the Mediterranean are the manifold human relations which have been woven across the Mediterranean at a personal level (mixed marriages for instance) and through the creation of numerous Franco-Maghrebi associations, creating what Henry terms a transnational Franco-Maghrebi community. The aspirations and strategies of such a community are at odds with Fortress Europe which leads, Henry argues, to a form of 'mild apartheid'.

---

du Rocher, 2003). For more on the question of rupture and continuity between Chirac and Sarkozy see M. Vaïsse, *La puissance ou l'influence? La France dans le monde depuis 1958* (Paris: Fayard, 2009).

15   I. Malkin, 'Introduction', *Mediterranean Historical Review*, 18/2 (2003), 1–8.

Migration across the Mediterranean, indeed, has become a major obsession on each side of the sea. On the southern shores, young people in particular are often desperate to cross the Mediterranean and are looking to Europe for the promise of a better life whereas on the northern shores, populist and extreme-right parties fan anti-immigrant feeling. One of the major changes which has taken place in terms of France's relationship with the Mediterranean in the postcolonial period is the shift in discourse on immigration, which has gone from being a technocratic issue in the 1960s and 1970s to one of the main themes structuring French and European political debate since the late 1980s. Focusing on Marseille's chequered history as a Mediterranean melting pot, Ed Naylor's chapter provides insights into the politics of postcolonial immigration in 1970s France. He illustrates how socialist mayor Gaston Defferre was a precursor in this politicization and the emergence of a racialized 'immigration question', likening immigration to 'a system which resembles both colonialism and the invasion of France'.

The extent to which France's colonial past on the southern shore of the Mediterranean explains the importance of anti-immigrant feeling today is discussed within the volume. Drawing on the work of Benjamin Stora, Bruno Levasseur explains the demonizing images of the 'ethnic other' as a consequence of French society failing to come to terms with its colonial past. On the other hand, Natalya Vince argues that discourses on immigration in France are the product of manifold contexts which cannot be reduced simply to the 'colonial fracture': other frames of reference – such as the 1979 Iranian revolution, the end of the Cold War, European integration, globalization and 11 September 2001 – are just as important for understanding the development of anti-immigrant discourses across Europe. Migrations across the Mediterranean have prompted the development of anti-immigrant discourses, but these are not necessarily grounded in a colonial past. Neither does such a colonial past provide the only framework to explain the nature of relations that (North African) colonies entertain with France and with their own memory. Vince shows that a generation of younger Algerians has 'a perception of the war [of independence, 1954–62] which is far removed from its anti-colonial context': on the contrary, they have a 'tendency to "rewrite" the Algerian War as a primarily religious, rather than anti-colonial, struggle'. In doing so, the focus within the Mediterranean shifts from its northern shores to the Middle East. So-called 'memory wars' – notably those which have re-emerged in France in the past decade over the events of the Algerian War – can be seen as less revealing of the impact of the colonial period on the post-independence relationship between two nation states and instead tell us more about internal Franco-French and Algero-Algerian tensions and divisions which are only partly linked to colonial history.

In his analysis of how French veterans of the Algerian War recount the conflict in the contemporary context, Andrea Brazzoduro explores the multiple and shifting frames of reference which shape their narratives. These include 'tendencies which seem common on a European if not worldwide scale' – such as informants' insistence on their youthfulness, and, it is implied, lack of responsibility, in the events they describe – and themes which can be located more precisely, 'linked to the recent French political mood' and social context – such as Sarkozy's insistence on a 'rupture' with colonial repentance.[16] Further, by drawing our attention to the growing importance attached to Islam in the way French veterans of the Algerian War today redefine their past struggles, Brazzoduro suggests that colonialism is not the only available frame of reference which can be used to decrypt the making and remaking of relations across the Mediterranean. This volume presents some material and ideas which it is hoped will enable readers to question some core concepts at the heart of postcolonial theory, such as memory transfer, the colonial fracture and hybridity.

---

16   With regards to Sarkozy's (often inconsistent) attitudes towards history and memory and his proclaimed rejection of the politics of repentance, see his controversial speech at the Université Cheikh Anta Diop, Dakar, Senegal, on 26 July 2007, available on the website of LDH Toulon http://www.ldh-toulon.net/spip.php?article2173 (accessed 1 October 2011), and his Tangiers discourse on the Mediterranean delivered on 23 October 2007, in which he declared: 'We will not construct the Union for the Mediterranean upon repentance, no more than Europe has been built upon atonement and repentance', http://www.ambafrance-uk.org/Discours-du-President-Sarkozy-sur.html (accessed 1 October 2011).

Culture and Postcoloniality

Hybridity is central to the argument developed by Mary Bonham, who, through an analysis of Marseille's Bibliothèque Municipale à Vocation Régionale, seeks to explain what a Mediterranean culture may be and how it can 'speak to multiple or merged cultural identities without becoming a pastiche of styles and symbols'. In doing so, Bonham explores the integrative – if not universal – quality of a Mediterranean culture. Conversely, Arboit argues that the multiplication of places where different cultures meet around the Mediterranean does not necessarily elucidate the nature of a Mediterranean culture: focusing on the Ateliers culturels méditerranéens as a tool of France's Mediterranean cultural policy, Arboit demonstrates that as a meeting point for – rather than of – different cultures of the Mediterranean basin, they do not fundamentally engage with the question of whether a Mediterranean culture actually exists either as a transcendent or hybrid culture: they are nevertheless politically useful to undermine both a Huntington-esque 'clash of civilizations' and Eurocentric concepts of culture. Likewise, in France, Levasseur argues, French left-wing *artistes engagés* have also relentlessly questioned the centrality of French republican universal values: their promotion of cultural hybridity instead provides a hospitable and convivial strategy to open the Republic to the ethnic other and to transcend the politics of race and ethnicity. Not everyone would agree on the value to ascribe to the politics of hybridity. Its radicalism may in fact mask the last avatar of a liberal position, a position which is not too different from Camus's Mediterranean humanism.[17] Margaret Majumdar points out that hybridity downplays the very experience of colonial domination:

---

17 N. Foxlee, 'Mediterranean Humanism or Colonialism with a Human Face? Contextualizing Albert Camus's "The New Mediterranean culture"', *Mediterranean Historical Review*, 21/1 (2006), 77–97.

> It is often the case that notions of hybridity have been assumed within an ideological stance, which would have us believe that there are no fundamental differences and opposition any more, that everything is on a par, of equal value and that the divisions between 'us' and 'them' are no longer credible, if they ever were. The history of the colonial period is rewritten to emphasize mutual influences and interactions and to downplay the binary dialectic of opposites as a figure of the colonial relations of domination and struggle.[18]

Yet hybridity is also a useful starting point to revisit and interrogate how binary divisions may or may not function. In her discussion of cinemas of the Maghreb, Patricia Caillé highlights the importance of interrogating our categories of analysis. 'Cinemas of the Maghreb', she argues, does not refer to a common source of film financing, production and circulation, nor does it represent a particular type of authorial voice or aesthetic style. Instead, 'Cinemas of the Maghreb' is used to describe films which engage in a certain number of themes, notably the condition of Maghrebi women, which in turn tends to attract commercial success on French and European shores. One example she uses is the reception of Tunisian director Moufida Tlatli's historical epic *The Silences of the Palace* (1994) in France. Clearly sensitive to present-day concerns about Islam, the 'Maghrebi family' and in particular the Muslim-Maghrebi woman in France, French audiences foreground the Tunisian female protagonists' battle against patriarchal oppression whilst marginalizing the ongoing struggle against colonial rule which is a key theme in the film. This gendered reading of Maghrebi society does not, as Caillé emphasizes, necessarily represent how female film-makers in these countries see themselves – many are resistant to the category of Maghrebi female film-makers, fearing 'being trapped in a gender and/or regional ghetto'. Thus, films produced in Tunisia, Morocco or Algeria are more likely to be seen in France than in a neighbouring Maghrebi country. In terms of a collective filmic identity, Pan-African cinema has a far greater online presence and film festival circuit than 'Cinemas of the Maghreb'. Between

---

18   M. Majumdar, *Postcoloniality: The French Dimension* (London and New York: Berghahn, 2007), 255.

European and African audiences, political, aesthetic and commercial demands, the southern shore of the Mediterranean finds it difficult to define its own cultural space.

The volume takes an multidisciplinary approach to question the multiple and complex set of relations across the Mediterranean and seeks to de-romanticize the subject by highlighting the ways in which the Mediterranean and France appear to be partly de-centred in a world where other spaces have become equally if not more important: Europe, Africa and the Arab-Muslim world. Furthermore, beyond relations between states, the changing set of relations between civil societies on each side of the Mediterranean are better understood at the local level and here, the two chapters on Marseille reveal both their complexity and ambiguity. The political reactions in France to the 2011 Arab Spring have highlighted that the Mediterranean region remains of strategic importance, but they also raise questions about the relevance of our categories of analysis. What differences are there, if at all, between an Arab and Mediterranean policy? Do countries with a border on the Mediterranean Sea have a definable 'Mediterranean character' or should we instead conceptualize the region as a series of concentric circles around the Middle East with the Mediterranean as a wider, and more diluted, category? Where is French 'Mediterranean' policy made and by whom? How has the relationship between the Elysée and the Quai d'Orsay shifted? How do the Elysée and Quai d'Orsay interact with other policy actors and complex networks, ranging from civil society to supra-national institutions such as the EU and the African Union within complex multilateral and bilateral relations? Is there more continuity than rupture between the policies pursued by Sarkozy in the region and those of his presidential predecessors? The Arab Spring has also demonstrated strong French interest in countries with whom it shares no colonial history, emphasizing the importance of analysing Franco-Mediterranean relations beyond the French colonial presence in the Maghreb, and notably beyond Franco-Algerian relations.

Bibliography

Aeschimann, E., and C. Boltanski, *Chirac d'Arabie* (Paris: Grasset, 2006).
Foxlee, N., 'Mediterranean Humanism or Colonialism with a Human Face? Contextualizing Albert Camus's "The New Mediterranean Culture"', *Mediterranean Historical Review*, 21/1 (2006), 77–97.
Hewlett, N., *The Sarkozy Phenomenom* (Exeter: Imprint Academic, 2011).
Majumdar, M., *Postcoloniality: The French Dimension* (London and New York: Berghahn, 2007).
Malkin, I., 'Introduction', *Mediterranean Historical Review*, 18/2 (2003), 1–8.
Vaïsse, M., *La puissance ou l'influence? La France dans le monde depuis 1958* (Paris: Fayard, 2009).
Youssef, A., *L'Orient de Jacques Chirac: la politique arabe de la France* (Paris: Editions du Rocher, 2003).

PART I

International Relations: France and the Mediterranean

JEAN-YVES MOISSERON AND MANAR EZZAT BAYOUMI

The Mediterranean: A Contested Concept

Over the past five decades, the Mediterranean has become the central reference for European public policy. The reinforced Mediterranean policy, the Barcelona Process, the New Neighbourhood Policy, and more recently the Union for the Mediterranean, all use and function within this geographical framework which appears to be obvious and completely organic.

However, in contrast with the overall scope and depth of integration achieved by the EU over the past fifty years, Euro-Mediterranean regional ambitions seem to have been thwarted by recurring difficulties and the results remain rather modest, if not outright disappointing. The reasons for this slow Euro-Mediterranean regional integration are widely documented[1] and it is not the objective of this chapter to discuss them any further. Rather, this chapter will seek to demonstrate that a partial responsibility for this failure finds its precise origins in the way in which the Mediterranean has been conceptualized, represented and used as a legitimate geographical framework to structure European public policy. It is our contention that the way the Mediterranean is conceptualized, as an obvious, given and organic geographical framework to determine and organize the parameters of European initiatives and policies, is based on a myth. As such, it acts as a veil, masking and obstructing the region's fundamental issues and making it difficult to find realistic and operative solutions to these challenges. In particular, to rely on such a Mediterranean myth often leads decision-makers to ignore the true dynamics of the area, upon which the construction of a common space between Europe and the Arab World could be possible. As Jean-Robert Henry argues: 'This myth

1   J.Y. Moisseron, *Le partenariat euroméditerranéen: l'échec d'une ambition régionale* (Grenoble: Presses Universitaires de Grenoble 2005).

needs to be put back into its true place and partially disassembled if we do not want to fall into the various traps of a virtual Mediterranean and an essentialist reading of the situation that risks clouding over the current issues at stake in the Mediterranean.'[2]

The Mediterranean as the Centre of the World: The Origins of a Concept

On the northern shore of the Mediterranean, one idea tends to dominate the writings and the arguments of those who are interested in Mediterranean public policy: whether unconsciously or not, their reasoning assumes that the region is defined by its 'unity'. The Mediterranean is seen, by virtue of its physical attributes, as an element of cohesion, a unifying player who is able not only to transcend the differences of the inhabitants of its shores, but also provide the original 'matrix' which defines the core parameters of all the civilizations concerned. This idea of the Mediterranean as the centre of civilization is the result of two earlier historic conceptions: first the 'Mare Nostrum' which expresses the idea of a sea in the middle of lands and second, the idea of the Mediterranean as a bridge which emerged with the discovery of the New World and later became central to the articulation of the European colonial project.

The Sea in the Middle of Lands: Mare Nostrum

The 'Mare Nostrum' conveys the idea of a peaceful world, with the Mediterranean as the centre of a relatively pacified and homogenous region dominated by a common civilization. In fact, it is above all Rome's cry of

---

[2] J.R. Henry, 'La nouvelle question méditerranéenne', *Questions internationales*, 31 (2008), 89.

victory over Carthage's empire after three Punic wars, and a cry of relief after avoiding defeat in the hands of Hannibal. Indeed, for more than a century from 264 BC to 146 BC the Mediterranean was a gigantic battlefield, with the northern shores and southern shores opposing each other. Rome and Carthage, the two main powers of the time, waged wars of an inconceivable violence. Such was the violence and resentment that for Rome only one radical solution could prevail: the total destruction of Carthage. It took a long time for the Mediterranean to become a peaceful and totally Roman sea. Piracy was at its zenith in the second century BC as a result of the development of the slave trade, and it was not until 67 BC that Emperor Pompey's maritime campaigns managed to rid the Mediterranean of piracy for several centuries.[3]

A 'Mediterranean lifestyle' progressively emerged, constructed upon the unity of common elements: an urban Roman civilization with its institutions and specific monuments, a network of roads and maritime channels between densely populated areas that reinforced the social fabric through religious unification, money, language, maritime communications and above all, an imperial organization. But this uniformity carefully hid fragmentary elements which were equally strong. 'Locals' expressed strong resistance and maintained their languages and social practices. The rhetorical opposition between the barbarian and civilized are the founding premises of 'Mare Nostrum'. History, as usual, is written by the victors, and here it serves up a false impression of unity. Thus, in his work *Geography*, the Greek geographer Strabon (64 BC–AD 24) evokes 'Mare Nostrum' ('Our Sea') to confirm the superiority of Roman civilization.[4] Writing at the beginning of our era, his Greek origins and the fact that he settled in Rome make him sensitive to the idea of the superiority of Roman civilization. It is for this very reason that he begins his geographical trip around the world with the Mediterranean Sea,[5] which from his point of view

---

[3] J. Carpentier, F. Lebrun and B. Bennassar, *Histoire de la Méditerranée* (Paris: Seuil, 1998), 97.
[4] F. Lasserre, *Strabon, Géographie* II (Paris: Les Belles lettres, 1967).
[5] Carpentier et al., *Histoire de la Méditerranée*, 100.

appears to be the centre. The very expression 'Our Sea' denotes an opposition between known and unknown worlds, an opposition which was to acquire a particular importance for geographers until the 'discovery' of the Americas. Thus, the Mare Nostrum is the sea of the known world which exists in opposition to the unknown sea, the outside sea, that is to say, the Ocean. Arab geographers would use the same binary opposition later on, differentiating the Sea of the East (*bahr al-shâm*) and the surrounding Ocean (*bahr al-muhît*).

The Mediterranean as a Marker of Difference

When Europeans 'discovered' the New World, the concept of a 'sea amid lands' started to be radically challenged: the world's scale changed drastically. Commercial maritime routes were opened to both the East and West Indies, and throughout the sixteenth and seventeenth centuries trade intensified but gradually the Mediterranean became marginalized: the international exchanges brought about by long distance trade no longer sailed principally through its waters. Thus, the discovery of the New World resulted in a strange regression in the idea of 'the Mediterranean, Centre of the World'. It ceased being the essential knot of maritime communications, and wealth did not necessarily have to travel through it. If Genoa could have claimed the status of world power until the beginning of the seventeenth century, it was ultimately outshone by Amsterdam, and then London. A modern capitalist economy may have emerged on the northern shores of the Mediterranean, but it largely developed itself outside its confines. The Mediterranean quickly ceased to be the locus for global economic development.

Thus, the Mediterranean became merely one sea among others, and not necessarily the most important one. What set it apart was no longer its centrality: indeed, it was then often described as a transitional space separating three different worlds: Asia, Africa and Europe. At the start of the nineteenth century this idea remained a central one. For instance, the Franco-Danish geographer Conrad Malte-Brun (1755–1826) certainly did not describe the Mediterranean in terms of its centrality: on the contrary, in his writing, not only did it have no more significance than any other

The Mediterranean: A Contested Concept

European sea, but its only distinctive feature was that it separated and isolated Africa from Europe, a prevailing theme among geographers during the two preceding centuries.[6] What lies beyond the Mediterranean is thus a 'far away' world, a world of 'otherness'. If Bonaparte's expedition in Egypt included such a plethora of scientists, it was to identify the unknown, those things that were foreign or strange.[7] Yet, the reactions of the botanists on Bonaparte's expedition are interesting: they were terribly disappointed. Searching for an 'elsewhere', they did not discover anything different about the flora. Many of the plants were common to both sides of the sea and were already well known. The grape vines, citrus, fig and olive trees that they recognized left little place for new plants or strange mutant varieties which could have in turn satisfied their search for the unusual, the Oriental exotica[8] which would eventually enable these learned biologists to leave their names tacked on to new species. The fauna was also familiar and thus disappointing. Anything which could draw attention to otherness or showcase the paradoxes and unfamiliarity of the place tended to be overemphasized: one was astonished, for instance, that 'grapevines abound here, despite a religion whose enemy is Bacchus'.[9] The historical theme of separation, therefore, remained very deep-rooted, and throughout the nineteenth century botanists would be very wary of using the adjective 'Mediterranean' to describe and thus provide a common identity to the regions around that sea. Malte-Brun may have spoken of a 'Mediterranean vegetation' but, as if demonstrating how intellectually uncomfortable such an idea was, he immediately added: 'if one can be allowed to use this expression'.[10] In the

6   C. Malte-Brun, *Précis de la géographie universelle: ou, Description de toutes les parties du monde sur un plan nouveau* (Berthot: Ode et Wodon, 1822).

7   P. Bret, *L'expédition d'Égypte, une entreprise des Lumières (1798–1801)*, Actes du colloque de Paris (8–10 juin 1998) (Paris: Académie des sciences, 1999).

8   E.W. Said, *Orientalism* (New York: Pantheon Books, 1978); translated into French as *L'Orientalisme: l'Orient créé par l'Occident* (Paris: Seuil, 1980).

9   M-N. Bourguet, B. Lepetit, D. Nordman and M. Sinarellis, *L'invention scientifique de la Méditerranée: Egypte, Morée, Algérie* (Paris: EHESS, 1998), 19.

10 F. Deprest, 'L'invention géographique de la Méditerranée: éléments de réflexion', *L'Espace Géographique*, 1 (2002), 73–92. Likewise, the botanist J.B. Bory de Saint-Vincent (1778–1846) was just as cautious: as the president of the Commission

early nineteenth century, the adjective 'Mediterranean' had no great currency and certainly a weak scientific legitimacy: it would take a long time to be integrated into learned and scientific papers.[11] It would also take some time for the Mediterranean to acquire a specific identity.

In this process, the Saint Simonians played an important, albeit ambiguous role: in the early 1830s, they envisaged a 'Mediterranean system' which, at first sight, suggests that the idea of Mediterranean unity was not altogether an unfamiliar theme. Saint-Simonian Michel Chevalier (1806–79) argued that the development of trade and industry would lead to the universal pacification of human relations and *in fine* the suppression of all wars. In 1832, as the director of the Saint Simonian newspaper *Le Globe*, Chevalier wrote a series of articles about peace between East (*Orient*) and West (*Occident*) which would then form the basis of his publication *Le système de la Méditerranée*. For Chevalier, the system went well beyond the Mediterranean Sea to include the Black Sea and the Caspian Sea. The area concerned was one of wars and multiple conflicts, real or mythologized, and the system ought to have eradicated such conflicts, not by imposing a 'European' model on other shores, but by insisting on the complementarity and interdependence of cultures. In his utopian project 'the Mediterranean will become the marriage bed of the East and the West',[12] each side was reconciled in the pursuit of interdependent economic and commercial objectives and each side understood the other better with the drastic reduction of distances that new technologies,[13] such as the railways, allowed.

> d'exploration scientifique d'Algérie (1839–42), he suggested that the vegetation in southern France and in northern Africa could be described as 'Mediterranean' but again demonstrated that this idea may be too farfetched when he added: 'if one is permitted to use this expression' (see J.-M. Drouin, 'Bory de Saint Vincent et l'invention de la botanique', in Bourguet et al. (eds), *L'invention scientifique de la Méditerranée*, 156).

11   Bourguet et al. (eds), *L'invention scientifique de la Méditerranée*, 153.
12   M. Chevalier, *Système de la Méditerranée* (Paris: Editions Mille et une nuits, 2006 [1832]), 15.
13   Railroads are the central element of Michel Chevalier's analysis. The theories that he exposed are fascinating when one considers the next 150 years of railroad expansion.

A common political destiny would emerged from economic cooperation, a Saint-Simonian utopia which yesterday was at the heart of the European construction and today underpins the Union for the Mediterranean, with, for instance, the development of a common technological project around solar energy. In the words of Chevalier: 'Industry is eminently pacific. Instinctively, it refuses war. That which creates cannot conciliate itself with that which kills.'[14] Four centuries after the discovery of the Americas, which made the Atlantic the centre of the world, Chevalier reintroduced the Mediterranean as the core element of Europe's vision of the world.[15] In his system, the Mediterranean remains a meeting place, a neutral ground between different cultures: it is not a unifier. The Mediterranean is only a first step to achieve a wider objective, what the author called 'the universal association'. One must wait until the apex of the colonial project and the opening of the Suez Canal to see the Mediterranean becoming a centre and a matrix of European civilization. This is particularly clear in the universal geography of Elisée Reclus.

In her penetrating study of the evolution of the *Universal Geographies*, Florence Deprest explains how the geographer Elisée Reclus (1830–1905) changed the paradigm through which the Mediterranean has come to be understood in Europe in the twentieth century.[16] For Reclus, time counted as much as space. Reclus' geography cannot be dissociated from

---

> As far back as 1832, the author was able to project through his writings a view of rail connections that were only to be constructed in the twentieth century, with the unique exception of the North African line, which is still not a reality. But Chevalier's project remains continental. At any rate, it appears very paradoxical to found a 'Mediterranean system' based on railroads. This is essentially explained by the fact that the Saint-Simonian project is above all a world project: it intended to construct a universal association that would unite mankind in a coherent system. But in order to achieve this objective, one would first have to overcome what seems to be the major schism of the time, that is, the opposition between the East and the West.

14   Chevalier, *Système de la Méditerranée*, 177.
15   J. Debrune, 'Le système de la Méditerranée de Michel Chevalier', *Confluences Méditerranée*, 36 (2000–1), 187–94.
16   Deprest, 'L'invention géographique de la Méditerranée'.

history. In his work, the Mediterranean is perceived as a space in which successive centres of civilization have developed: Egyptians, Phoenicians, Greeks and then Romans. Departing from the southern banks of the Mediterranean, passing through its oriental shore, turning north and ultimately west, the centres of civilization define a circle with its apex in Europe and, more precisely, Reclus argued without much humility, in France. The progress of maritime technologies, the end of piracy and peace with Great Britain had now turned the Mediterranean into a space 'beginning to be tamed' throughout the nineteenth century. In Reclus' words, the Mediterranean no longer 'swallows up' boats; it 'carries' them. It becomes a sea of development, progress and wealth. A common metaphor that describes the region as 'an interlacing of islands and peninsulas resembling the folds of the human brain' is sufficient to understand how it had come to stand for the matrix all civilizations. The Mediterranean became 'this huge mediator without which all of us in the West would have remained in a primitive, barbaric state'.[17] The Mediterranean is elevated to a historical status, *un principe créateur*, which also justified the colonial project. For Reclus, nothing was more 'natural' than to bring civilization to those who transmitted it to us. Civilization must continue its voyage and turn southward to complete its circle. In this intellectual framework, France's colonial 'civilizing mission' is nothing other than the necessary historical response to a civilizing movement which demands that France restores a civilization which has its roots in the southern, barbaric shore. With Elisée Reclus, the Mediterranean again acquired a magnificent aura. It became 'a machine that manufactures civilizations', as famously described by the poet Paul Valéry.[18] As a living organism compared to a thinking brain, a creative matrix, the Mediterranean evolved into a myth that justified colonization, occupation of territories and exploitation of resources. As foreseen by the Saint-Simonians, the intensification of trade and the development of

---

17   E. Reclus, *Nouvelle géographie universelle: La terre et les hommes* (Paris: Hachette, 1876), 33.
18   P. Valery, *Regards sur le monde actuel* (Paris: Gallimard, 1951), 317.

technologies narrowed the distances between the north and south shores of the sea. Proximity had to replace 'otherness': proximity of resources and markets, proximity of peoples who must assimilate into the new civilization and be integrated into the logic of a global economy, unless their otherness, their differences, were deemed too strong or too absolute to be part of it. These contradictions between 'differences' and 'proximity', which structured the various representations of the Mediterranean, were at the heart of the colonial project.

The 'Mare Nostrum' which had been the Romans' sea became the sea of all the peoples surrounding the Mediterranean. The mobile frontiers of this 'our' open the doors to many ambiguities. If 'we', the inhabitants of the two shores, are completely similar, the civilizing action of the West can better justify itself than if the differences between the East and West were absolute. But at the same time, if the West is not different, that is to say, more advanced in the historic evolution of progress, there is no justification to colonize other peoples or to 'bring them civilization'. In this case the colonial project can no longer disguise itself behind the mask of a legitimate historic role. The Mediterranean, with the wealth displayed in its various representations, permits the conception and articulation of the contradictions between 'difference' and the 'proximity' necessary to realize the colonial project.

The Mediterranean: The Diversity of its Representation

Representations of the Mediterranean are both shifting and heterogeneous. Such representations vary in the extreme depending on the specific histories of the peoples who inhabit its shores. Thierry Fabre and Robert Ilbert's research has sought to map out the different representations of the Mediterranean, but the enlightening ten volumes which resulted from their project do not yet provide a definitive account of all the reasons explaining the diversity of such representations.¹⁹ The following part of this chapter will try to illustrate this diversity of representation through the selection of some significant examples. If this is a substantial reduction of the diversity of representations which were so valued by Fabre, Ilbert and their research team, it will nevertheless serve to illustrate how the Mediterranean is a contested concept and how its representations have a real impact on the way recent policies have been framed.

In Europe, the concept of the Mediterranean has been extremely malleable: Mediterranean heritage, imaginary representations and the interpretation of historical events have been used to fulfill the needs of the moment or to justify a specific ideological project. For instance, the Mediterranean as Mare Nostrum was exploited by fascist Italy, whose identification with the Roman Empire was one of its most consistent narratives. After the pacification of Libya and the conquest of Ethiopia, Rome naturally became the capital of the 'Empire' and used Roman scenography[20] to contest the legitimacy of other colonial powers in the Mediterranean, notably that of France. After the war, Italy turned its back on the Mediterranean which had become too closely associated with the vagaries of fascism and its delirious dream of a new Roman empire. Post-war Italy was chiefly concerned with economic reconstruction and modernizing the country. The process of the European integration project became paramount to realize these objectives and Italy looked towards its northern neighbours for support and collaboration. The Mediterranean faded away and has only become an important part of Italy's foreign policy through EU initiatives, notably the instigation of the Barcelona Process. The case of Spain is also interesting: modern Spain was built during the *reconquista* in opposition to the Arab world.

---

19   T. Fabre and R. Ilbert, *Les représentations de la Méditerranée* (Paris: Maisonneuve & Larose, 2000).
20   East of Tripoli, the ruins of Leptis Magna, a city of the Roman Empire, was exploited by the regime to strengthen the fascist colonial claim that the Mediterranean was part of Italy's natural heritage, notably during Mussolini's visits in 1926 and 1937. See M. Galaty and C.S. Watkinson, *Archeology under Dictatorship* (New York: Plenum, 2004).

The recent conflicts concerning Ceuta and Mellila, and also the acute dispute concerning the Perejil Island, show how the distant past still informs representations and policies. A more recent past also has a great bearing on Spanish diplomacy in the Mediterranean: Franco's politics and notably its pro-Arab orientation still informs Spain's diplomacy. At present, Spain is a key actor in the Euro-Mediterranean partnership, and positions itself as a natural intermediary between both shores. It would not be difficult to provide a large number of examples illustrating the multiple representations of the Mediterranean found on the northern shores of the sea. What is most striking is nevertheless the range of different Mediterranean representations, their changing significance with time and their varying centrality within the national narrative. If this diversity of representations is the rule on the north shores of the Mediterranean, it is also the rule on its southern shores.

A Western perspective may assume too quickly that countries of the southern shores of the sea would naturally define themselves as Mediterranean. This is very questionable, not only because they display different representations of the Mediterranean, but also because they may not conceive themselves as Mediterranean. For instance, for the heir to the Ottoman Empire, the idea of a Mediterranean Turkey is empty of meaning. The sea is a symbol of defeat and of the very long decline of the Ottoman Empire. Modern Turkey has reconstructed itself as a continental power: its modern identity turns its back on the Mediterranean Sea. This is equally the case with Egypt: the Nasserian revolution sought to connect Egypt with the Arab, Islamic and African worlds, not with a Mediterranean one. This is not to deny that there are no Egyptian voices that seek to anchor the country to a Mediterranean identity; the prominent Egyptian writer Taha Hussein, who studied in Montpellier and had a French wife, was able to defend the idea that Egyptian identity was partly Mediterranean, but he never had a large following. Alexandria, which used to be a cosmopolitan city open to the West, is today mostly an Arab city. The reflections of the Egyptian intellectual Mohammed Afifiare are interesting in this respect:

> The Mediterranean came back into the spotlight during the 1990s in Egypt but this time in a different context. For the first time in the history of Egypt, the idea was presented by an initiative of the Government and not by certain intellectuals, as was the case before. This initiative is probably due to the new 'pragmatic' political theory adopted by the State in its foreign policy. [...] If the State adopted this initiative, it was to attempt to profit economically from its European neighbours in the context of worldwide economic mutations. [...] The majority of political opposition parties refused this 'new return' to the idea of a Mediterranean-based partnership.[21]

One should be cautious when considering the so-called Mediterranean character of the countries on the southern shores of the Mediterranean. There is a wide gap between the energy deployed by some NGOs who make their voices heard in various Euro-Mediterranean forums and the relative indifference of local populations about what appears to be a somewhat abstract and remote idea of a Mediterranean identity. Several Mediterranean-based NGOs survive essentially through European financial arrangements engaged in the framework of the Barcelona Process. If 'Orientalist' was a profession in the past, today one can make a career out of 'Euro-Mediterranean' business. Speaking the 'Euro-Mediterranean' language can increase the visibility of a southern Mediterranean NGO, even if this language remains largely meaningless for the communities in which the NGO is supposed to be embedded. Too often, the objective seems to be to secure a position of valid interlocutors to the European Union: it is fascinating to watch Europe 'speak to itself' during international forums, using concepts such as democratization, human rights, gender equality and individual liberties whose meaning for the southern Mediterranean populations remain obscure.

In fact, the 'Mediterranean' has so little meaning in the representations of the southern shore that there is not even a specific word in the Arab language to designate this sea. In ancient texts, the expressions *bahr al-shâm* (the Sea of the Eastern Edge, that is, the Syrian (*al-shâm*) Sea) and *bahr al-rum* (the Byzantine/Christian Sea, that is, the 'Roman' (*al-rum*)

---

21 M. Afifi, R. Ilbert and T. Fabre, *La Méditerranée égyptienne* (Paris: Maisonneuve & Larose, 2000), 32.

The Mediterranean: A Contested Concept

Sea) were commonly used, whereas today it is referred to as *al-bahr al-abyad al-mutawassat*: 'the White Sea in the Middle'. At first glance, this expression may seem enigmatic; why add an adjective of colour? The Turkish expression *Akdenizuswedto* designating the Mediterranean Sea means the 'White Sea' in contrast to another sea – '*Karadeniz*', the Black Sea. This difference of colour does not point to the centrality of the Mediterranean which is defined by contrast rather than rank. Thus, the modern Arab expression – *al-bahr al-abyad al-mutawassat* – is borrowed from Turkish, indicating that there is neither a specific Arab representation of the Mediterranean Sea, nor has the Mediterranean acquired the rich and varied connotations that are so prevalent on its northern shores.

How a Vision can Obstruct Action

The study of the different representations of the Mediterranean and of the different words used to name it allows us to appreciate that it does not generate the same images, the same passion, the same values on its northern and southern shores. Such differences have a real impact on how public action is constructed, projects and objectives are defined and potential blueprints are developed; in the end, they underpin diverging ways to formulate public policies. This perspective partly explains the semi-failure of the Barcelona Process (see the following two chapters in this volume), and points to the forthcoming challenges that the Union for the Mediterranean will inevitably meet. It is thus important to analyse further the Mediterranean as a concept to understand how this very concept has today become a major hindrance to meaningful and effective public action.

When referring to the Mediterranean, it is not unusual for European political leaders to suffuse their discourses with mythical images which underpin the unity of the region. Jean-François Daguzan reminds of some very striking examples.[22] Thus, Jacques Chirac's former foreign secretary (2002–4) and prime minister (2005–7), Dominique de Villepin, who was born in Rabat, Morocco, in 1953, declared in a 2002 speech entitled 'The Dream of Two Shores' (*Le Rêve des deux rives*):

> The two shores of the Mediterranean have constantly borrowed, held dialogues with and exchanged with one another, thus forging our identity and our destiny. The shores are like the two lips of the same mouth that only speak when reunited. The Mediterranean: sea of commerce and exchanges, where oil circulates in ceramic urns and woven fabrics under the decks of ships, where the Greek warrior alights in Africa, where the dreams of Ulysses give voice to grottos and rocks.[23]

This speech is a vibrant and passionate call for a union of the peoples around the Mediterranean Sea. It mixes poetry, references to Greek mythologies and ancient history and assumes a shared and common heritage, similarly conceived, understood and valued on each shore of the sea. But when de Villepin asserts that what added force to this constant dialogue was 'our attachment to the values of democracy and liberty that spread beyond our own frontiers', he not only reformulated France's traditional *mission civilisatrice*; he also underlined how it has become a central theme of the Barcelona Process, the European Neighbourhood Policy and the Union for the Mediterranean.

A dialogue between the two shores is perceived to be even more urgent and justified if a 'clash of civilizations' is to be avoided: if the Mediterranean is the 'source of all civilizations', then it also bears the promise of mutual understanding, progress and peace.[24] Here, there is an astonishing continuity between the visions of the Saint-Simonians and the justifications for the Union for the Mediterranean, as articulated by France: once again, all the various representations of the Mediterranean that the north has traditionally developed are mobilized to avert a potential clash of civilizations.

---

22 J.F. Daguzan, 'French Policy in the Mediterranean: Between Myth and Strategy', *Journal of Contemporary European Studies*, 17/3 (2009), 387–400.
23 D. de Villepin, 'Le Rêve des deux rives', speech given at the Université Mohamed V, Rabat (31 October 2002).
24 Daguzan, 'French Policy in the Mediterranean'.

But are these representations and values of any real use for the south? To see the Mediterranean as a source of all civilizations is a difficult thing to understand and indeed to accept on the southern shores of the sea. The civilizational matrix with which Arab communities identify themselves is above all delineated by the use of Arabic and the practice of Islam. Mecca and Medina cannot easily be included on a Mediterranean map, even an overstretched one: indeed, the vast majority of the *mihrabs*[25] in Mediterranean mosques turn their back to the sea. To use the Mediterranean as an analytical category is thus fundamentally problematic as it often marginalizes, if not denies, the cultural specificities of those who are born on its southern shores. Indeed, those who defend a 'Mediterranean civilization' or a 'Mediterranean culture' are usually ill at ease when asked to define the role of Islam in this 'common civilization'. The basis for so many speeches about the Mediterranean, such as the one made by Dominique de Villepin, may be understood by the elites of the Arab World because they can decipher and understand, make sense of such values, or by some Euro-Mediterranean NGOs, but what can they mean for the vast majority of the population on the southern shores of the sea? The very idea of a clash of civilizations is also puzzling: apart from fundamentalist movements who preach a radical opposition to the West, one is hard pressed in the south to find any evidence that the potential clash of civilizations informs either the formulation of policy objectives or indeed is a preoccupation of local populations. For instance, most of the youth on the southern shores of the Mediterranean, notably in Algeria, are not seeking to defy the West, but rather to find a place there, sometimes at terrifying risks. To accede to a consumer society remains a desirable objective and the effects of consumerism have already had a strong impact on southern Mediterranean countries. Islam is not necessarily opposed to the diffusion of this social model even if it may criticize some of its overindulgence, as would any spiritual leaders in the face of excessive materialism.[26]

25 The niche in a mosque showing the direction of Mecca.
26 P. Haenni, *L'islam de marché: l'autre révolution conservatrice* (Paris: Seuil, 2005).

Globalization and its economic and social dynamics – new modes of production, migration, access to information and education, to name a few – have a profound effect on societies on both sides of the Mediterranean. If a successful partnership is to emerge between the two sides of the Mediterranean, public policies, informed by social scientists rather than grand narratives derived from the readings of the 'Odyssey', should address such practical concerns and bear in mind that the Mediterranean might be considered a concept which has more resonance in the north than on the southern shores of the sea. A genuine union with the south would seek to understand, recognize and respect the particularities of each partner. In this volume, Jean-Robert Henry demonstrates how culturally loaded references to the Mediterranean often block rather than enhance the Euro-Mediterranean dynamics on the ground. Co-operation is at its most effective when grand references to the 'Mediterranean myth' are left aside: on the ground, beyond the Barcelona Process, there exists a dense network of exchanges between the two shores of the Sea: Euro-Mediterranean NGOs may tactically use for their own profit the language EuroMed, but by virtue of the sheer number and the intensity of their work, they enhance co-operation between the northern and southern shores: for instance, student exchanges and scientific co-operation do create a sense of proximity and lead to the implementation of specific projects. More important is the role played by immigrant populations and the creation of a 'grassroots' melting pot: mixed marriages, dual nationals, participation of immigrants in economic, social, cultural and political life. This does not mean that discrimination will subside or that fear of Islam will regress, but the dynamics are much more complex and richer than the Manichean alternative which limits the choice between a 'clash of civilizations' or a mythical 'Mediterranean unity'.

Conclusion

This chapter has sought to give an insight into how representations of the Mediterranean have developed in the north and continue to inform French and EU policy towards countries on its southern shores. One could certainly argue that the choice of 'Mediterranean' as a category of analysis and the meanings with which this term has been charged could partially explain the origins of the difficulties encountered by the Barcelona Process. What is most interesting is precisely how such a representation is able to become an obstacle to efficient action in a given area or in the world.

We have pointed out that the 'Mediterranean' as a concept and as a representation plays an essential role while other references, such as the 'Arab World', are marginalized. The rhetoric of thought concerning the Mediterranean is not a mobilizing force for the south because it is a construction of the north that is articulated as a function of its colonial history. If it finds any echo at all in the south, it is principally because of the advantages that the southern countries can find in it. To agree with a 'Mediterranean' political action or to be inserted into one of the many Euro-Mediterranean partnerships allows countries and NGOs to benefit from an array of budgetary funds and financing offered by the European Union, demonstrating another form of the instrumentalization of the concept of the Mediterranean.

Bibliography

Afifi, M., Ilbert, R., and Fabre, T., *La Méditerranée égyptienne* (Paris: Maisonneuve & Larose, 2000).

Ahmed, A.S., and Hart, D.M., *Islam in Tribal Societies: From the Atlas to the Indus* (London: Routledge/Thoemms Press, 1984).

Arrault, J.B., 'A propos du concept de méditerranée: l'expérience géographique du monde et mondialisation', *Cybergeo* (January 2006), 332.

Bourguet, M.-N., Lepetit, B., Nordman, D., and Sinarellis, M., *L'invention scientifique de la Méditerranée: Egypte, Morée, Algérie* (Paris: EHESS, 1998).

Bret, P., *L'expédition d'Égypte, une entreprise des Lumières (1798–1801)*, actes du colloque de Paris (8–10 juin 1998) (Paris: Académie des sciences, 1999).

Carpentier, J., Lebrun, F., and Bennassar, B., *Histoire de la Méditerranée* (Paris: Seuil, 1998).

Chevalier, M., *Système de la Méditerranée* (Paris: Editions Mille et une nuits, 1998).

Daguzan, J.F., 'French Policy in the Mediterranean: Between Myth and Strategy', *Journal of Contemporary European Studies*, 17/3 (2009), 387–400.

Debrune, J., 'Le système de la Méditerranée de Michel Chevalier', *Confluences Méditerranée*, 6 (2000–1), 187–94.

Deprest, F., 'L'invention géographique de la Méditerranée: éléments de réflexion', *L'Espace Géographique*, 1 (2002), 73–92.

Drouin, J.M., 'Bory de Saint Vincent et l'invention de la botanique', in M.N. Bourguet, B. Lepetit D. Nordman and M. Sinarelli (eds), *L'invention scientifique de la Méditerranée* (Paris: EHESS, 1988).

Fabre, T., and Ilbert, R., *Les représentations de la Méditerranée* (Paris: Maisonneuve & Larose, 2000).

Galaty, M., and Watkinson, C., *Archaeology under Dictatorship* (New York: Plenum, 2004).

Haenni, P., *L'islam de marché: l'autre révolution conservatrice* (Paris: Seuil 2005).

Henry, J.R., 'La nouvelle question méditerranéenne', *Questions internationales*, 31 (May–June 2008), 82–93.

Lasserre, F., *Strabon, Géographie II* (Paris: Les Belles lettres, 1967).

Malte-Brun, C., *Précis de la géographie universelle: ou, Description de toutes les parties du monde sur un plan nouveau* (Berthot: Ode et Wodon, 1822).

Moisseron, J.Y., *Le partenariat euroméditerranéen: l'échec d'une ambition régionale* (Grenoble: Presses universitaires de Grenoble, 2005).

Rabelais, *Gargantua* (Paris: Editions R. Calder & M.A. Screech, 1534).

Reclus, E., *Nouvelle géographie universelle: la terre et les hommes* (Paris: Hachette, 1876).

Saïd, E.W., *L'Orientalisme: l'Orient créé par l'Occident* (Paris: Seuil, 1980).

Valéry, P., *Regards sur le monde actuel* (Paris: Gallimard, 1951).

Villepin, D. de, 'Le Rêve des deux rives', speech given by Dominique de Villepin, Université Mohamed V, Rabat (31 October 2002).

JEAN-ROBERT HENRY

France's Mediterranean Policy: Between National Ambitions and European Uncertainties

France's Mediterranean initiative, announced by Nicolas Sarkozy in February 2007 as the great diplomatic ambition of his presidency, upset the routine of Euro-Mediterranean policy for two years and finally led to a desultory result: since early 2009, the partnership has been put on ice in practical terms. Even in Paris, Mediterranean activism has eased off, as demonstrated by the very weak place accorded to it by the President of the Republic in his speech of 26 August 2009 to France's ambassadors: he proposed only holding a summit for the Union for the Mediterranean the following autumn on the Middle East peace process. This summit did not take place; neither did the one which the Spanish presidency of Europe had planned to organize in June 2010 in Barcelona, which now seems to have been postponed *sine die*. Despite this, what has happened since 2007 on the Mediterranean scene must not be ignored. The episode is doubly revealing: on the one hand it highlights the constancies and inconstancies, the inflections and contradictions of France's Mediterranean policy; on the other – and above all – it sheds light on Europe's current difficulties in thinking about its relations with countries on the other side of the Mediterranean.

The Changing Face of France's Mediterranean Policy

France's Mediterranean action – along with that of the UK and Spain, notably – is not limited to community policy. These countries carry out their own policies on the Mediterranean stage, which each have their own history and constitute specific aspects of the Euro-Mediterranean relationship, adding to or connecting to, with varying results, the attempt to develop a common policy. Like the other 'great powers' of the European Union, France is reactivating its assets, practices and networks, which are kept alive through its rich diplomatic capital accumulated in the course of centuries in the Mediterranean, in order to re-invest parts of them in the future of Europe. In return – just like others – she is interpreting and using to her own advantage the objectives of the Barcelona Process according to her tradition, her interests and her own preoccupations. One cannot therefore really talk of a lack of continuity between past and present in France's diplomatic action in the Mediterranean, but what it at stake is rather how France, as a regional power, is recalibrating her strategy to fit new contexts. One must bear this continuity in mind in order to comprehend the current diversity – and, at times, the inconsistencies – of 'France's Mediterranean policy'. The latter is the complex and changing result of multiple factors which were formed through a long history of relations with the Arab-Muslim East, a shorter and more recent history of relations with the south within the European context, and the mobilization of various cultural, economic and political assets, amongst which is the relative goodwill towards France on the part of various partners in the south.

The Advantages of a Long History of Relations with the East

For centuries the Mediterranean has been a major scene in France's foreign policy and has witnessed the worst excesses of wars of conquest as well as relations more respectful of the personality of societies on the other side of the Mediterranean. Even the heritage of the crusades is ambivalent. A traumatizing exercise for the societies which were attacked and sometimes for the victors, too; but for French society and its young monarchy they were also an opening to the East, the traces of which proved enduring.

In the sixteenth century, diplomatic relations were established between France and Turkey; this meant that the Mediterranean was no longer a space primarily structured by religious solidarity and conflicts. As a result, formal diplomatic relations between the two powers contributed to the emergence of rules governing the functioning of the modern international system. Franco-Ottoman relations remained active for more than three centuries, despite periods of confrontation or tension. In the eighteenth century France appeared to be a major European power next to the Turkish and Spanish protagonists, who were weakening. France's diplomatic interest in this region even manifested itself in the creation of a office specifically for foreign affairs.[1] Early on, French specialists were trained in the understanding of 'eastern' languages and societies.

In this context, Bonaparte's Egyptian expedition was not a simple, unfortunate military adventure. It had a determining influence on the modernization of Egypt and contributed, even in France, to the forging of a different vision of the East–West relationship which would then be developed by the Saint-Simonians. From 1832 onwards, the latter imagined a 'system of the Mediterranean' which claimed to be a central element of a 'universal association of peoples'.[2] This was the dream of social inventors who were anchored in the scientific culture of their time (many of them had come from the École Polytechnique) and were determined to articulate ideas about Utopia and action. As a major pressure group with direct access to decision-makers, they influenced major political projects in the 1860s such as Napoleon III's 'Arab kingdom' in Algeria, and they were able to carry off large technical projects such as the Suez Canal before being caught up by various business scandals. Lyautey's Morrocan policy has at times been linked to their influence, even if he was, above all, a conservative with ambitions to modernize the Moroccan monarchy. In the 1920s he was part of those who wanted to make a 'great Muslim power' out of France, according to an ambiguous expression which expressed both concerns about opening up to the 'other' and the desire to dominate and exploit it, within the context of a power struggle with Germany on the European stage.

1   J. Carpentier, F. Lebrun and B. Bennassar (eds), *Histoire de la Méditerranée* (Paris: Seuil, 1980).
2   M. Chevalier, *Le système méditerranéen* (Paris: Editions Mille et une nuits, 1986 [1832]).

Closer to home, some saw a renewal of the Saint-Simonian way of thinking in certain ideas regarding postcolonial co-operation.[3] In reality, the latter mixed determinedly neo-colonial concerns with the promotion of a new vision of North–South relations, defined by the Jeanneney Report under General de Gaulle's presidency.[4] In the Mediterranean in particular it was important to move away from the dramatic experience of decolonization in an attempt to write a new chapter of shared history with the once-colonized societies. The Algerian case – then considered exemplary – justified the implementation of a huge policy of co-operation working towards a strategy of reconciliation reminiscent of that which was sealed with Germany in 1961 on the occasion of the 'Franco-German summer'. The conditions were not, however, the same. The restructuring of the colonial relationship required France to come to terms with its Algerian colonial past. As a result of not being able to manage this, relations between the two countries went from one misunderstanding to another, and despite a solemn state reunion in 2003, the project of signing a treaty of friendship and co-operation between the two countries misfired, placed in jeopardy by historical controversy in Algeria and in France.[5] On an economic level too, the ties between France and the Maghreb are considerable: around a quarter of the Maghreb's international trade are with France, which is supplied with hydrocarbons by Algeria and Libya. Nicolas Sarkozy has sought to strengthen these economic ties by proposing selling high-technology commodities to the countries of the southern Mediterranean: nuclear

3   M. Levallois, 'Preface', in I. Urbain, *L'Algérie française: indigènes et immigrants* (Paris: Séguier, 2002 [1862]).
4   La politique de coopération avec les pays en voie de développement, Rapport de la Commission d'études, set up by the decree of 12 March 1963 (Paris: Ministère chargé de la réforme administrative, July 1963).
5   The conditions in which those responsible in France became bogged down in the handling of this story have not been totally clarified. It seems they wanted to satisfy an Algerian demand in recognizing the 'inexcusable tragedy' of 8 May 1945 at the same time as paying lip service to those who were repatriated from Algeria by accepting that in February 2005 a bill was passed which recognized the positive aspects of colonization. Whether it was intentional or not, the parallelism of these two gestures has been perceived in Algeria as unacceptable duplicity.

power stations (five agreements signed in 2008), aeroplanes, high-speed trains, arms and large building projects.

However, it is largely due to the long-standing nature of its relations with civil societies on the other side of the Mediterranean that France's Mediterranean policy has the assets which she is mobilizing today: it is particularly in the cultural domain that the most specific assets can be seen. France's foreign policy in the Mediterranean is based on what remains from its huge heritage from the past. If the French language is no longer the dominant language in the Mediterranean region and a cultural tool used to modernize the Ottoman Empire,[6] as it was in the nineteenth century, French influence continues to be based on a network of several hundreds of public or private high schools, cultural centres, Alliances Françaises and research centres, the importance of which at times seems disproportionate in relation to the Foreign Ministry's reduced financial means. But this cultural influence now only reaches a small section of the elite in most of the countries, except in the Maghreb (which backs onto a 'French-speaking' west Africa), where it is powerfully reinforced by the fact that the societies there have adapted the French language to their own needs in a reasonably easy-going co-habitation with Arabic: the French-speaking element remains a strong cultural fact, whilst Islam has become a cultural reality for French society. This reality is increasingly recognized by the political authorities as President Sarkozy's message on 12 December 2008 to the Muslim community in the Seine Saint-Denis region demonstrates: 'Today, and in a definitive manner, Islam is an integral part of our cultural, economic and urban environment [...] Going by numbers, France is home to Europe's largest Muslim community.'[7] Moreover, an institution such as the Institut du Monde Arabe stresses the exceptional nature of Franco-Arab cultural relations.

---

6   At the start of the twentieth century the Ottoman Empire had several hundred newspapers in French. See G. Groc, *La presse française de Turquie* (Istanbul: Isis, 1985).
7   N. Sarkozy, 'Message à la communauté musulmane de la Seine-Saint-Denis' (Union des Associations Musulmanes de la Seine-Saint-Denis, 12 December 2008), http:// www.uam93.com (accessed 12 February 2010).

These linguistic and religious elements aside, French policy equally knows how to mobilize – and this was again apparent in the argumentation developed by Sarkozy's project of Mediterranean union – an entire mythology based on the Mediterranean, fuelled since the 1930s by the works of leading figures such as Gabriel Audisio, Fernand Braudel, Jacques Berque and Edgar Morin. For a long time it has also been mobilizing actors who are not linked with the state: after the religious orders and lay movements, who played a large role in 'France's cultural presence' in the Middle East, hundreds of associations today claim to have an interest – which is predominantly cultural – in the Mediterranean.

One specific asset in 'Franco-Mediterranean' relations is ultimately due to the importance of their human dimension on an individual and collective scale. Many members of the southern Mediterranean elite have been through the French university system, whilst countless French politicians have had their own personal experiences of the Maghreb in the course of their careers. However, this generation has now left power and the personal link between new French leaders and societies on the other side of the Mediterranean generates less interest today than it previously did.

On the other hand, one fact which today heavily influences France's Mediterranean policy is the interweaving of the human and cultural dimensions found in their societies. Between eight and nine million people living in France have a relationship of varying degrees of complexity with the Maghreb; similarly, millions of Maghrebi people have had a direct experience of France. This human Franco-Maghrebi fabric is also characterized by the existence of a Franco-Maghrebi society (often of dual nationality, legally) made up of several million people whose expectations and strategies find a voice in individual strategies and in many different associations which have a transnational interest. Silent for a long time, this frontier society is today speaking out in order to make its existence known and to highlight its claims using various means (for example, the Indigènes de la République movement, the film *Indigènes*, the question of the veil, the

Year of Algeria in France, literary and scientific works).[8] Their aspirations are increasingly taken into account as a major factor in France's relations with the Maghreb and within the Islamic-Mediterranean space. Above all, this transnational human and civil reality helps to question both the idea of a Mediterranean frontier in Europe and the policy of shutting down the European human space which are a part of this.[9]

France's foreign policy has to manage the consequences of this transnational dimension as well as more traditional international relations. But such an interface does not easily give rise to either the formulation of public policies or even to the development of research agenda.[10] On the other hand, the interface between the national dimension of France's Mediterranean policy and its European dimension has received more attention.

Between a European Approach and a National Approach: The Two Registers of France's Mediterranean Policy

The heritage of the past and its resulting assets in the region explain France's ambitions and strategy in the Mediterranean. France has used Europe's

---

8   For example the novels or short stories by Leïla Sebbar, such as *La Seine était rouge* (Arles: Actes Sud, 2009) or larger works such as M. Arkoun et al., *Histoire de l'islam et des musulmans en France* (Paris: Albin Michel, 2006). Moreover, a 'Franco-Algerian space' was created in 2007 in order to form an association and represent the tens of associations claiming a mixed identity.

9   This reality is now recognized by certain politicians; Michel Vauzelle, president of the PACA region and defender of an active solidarity between the peripheral regions of the Mediterranean, does not hesitate to affirm that the notion of neighbourhood is inappropriate to the transnational human reality which characterizes the links of Mediterranean proximity and notably Franco-Maghrebi ties: 'We do not have a neighbourly relationship with the Maghreb and the southern and eastern Mediterranean. This spaced is linked by a common destiny' M, Vauzelle, 'Ouverture du Forum des autorités locales et régionales de la Méditerranée' Forum des autorités locales et régionales de la Méditerranée (Marseille, 2008).

10 C. Visier, *L'Etat et la coopération, la fin d'un monopole: l'action culturelle française au Maghreb* (Paris: L'Harmattan, 2004).

policy in Africa and the Mediterranean as part of its strategy to remain a decisive and powerful actor in the region, and through this, to make its presence felt on the international stage. As Hayète Chérigui has shown, in using this strategy, France kept two irons in the fire on the Mediterranean stage: on the one hand, she has been heavily involved in Europe's Mediterranean policy with ambitions of exercising a leading role on this stage – or to fight over it with her partners – on the other, she has pursued and developed an 'Arab policy', a singular policy which claims to be essential in the defence of her national interest, and which displays pro-Arab and rather anti-American tendencies in its support for what used to be referred as 'Third world' countries.[11]

On the one hand, France played a major role in shaping Europe's Mediterranean policy, seeking to ensure the prominence of her positions on that stage and limiting the ambitions of other European partners (Italy, and later the UK and Spain). After all, in the early years of the then European Economic community, it was France who under the leadership of prestigious commissioners, such as Edgar Pisani and Claude Cheysson, ensured that the EEC would build up positive relations with Mediterranean countries, notably by developing cooperation agreements with her former colonies.

Spain and Portugal's entry into the European Community modified the deal as shown by the avatars of the first 'French initiative into the Mediterranean', which was launched in 1988 following an idea of François Mitterrand. With its aim of bringing together countries in the western Mediterranean, this initiative was the start of the '5+5 dialogue' (Portugal, Spain, France, Italy and Malta in the north, with Mauritania, Morocco, Algeria, Tunisia and Libya in the south). But the fact that this initiative was heavily directed by Paris and far too linked to its own interests sparked off a reaction from its various partners. Finally, the project to organize a conference for the foreign affairs ministers of the western Mediterranean was made possible following an Italian proposal with Spanish support. But

---

11   H. Chérigui, *La politique méditerranéenne de la France: entre diplomatie collective et leadership* (Paris: L'Harmattan, 1997).

the process had no direct result because in the meantime the European Union had revived a project on a Euro-Maghrebi partnership in 1992 which received strong support from the French diplomatic authorities.[12] When this Euro-Maghrebi project was, in turn, converted into a Euro-Mediterranean project hard on the heels of the Middle East peace process, it was Spain, rather than France, who managed to finalize the enterprise by hosting the conference launching the Euro-Mediterranean partnership in Barcelona in November 1995.

Nevertheless, French diplomacy did its best to make its presence felt within this new Euro-Mediterranean space and to launch with varying degrees of success a series on initiatives in the field. Hayète Chérigui has made a good analysis of the modes of action and the instruments of France's Mediterranean policy which it used to maintain an eminent position on the Euro-Mediterranean scene.[13] The main instrument used in this game of influence was the 'Mediterranean Forum' comprising eleven members launched in 1994 by France and Egypt from what remained of the Mediterranean Forum of 1988. Whereas the latter was based on the western Mediterranean, the new Forum did not include Libya (under embargo) or Mauritania, but increased in size to include Egypt, Turkey and Greece. This Mediterranean Forum, whose strength resides in the informal character of the rules by which it is run, took care not to enter into competition with the Barcelona Process but instead to act within it as a diplomatic lobby or as a 'laboratory of ideas for the Euro-Mediterranean partnership', aiming to maintain the Mediterranean as a priority within the enlarged EU. One of its assets was not to have amongst its members any countries which were directly linked to the conflict in the Middle East.

The landscape became even more complicated with the resurrection of the '5+5 dialogue' in 2001 after a ten-year break; its priority was to reinforce partnership with the Maghreb. This group – also informal – could

---

12   On this Euro-Maghrebi dynamic, see J.-R. Henry, 'L'Europe du Sud et le Maghreb: le rêve andalou à l'épreuve', in J. Bourrinet (ed.), *La Méditerranée, espace de coopération: en l'honneur de Maurice Flory* (Paris: Economica, 1994).
13   J. Carpentier, F. Lebrun and B. Bennassar (eds), *Histoire de la Méditerranée* (Paris: Seuil, 1980).

boast of some success with the meeting in December 2003 in Tunisia of the heads of state or government of the '5+5', and above all with the establishment of co-operation amongst various offices (defence, civil security, national security, education ...) that the Euro-Mediterranean partnership had somewhat failed to address. The '5+5 dialogue' was recognized by the Euro-Mediterranean conference in Naples in 2003 as complementary to the Barcelona Process, within the context of encouraging co-operation at a sub-regional level. Economists saw here a good opportunity for regional economic development.[14] In 2007, at the end of Jacques Chirac's presidency, enlarging the group to become a 'dialogue 6+6' had been envisaged; this would have included Greece and Turkey whilst at the same time remaining outside the conflict in the Middle East. But this perspective, which would have in fact led to merge the two sub-regional forums ('5+5 dialogue' and the 'Mediterranean Forum *à 11*') was rendered obsolete by Nicolas Sarkozy's Mediterranean initiative. After that, the '5+5 dialogue' survived in a discrete – which is not to say inefficient – way as a modest alternative to the fiasco of the Union for the Mediterranean – away, always, from the conflict in the Middle East.

Alongside its singular contribution to Europe's policy in the Mediterranean, France continued to play – depending on any occasions provided by the international context – its Arab policy card which, in the Gaullist tradition, claimed to be supportive of Third World countries, pro-Arab, relatively anti-American and anti-Israeli, at least until Nicolas Sarkozy's accession to the presidency of the Republic. Above all, it claims to be a manifestation of independence and of diplomatic 'grandeur'.

Paradoxically, the idea of 'France's Arab policy' came after de Gaulle. It came into general use following a publication by Paul Balta and Claudine Roulleau – journalists who were close to Pompidou – which was published in 1973,[15] that is several years after the cooling in relations with

---

14 J.-M. Chevalier and O. Pastré, '5+5 = 32: ambition d'une association renforcée feuille de route pour une Union Méditerranéenne', *Les Cahiers du Cercle des Economistes*, 4 (Paris: Editions Perrin, 2003).
15 P. Balta and C. Roulleau, *La politique arabe de la France: de de Gaulle à Pompidou* (Paris: Sindbad, 1973).

Israel, under de Gaulle, following the Israeli-Arab war of 1967. It was the defence and rationalization of a Gaullist vision of France's action in the Mediterranean. This 'Arab policy' was supposed to be part of a wider Francophone project: or, according to the famous expression by Boutros Boutros-Ghali,[16] Francophonie, both as a cultural project and a set of institutions, is part of an 'objective alliance' with the Arab world.

In reality, the origins of France's Arab policy can be traced back to the end of the war in Algeria; this policy was, at the time, breaking away from the colonial, pro-Israeli policy of the majority of French socialists, which had led to the Suez fiasco in 1956. As a policy of co-operation, 'France's Arab policy' displays all the ambiguities of a postcolonial diplomacy: it claims to be an good example of French colonialism's conversion to becoming more supportive of Third World countries, whilst at the same time taking care to defend its national interests which have been inherited from this colonial past.[17]

France's Arab policy became established amongst different successors of General de Gaulle like an essential rhetoric: even Nicolas Sarkozy felt compelled to declare in 2007 that his project of Mediterranean union was 'the new name for France's Arab policy'. The idea of France having a clear policy regarding the Arab world is entirely in line with that of the Euro-Arab dialogue, the concept of which France had defended in the mid 1970s. Following the Yon Kippur War in 1973, and the first oil crisis, President Pompidou and his foreign minister, Michel Jobert, sought to formalize a dialogue between the EEC and the Arab League. Enhanced cooperation

---

16 Egyptian diplomat, UN secretary general (1991–6) and secretary general of Francophonie (1997–2002).

17 Not only was this Arab policy – a leading proponent of which was Michel Jobert, foreign minister under President Pompidou – not met with unanimity amongst politicians but it also transgressed traditional political divides. During the first Gulf War one could thus perceive a very clear opposition between the Mitterrandists and the neo-Gaullists of the left and the right. The high point of the crisis was the resignation of the socialist defence minister, Jean-Pierre Chevènement, in 1991. Chevènement openly criticized President Mitterrand's policy in support of the US position in the Gulf. Chevènement's pro-Arab stance is well illustrated in J.-P. Chevènement, *Une certain idée de la République m'amène à* … (Paris: Albin Michel, 1992).

between Europe and the Arab world could be perceived, of course, as running against US interests in the Middle East. The Euro-Arab dialogue never really took off, but France made several attempts to relaunch it: one of the last fleeting attempts took place in April 2006 with the support of President Chirac, but with no representation of the European Union.[18] Likewise, Paris's desire to bring the Arab League back into the Euro-Mediterranean partnership, during the Euro-Mediterranean conference held in Marseille in November 2008, was not unrelated to the idea of re-introducing the principle of Euro-Arab dialogue into the Euro-Mediterranean game. Finally, it should be emphasized that as France seeks to reinforce its bilateral links with each of her southern Mediterranean partners, the notion of 'France's Arab policy' conveys a range of ideas and ideals which are far more convincing than the notion of 'Euro-Mediterranean policy'. Given that bilateral relations remain one of the great modes of action in France's diplomacy in the Mediterranean, the way France's chooses to define and characterize her policy in the region is not to be underestimated. Indeed, when we look at the body of France's Mediterranean policy in its entirety – which at times promotes European interest, and at others, national interest – we can put forward two thoughts. The first is that the title given to this policy is also a question of how it is perceived and received by France's partners: the Mediterranean reference 'sells better' to European partners, who are alarmed by 'Arab policy'; whereas for Arabs the equation has long been: Arab policy = pro-Arab policy = Gaullist policy. A second observation is that the bilateral Franco-Arab logic clearly tends to prevail over the multilateral Euro-Mediterranean logic. On the one hand, it claims to be more ambitious and daring than the latter, and does not hesitate to upset the inevitably moderate nature of European consensus. On the other hand, the important moments of French *rapprochement* with the south are largely of a bilateral nature (with Algeria in 2003, for example). In a general way,

18 The 'Europeans' absence and lack of involvement' was deplored and was included in the final declaration which mentions the European Commission's 'unjustified absence'. For further details, see C. Lager, 'Euro-Arab Dialogue Forum, a new instrument to reinforce Euro-Arab relations' CERMAM (November 2005), http://www.cermam.org/en/logs/zoom/1st_euroarab_dialogue_forum_a/ (accessed 13 January 2010).

France maintains strong bilateral relations with each of the states on the other side of the Mediterranean and these relations tend to prevail over multilateral approaches.

All things considered, therefore, France's Mediterranean policy uses European objectives but internalizes them very little, with the exception of what happened in 2003 during the confrontation between the Franco-German 'old Europe' and George W. Bush's policy in the Middle East. But this episode is atypical and Germany's support for what might have appeared as France's 'Arab policy' was, in the end, a mere flash in the pan. What is rather more interesting is that the 2003 confrontation reveals France's *modus operandi* regarding her action in the Mediterranean and Arab world: to seize all opportunities to promote publicly France's position on the international scene. Alongside unilateralism, the exhibitionist style which characterized Nicolas Sarkozy's diplomacy, and in particular the way he initiated and presented his project of a Mediterranean Union should not mask the fact that France's new policy in the Mediterranean has, in essence, moved away from its Gaullist heritage.

The Difficulties of Diplomatic Ambitions

Nicolas Sarkozy's Mediterranean initiative intended to restructure the Euro-Mediterranean partnership whilst at the same time promoting France's Mediterranean policy. But at the end of his first term in office, all emerge from it very weakened. Far from attaining its objectives, the initial project of a Mediterranean union, or more precisely a 'Union *of* the Mediterranean', has even contributed, in the course of the difficulties that arose from its implementation, to the further weakening of the Barcelona Process. This result is quite paradoxical; it would be quite legitimate to describe it today as a fiasco. What factors led to such a failure?

When Nicolas Sarkozy outlined his project for a Mediterranean union in February 2007 in Toulon, he was not really breaking with the initiatives which had tried – in France, notably – to prevent the Barcelona Process from gradually reaching a position of stalemate following the failure of the summit held in November 2005 on the tenth anniversary of the Barcelona

Declaration. Progressively, the idea that this process could be reformed from within seemed less and less credible: this led to different means of overcoming the malfunctioning of the Euro-Mediterranean partnership being sought out. It was already from this perspective that the European countries in the western Mediterranean had, from 2001 onwards, revived the '5+5 dialogue' by highlighting the ways in which the economies and cultures particular to this region – which had been spared the conflicts in the Middle East – complemented each other.

In France, attempts to respond to the failures of Barcelona followed one another at a rapid rate during the last year of Jacques Chirac's presidency. Within the space of a few months in 2006, French officials relaunched the Euro-Arab dialogue, celebrated the Franco-Maghrebi partnership and promoted Mediterranean cultural exchange within the 'Alliance des civilisations' framework. The desire to reform the partnership was therefore clearly on display, even if the coherence and efficiency of these actions were uncertain.

Sarkozy's contribution is to have placed the Mediterranean (and Euro-Arab) question firmly back in the centre of regional and European debate. His speech in Toulon in February 2007 deals wholly with this. The argumentation given by Henri Guaino, the president's special adviser, often described as a *gaulliste social*[19] was persuasive in its long introduction which borrowed from Braudel, Valéry, Camus, Morin. Reflecting that 'Europe's future will be played out in the Mediterranean', he developed the vision of a Mediterranean union to be built 'using the European Union as a model'. Nevertheless, one soon realizes that a principal reason for creating this new group is the need to give Turkey – whose European membership Sarkozy objects to – an important, compensating place in the regional equilibrium. In the same way, the generosity of the initial proposal was seriously weakened by the project's security objectives: above all, it was a question

---

19   Henri Guaino, a declared Euro-sceptic once close to Philippe Seguin and Charles Pasqua, has largely inspired Sarkozy's fateful 'Discours de Dakar' (26 July 2007) where he claimed that 'the tragedy of Africa is that Africans have not fully entered into history'.

of reinforcing – with the help of neighbouring countries – control over the influx of immigrants.

Beyond its ambiguities, the aim of the Mediterranean union is to install on Europe's southern doorstep a sub-regionalism which is benevolent with regard to the European project without having the extent of its competence or competing with it. On an institutional level, it is obvious that the concept of Mediterranean union takes its inspiration from two experiences of organizing relations between the north and the south of the Mediterranean: the '5+5 dialogue' and the Mediterranean Forum. The Union of the Mediterranean could have contented itself with increasing the extent of co-operation between neighbours within the framework of a '6+6'. This would have been readily accepted by Europe and the countries concerned if the initiative had been undertaken in a less ostentatious, cavalier way with regard to its European partners.

On the contrary, on the very night of his electoral victory the new president chose to launch a solemn appeal to France's partners in the southern Mediterranean in order to propose his project of Union *of* the Mediterranean to them, which he introduced as an innovative vision of how to organize the Mediterranean space. From July he went to the Maghreb to defend the project but his trip turned out to be rather disappointing. As far as Europe was concerned, Italy and Spain showed polite interest but it did not take long for other European partners such as Germany and the UK to express their irritation. The increase in the strength of this reluctance explains the successive correctives made to the project for Mediterranean union in the speech the president gave to France's ambassadors on 27 August, then in his speech in Tangiers in November, and finally during interviews in Rome at the end of December with his Italian and Spanish counterparts, before the ultimate Franco-German compromise which was endorsed in March 2008 by the European Council.[20]

20   On the twists and turns and revisions of the concept of Mediterranean union, see J.-R. Henry, 'Initiative française en Méditerranée: un retour à la case départ?', *Med.2008: L'année 2007 dans l'espace euro-méditerranéen* (Barcelona: IEMed & Cidob, 2008), 38–45, and J.-R. Henry and I. Schäfer (eds), *Mediterranean Policies from Above and from Below* (Baden-Baden: Nomos, 2009).

It was on 20 December 2007 that the French, Italian and Spanish leaders, meeting in Rome, launched an 'appeal for the Mediterranean', claiming that 'the Union *for* the Mediterranean will have the task of reuniting Europe and Africa around the countries bordering the Mediterranean'.[21] The French were hoping that the UfM would maintain its own political identity, quite autonomous from existing European mechanisms. Following a final bout of Franco-German haggling, a general consensus during the European Council of 13–14 March 2008 emerged and the union acquired officially its title. The Union *for* the Mediterranean (it is not clear what *for* means in this case) was immediately emptied of its (French) substance and reduced to the rank of auxiliary to the 'Barcelona Process'. Henceforth the idea of a Union *of* the Mediterranean lost most of its relevance and with it a good part of the French initiative. The only political element which survives from the project is the joint presidency of the new union, awarded to two countries bordering the Mediterranean on the north and south sides, and the creation of a permanent secretariat.

But the 'normalization' of the Mediterranean union project pushed Paris to question its orientation. As France was presiding over the European Union from July to December 2008, she used the first, solemn meeting of UfM on 13 July 2008 in Paris, on the eve of the Bastille Day, to assert her diplomatic position. Inviting the Syrian President Bashar al-Assad to the national celebrations was interpreted as an unequivocal sign of a desire for autonomy on the part of French diplomacy in relation to European expectations. This determinedness was confirmed a few months later in Marseille during the Euro-Mediterranean conference for foreign ministers on 3–4 November 2008.[22] There, French diplomacy tried very hard to

21 The Union for the Mediterranean encompasses the twenty-seven EU member states, the European Commission and sixteen Mediterranean countries (Albania, Algeria, Bosnia and Herzegovina, Croatia, Egypt, Israel, Jordan, Lebanon, Mauritania, Monaco, Montenegro, Morocco, the Palestinian Authority, Tunisia, Syria, Turkey). The Arab League also participates in all meetings at all levels of the UfM.
22 During this Euro-Mediterranean conference in Marseille, France did not manage to convince her European partners that the UfM should have its headquarters in Marseille. Tellingly, they are located in Palacio de Pedralbes in Barcelona,

bring both Arabs and Israelis into the UfM, as well as to reintroduce the Arab League into the system. The UfM's main failure was to prevent the renewed violence between Gaza and Israel and very quickly these tensions paralysed the newborn institution: the fact that the Palestinian authority and Israel are both members of the UfM did not make any difference. It is unthinkable that the French president had not been kept informed in December 2008 in Paris by the Israeli foreign minister of the imminence of the Israeli offensive on Gaza. Without doubt, it is with good reason that Sarkozy has been suspected by the Arab leaders of duplicity in favour of Israel in this matter, who saw it as confirmation of a major change in direction in France's Mediterranean policy.

On a national level as well as on the Euro-Mediterranean level, an assessment of France's Mediterranean activism since 2007 thus proves disappointing. It is true that the way in which the operation was undertaken justifies the criticisms. Not enough coherence, too many improvisations and too much media hype have devalued the French initiative, damaged the country's external image and aroused in France reservation from traditional diplomatic quarters which have been largely kept out of this initiative by the Elysée. It is debatable whether the centralization of the Mediterranean policy at the Elysée rather than the Ministry for Foreign Affairs has created a sense of cohesion: what it certainly did was to make the President to bear the mistakes and blunders of his own policy.

Indeed, French policy makers, including the president, seem to have been held captive by a hastily thought-out formula, marred by too many contradictions. The first contradiction was the fact that they pitted Mediterranean space against European space, whereas they should have thought about how the two intersected in order to propose a more coherent policy; on the contrary, it was surprising to hear how France claimed it belonged both to the European and Mediterranean unions whilst at the same time refusing this to Turkey. Another contradiction concerned the functional linkage between the European Union and the UfM, and

---

signalling clearly that the UfM is one part of the Euro-Mediterranean partnership (EUROMED).

the distribution of their responsibilities. Those who promoted the project have, for a long time, been reluctant to involve the institutions and other members of the EU during the elaboration and development of the new project. Likewise, the initial project was ill-conceived because there could never be any symmetry between the two unions, the European and the Mediterranean unions, but only complementarity and subsidiarity: indeed the EU countries bordering the Mediterranean could not delegate to a new Mediterranean union the competences that had already been given to the European Union. Finally, the most important contradiction remains the discrepancy between the project's idealism and the modesty of its actual proposals. Rhetoric about the Mediterranean has not only been instrumentalized but submerged by security objectives. And if the European model has been much sought out in order to invent a Mediterranean union, this has come about by forgetting what gives it its substance and purpose – its human dimension: there has never been a question of following the European Union's example and creating a common human space within the Mediterranean union.

Thus, a more serious criticism that can be levelled at this project is that it has served – whether unwittingly or not – as a smoke screen and as a cover for fundamental changes made to France's policy in the Mediterranean: the different institutional variations of the Mediterranean union and the ambiguous or vague discourses which justify its existence did not conceal a drift towards Atlanticism, which Hubert Védrine had warned against in his report to the President of the Republic.[23]

In the first months of Sarkozy's presidency, different declarations were made which confirmed a general shift towards Atlanticism and generally dented a several-decades-old diplomatic heritage in the Mediterranean. France's new foreign policy seemed more concerned with improving its status as the United States' preferred ally in the region and, therefore, with supporting George Bush's diplomacy. Thus, France's traditional position in favour of a global 'denuclearization' of the Middle East was replaced

---

23 H. Védrine, *Rapport pour le Président de la République sur la France et la mondialisation* (Paris: Fayard, 2007), 63.

with a strong criticism regarding the nuclear risk posed by Iran, with no reference to Israel's possession of nuclear weapons, except to reiterate the right and the need for Israel to ensure its security through nuclear deterrence. In this context, to deplore officially the colonization of Palestinian territories by Israel appears as a light rebuke.

The weakening of American diplomacy in a period of presidential elections did, however, lead French policy to withdraw from George Bush's unpopular international policy and go it alone. At the same time, the French presidency of the European Union (2008) was a great opportunity for France to make its foreign actions more visible and prestigious. This concurrence of favourable factors meant that the Paris Summit on 13 July 2008 and the Euro-Mediterranean conference in Marseille in November both ran smoothly, before the bombing of Gaza by Israel undermined their success.

Beyond these particular set of events, one worrying element of France's new policy in the Mediterranean region nevertheless remained the question of its ideological re-orientation. Nicolas Sarkozy's insistent reference in August 2007 to the 'confrontation between Islam and the West', as the 'major challenge' to current international relations – using the purest Huntingtonian vocabulary – has certainly introduced some new ideas which one would be hard pressed to find in the discourse of previous French presidents. Since de Gaulle, French presidents have always been careful to avoid presenting Mediterranean conflicts in crude cultural and political terms and sought to maintain an alternative position to the one usually put forward by the 'West' or the US and as such benefitted from a rather positive response from southern Mediterranean public opinion and states. In this respect, the contrast between France's new policy in the Mediterranean and that which was defended in 2003 seems huge.[24]

---

24 Perhaps one should not attach such huge importance to these variations in discourse, which are partly due to new methods of making French foreign policy. Compared to Gaullist diplomacy (in which the president defined grand guiding principles, that brilliant diplomats then made more specific and put into place), the vision of the world which inspires Sarkozy seems very dependent on that of his advisers, particularly Henri Guaino, whose departure has been announced several times but

It is true that the response of other European countries to the challenges posed by the region is barely better: too much economic and cultural – and not enough human – realism. Behind the formal quarrels, personal incompatibilities and power struggles between European leaders, there are also many points of convergence: the same Atlanticism, the same support for Israel despite some cautious reservation, the same reluctance about opening up Europe to Turkey and the Muslim world, and, above all, the same difficulty in accepting the human proximity with the south – that is, in thinking and encouraging the development of partnerships between civil societies on each side of the Mediterranean.

The Crisis of Europe's Mediterranean Policy

The way in which the French initiative manifests itself is only the tip of the iceberg in the whole matter. In reality, the whole of Europe is responsible for the stalemate in Euro-Mediterranean relations, and there are several reasons for this.

On the one hand, the way in which the European system currently works is not favourable for an imaginative and substantial overhaul of Euro-Mediterranean relations. The absence of a common, clear foreign policy as well as confusion in the decision-making process ultimately favours the fragmentation and diversification of Mediterranean policies developed by individual member states, or even encourages competition between

---

who was confirmed as a prominent adviser by being named head of the Mission interministérielle pour la Méditerranée in November 2008. In reality, it seems that President Sarkozy attaches less importance to any doctrinal referents (which he changes without any hang-ups) than he does to acting in a tactical and opportunist manner as he seems more anxious to impose his own image – more than France's – on the international scene.

them.[25] The return to national strategies – or even nationalist reflexes – amongst the big European actors is a major trend which contrasts the way in which some small countries continue to defend policies they have in common. The Finnish presidency had thus been able to reactivate in 2006 Euro-Mediterranean co-operation, which had been mishandled by the failure of the tenth anniversary of the Barcelona Conference. At the same time, following a deficient Czech presidency, a great deal had been expected from the Swedish presidency. But it is not realistic to count on the rare member states that are still passionate about and wholly committed to the European project to compensate by their effort the lack of community resilience.

In the second instance, we have mentioned above the obvious role played by the Gaza crisis as an immediate cause for the paralysis which gripped the UfM from 2008 onward, that is from its inception. The impossibility of carrying out the decisions adopted by the Euro-Mediterranean conference in Marseille in November 2008 is largely due to the fact that the Europeans *in their entirety* – and not just France – did not really try to dissuade the Israelis from launching their offensive on Gaza, in contrast to the commitment that had just been made at the conference not to resort to force as a means of resolving conflict. Their position in this matter – as well as their inability to adopt a clear position regarding the conflict in the Middle East – had a negative, long-lasting effect on the relationship of trust which had begun to take shape in Marseille.

In the third instance – and most importantly – many European countries and leaders, including the German chancellor, shared Sarkozy's reluctance with regard to Turkey's accession to the EU: it is 'not part of Europe'. This reluctance more or less explicitly expresses a fear and rejection of the Muslim world, of a feeling that Islam would be incompatible with the European project or, at least, does not really belong in the 'European

---

25 J.-R. Henry and C. Groc (eds), *Politiques méditerranéennes entre espace étatique et espace civil* (Paris: Karthala, 2000).

home'.[26] As with the influx of immigrants, this fear is attributed to public opinion which, however, most politicians shy away from enlightening on this matter. We are far from the moral authority which post-war European leaders had to use in addition to their own opinions when restructuring a shattered Europe on the basis of values that were humanist and universalist. Today the European project seems to focus essentially on the search for a European identity; an inward-looking obsession open to the wildest interpretations. Trapped in its populist temptations about identity, Europe struggles to think through its relationship with the south and with the Muslim world beyond using it – in a passive, implicit way – as a foil to European identity, a 'Muslim frontier' in and for Europe. Over the last few years the stigmatization of Islam has repeatedly been used in response to the question 'What is Europe?' This could be seen with the recurrent question about Turkey's accession, with the Muhammad caricatures in Denmark and the minaret caricatures in Switzerland, or even during the French debate on national identity organized by the minister of immigration, integration and national identity. Issues related to religious dress code, such as the use of the Muslim veil or the burqa, have spread across Europe, and Islamophobic rhetoric is often found in conservative party political broadcasts during the national or European elections. Modern-day Europe's propensity to make out of Islam its external (and internal) 'other' is relatively recent: the 'little Europe' of the 1950s, which encompassed the French administrative departments in Algeria, did not make a natural or cultural frontier out of the Mediterranean. This tendency contrasts above all with the more open, generous vision of relations with the Muslim world which was elaborated upon by President Obama in his speech of 4 June 2009 at the University of Cairo.

---

26 Indirectly, this is what can be deduced from the speech given by the pope in Prague on 26 September 2009: 'Europe is more than a continent. It is a home!', a 'spiritual homeland', he says a few lines later, unwittingly misusing the vocabulary of Gabriel Audisio from the 1930s on a 'Mediterranean homeland' – open to all. Without making explicit reference to Islam, the pope stresses 'Christianity's irreplaceable role' – notably its distinction between the political and religious domains – for the 'promotion of a fundamental consensus on ethics' in the European home.

But relations with the Mediterranean are much more crucial for Europe than they are for the United States – not only for reasons of strategy, energy resources and economic growth but because the nature and the future of the European project partly depends on how the human element of the Mediterranean space is managed. This *human dimension to Euro-Mediterranean relations* – repressed for a long time – has reappeared over the last few years as a major issue about how to build new relations with civil societies on the other side of the Mediterranean: the need to manage better the *human and cultural proximity* with the south, taking into account the weight of the past as much as the opportunities that the future may offer, is gradually being rediscovered.

For those who reject the idea that there is a 'Mediterranean fracture' based on a 'East–West confrontation', the word 'proximity' defines fairly well the links existing between Europe and southern Mediterranean societies: an obvious geographic and historical proximity, a certain economic and environmental proximity but also a not-inconsiderable human and cultural proximity. To talk of a 'Mediterranean melting pot' may be an hackneyed expression but it is a fact that the last two centuries have shuffled around the populations of the Mediterranean region at least as much as they did with those in Europe. The overlapping of human and cultural relations which resulted from this shuffling has created a network of solidarities and shared cultural practices beyond and across political borders. For example in terms of matrimonial practices, with equal populations, there are far more mixed Franco-Maghrebi marriages than Franco-German ones, and 30 per cent of Marseille's population – according to its mayor's office – is made up by Muslims, whose vast majority originate from the Maghreb.

Until now the European project has not managed this human reality very well. Since 1986 and the Schengen Agreement, it has striven to reinforce Europe's human frontier in the Mediterranean with the development of more stringent immigration policies – without having any long-term vision or evaluating the effects of such policies. The Barcelona Process did not seek to question fundamentally this general trend, in spite of its concerns about reinforcing partnerships with countries on the other side of the Mediterranean: in the end it confirms the divorce between economic space and human space whilst at the same time relying on cultural dialogue between civil societies to lessen its effects.

This policy of separate human development in the Mediterranean – which can be described as a *mild apartheid* – has not resolved the question about the human dimension to trans-Mediterranean relations, but, on the contrary, has had some unexpected and tragic effects, such as the development of illegal migration and its thousands of victims.[27] Another, unexpected consequence of European policy is the sharp increase in number of individuals who are in the position to claim – legally – a dual nationality, thus bypassing the borders that Europe has sought to reinforced between the northern and southern shores of the Mediterranean. Cases of dual nationality are no longer isolated but are among the millions today; they testify to the emergence of a genuine *trans-Mediterranean individualism*, which reinforces the establishment in Europe of Mediterranean societies that are at an *interface* with each other.

Alongside individual initiatives, organized civil societies with an interest in the Mediterranean have gained a great deal of consistency and credibility since the Barcelona Process and have acquired a certain degree of autonomy from the objectives defined at intergovernmental level. Humanist (such as the human rights leagues), religious (San Egidio in Italy, CIMADE in France) or political organizations (social forums, ATTAC) have taken upon themselves the task of criticizing the human consequences of the closing down of European borders. Today, a genuine public awareness of such issues is emerging, helped by the media.[28] This awareness helps European citizens to understand that the European project can have dramatic consequences, and that illegal immigrants are also human beings who aspire to move freely within the regional space. The civil Euro-Mediterranean Forum in Marrakech in November 2006 endorsed this awareness but it took a long time for it to filter through to the political leaders and policy makers.

---

27 Taking a low hypothesis of 2,000 people who drown or disappear each year amongst those who attempt to reach the European coast, that makes nearly 40,000 victims in twenty years created by the closing of the human frontiers in Europe.
28 For example, the films *Eden à l'Ouest* by Costa-Gavras and *Welcome* by Jerôme Lindon, which came out in France in 2009.

Nevertheless, during the last European elections some lists – including those of the Green parties – denounced the negative human effects of 'Fortress Europe'.

Despite a greater involvement – or at least a greater visibility and a greater activism – of non-governmental organizations (NGOs), these only played a minor role in the diplomatic debates and negotiations which resulted in the creation of the UfM. Even more than in Barcelona in 1995, NGOs were marginalized in the process of setting up the UfM. On the issue of human mobility, the decisions taken by the UfM founding summit in Paris in July 2008 and developed further by the UfM conference in Marseille the following November expressed a strong consensus on security issues between the northern and southern states: but they did not take into account the demands or expectations formulated by the civil societies on each side of the Mediterranean. The Migration Pact which was adopted under the French presidency of the European Union in 2008 demonstrated how the EU linked immigration and security issues and tightened further migration flows from the southern shore of the Mediterranean into the EU. To oppose a 'chosen' immigration to one which would be inflicted is a short-sighted strategy and it has become criticized by an increased number of economists and geographers as well as by European Community leaders.[29]

If the hiatus between the inter-state agreement which was cemented in Paris and Marseille, and the unfulfilled expectations of civil society characterizes the new, institutional, Euro-Mediterranean relationships which have resulted from the UfM, the tension between the objectives of European States and the aspirations of civil societies may also – paradoxically – constitute a source of dynamism. Today, issues related to trans-Mediterranean human relations only seem able to change as a result of demands, power struggles and conflicts. Today, NGOs behave less and less

---

29   P. Fargues and H. Le Bras, 'Migrants et migrations dans le bassin de la Méditerranée', *Note IPEMED*, 1 (September 2009), http://www.ipemed.coop (accessed 13 January 2010); see also Y. Courbage, 'Vers une convergence démographique entre les rives Nord et Sud', *Questions internationales*, 36 (March–April 2009).

like passive auxiliaries to the public powers, and expect to play an exacting role in monitoring respect for human rights. For example, CIMADE in France has been involved in a fierce dispute since 2008 with the Ministry for Immigration and National Identity about how detention centres for illegal immigrants are run. In the same period of time, alternative voices from very different sources – including Christian and Muslim religious groups – have been making themselves heard and gathering together to contest or denounce certain aspects of the Euro-Mediterranean approach.[30] Researchers themselves seem affected by this development: they are less tempted by expertise, often financed directly or indirectly by states, and more concerned, as a citizen would, about promoting an independent analysis of Euro-Mediterranean relations.

Of equal significance is the tension between governmental and non-governmental actors which can be seen within the Anna Lindh Foundation,[31] one of the rare examples of the Euro-Mediterranean system in which the linkage between the two types of actors is organized explicitly. Contrary to the recommendations of the *Rapport des Sages sur le dialogue des peuples et des cultures en Méditerranée* (2004), which recommended turning it into an independent body, the foundation was created under the bizarre and comical form of an 'intergovernmental organization of civil society', which explains why the results of its first few years of operation were not particularly impressive. The criticism provoked a change in policy, structure and team which resulted in the adoption, in November 2008, of a three-year programme in which 'rank and file' demands about human mobility are relayed. But, in turn, this ambitious policy clashed with the states' political

---

30   For example, a conference organized by SODEPAU in Barcelona in November 2007, which intended to join the tradition of the alternative Euro-Mediterranean forum which was held in Barcelona in November 1995.
31   To quote the Foundation's own website: 'Our purpose is to bring people together from across the Mediterranean to improve mutual respect between cultures. Since its launch in 2005, the Anna Lindh Foundation has launched and supported action across fields impacting on mutual perceptions among people of different cultures and beliefs, as well as developing a region-wide Network of over 3000 civil society organisations.' See http://www.euromedalex.org/ (accessed 1 January 2011).

reluctance and financial unwillingness to support such initiatives. Despite the uncertainty which hangs over the foundation's future, this example does show that within a context that is globally pessimistic vis-à-vis the future of Euro-Mediterranean relations, NGOs have a capacity to become a critical, but essential actor. That is why some political actors such as Risto Weltheim, Finnish ambassador to the Euro-Mediterranean process, believe that any serious relaunch of the Euro-mediterranean partnership must rely primarily on a greater and better involvement of NGOs.[32]

One other positive factor – albeit a very different one – which deserves to be pointed out is that over the last two years the various debates have provided the opportunity, despite disappointed hopes, for a big 'brainstorming' session on the Euro-Mediterranean future. Before opting for the ambiguous and slightly meaningless concept of 'Union for the Mediterranean', other avenues were explored in thinking about the Euro-Mediterranean future. The most interesting title was that of 'Euro-Mediterranean Union', which was proposed at the end of 2007 by the Spanish Foreign Affairs Minister, Angel Moratinos, and which had already been suggested previously by other politicians (Dominique Strauss-Kahn, for example, in his *Cinquante propositions pour l'Europe* from 2004). The formula defended by Angel Moratinos at the end of 2007 possibly came close to a conjectural choice, but its advantage was that it was polysemous. It could, in effect, take on a meaning that could be both maximalist (the enlarging of the European Union to include partners from the south) and minimalist (the creation of a structure which encompassed the European Union and the countries on the other side of the Mediterranean). In both cases, it left the door open for a better management of the human European space and above all the development of a common destiny, whereas the Union for the Mediterranean can by no means act as a mobilizing common Utopia for people who are unable to move freely within the Mediterranean space. Beyond the name games, the debates over the last three years have also been the occasion for some politicians to take a position in favour of a Europe which is more open towards the south with regards to the movement of

---

32   Memo from 6 May 2009 which was distributed by the Euromed authorities.

people. In France, for example, politicians from different sides are now declaring their concern regarding the perverse effects of closing the human frontiers in Europe.[33] It is true that pleading for the return to freedom of movement and for people's mobility within the Mediterranean requires political courage. 'It is an issue which paralyses politicians, even when they are aware of its importance', admitted one diplomat who is involved in the French initiative and who is well aware of how the issue has been instrumentalized for populist gain.

In the short or medium term, it is unrealistic to think that a conflict-free zone can be created in and across the Mediterranean without accepting the human proximity and the continuity of human space between the societies on its shores; this is why one should take into consideration the methods which have been successful in harmonizing the human space in post-war Europe.

The first of these methods was to link – as was done from the Schuman Declaration of 1950 onwards – an ambitious, credible Utopia for the populations with the policies of what was possible. This is still only a distant possibility as far as the Mediterranean is concerned. Yet the progressive enlargement of the Council of Europe to the Mediterranean region would be a measure within easy reach and would have a strong symbolic resonance. Another way would involve cultivating the human dimension of inter-Mediterranean relations by giving priority to policies aimed at young people (a Euro-Mediterranean office for youth, a Mediterranean ERASMUS, a harmonization of education systems). All this would, of course, require a considerable, significant return to the free movement of populations. Similarly, further co-operation (including economic cooperation) between the regions bordering the Mediterranean – particularly important for southern European countries – is not compatible with the thinking behind Schengen.

---

33 This was the collective position of the Greens during the European elections, but it is also the personal position defended by Jean-Luc Benahmias, vice-president of MODEM, and of Michel Rocard, former socialist prime minister.

In conclusion, it is certain that the human dimension to Euro-Mediterranean relations seriously questions the European project itself. The future of the European project depends on how human relations with the south are managed: either one withdraws to a pseudo-European identity limited by the Mediterranean: Europe, as an exclusive club, but a club under the permanent threats of southern barbarians and invading Muslim hordes; or one returns to the gamble of a *Europe sans rivages*[34] – that is, to that of a virtuous circle which wholly accepts its human and cultural proximity with the other shores of the Mediterranean. The issue would therefore not be about the Europeanization of the Mediterranean but, on the contrary, about the re-Mediterraneanization of Europe; about reconciling Europe with its Mediterranean roots.

Aiming for such an objective would not be an easy task: the French initiative, despite its shortcomings, highlighted clearly a practical issue: how does one take on the diversity of interests in the Mediterranean and out of different tropisms, different assets and objectives as they exist between European states and societies, how does one build a common Mediterranean project? It is obvious, for example, that the political relations between Spain or France – as well as the cultural and human ones – and a country such as Morocco are much closer than the relations they have with many of their European partners. How can the diversity of relations across the Mediterranean becomes the basis of for a common European policy? It is a question which the European Union has never seriously asked, apart from within the small context of 'reinforced co-operation'. The disappointing results over the last three years show that managing Europe's relations with its south cannot be exclusively based either on a hypothetical, common foreign policy or on competing, uncoordinated national initiatives. Harmonization of the levels of Europe's Mediterranean policy is something desirable; but is it achievable without being hampered by the shortcomings and lack of serious discussion about the European project?

---

34 'Europe without shores', from the title of a book by F. Perroux, *L'Europe sans rivages* (Paris: Presses Universitaires de France, 1954).

Bibliography

Arkoun, M., et al. (eds), *Histoire de l'islam et des musulmans en France: du Moyen-Age à nos jours* (Paris: Albin Michel, 2006).

Balta, P., and Roulleau, C., *La politique arabe de la France: de de Gaulle à Pompidou* (Paris: Sindbad, 1973).

Carpentier, F., Lebrun, F., and Bennassar B. (eds), *Histoire de la Méditerranée* (Paris: Seuil, 1980).

Chérigui, H., *La politique méditerranéenne de la France, entre diplomatie collective et leadership* (Paris: L'Harmattan, 1997).

Chevalier, J.-M., and Pastré, O., '5+5 = 32: ambition d'une association renforcée feuille de route pour une Union Méditerranéenne', *Les Cahiers du Cercle des Economistes*, 4 (Paris: Editions Perrin, 2003).

Chevalier, M., *Le système méditerranéen* (Paris: Editions Mille et une nuits, 1986 [1832]).

Chevènement, J.-P., *Une certain idée de la République m'amène à ...* (Paris: Albin Michel, 1992).

Courbage, Y., 'Vers une convergence démographique entre les rives Nord et Sud', *Questions internationales*, 36 (March–April 2009).

Fargues, P., and Le Bras, H., 'Migrants et migrations dans le bassin de la Méditerranée', *Note IPEMED*, 1 (September 2009), http://www.ipemed.coop (accessed 13 January 2010).

Groc, G., *La presse française de Turquie* (Istanbul: Isis, 1985).

Henry, J.R., 'Initiative française en Méditerranée: un retour à la case départ?' *Med.2008: L'année 2007 dans l'espace euro-méditerranéen* (Barcelona: IEMed & Cidob, 2008), 38–45.

Henry, J.R., 'L'Europe du Sud et le Maghreb: le rêve andalou à l'épreuve', in J. Bourrinet (ed.), *La Méditerranée, espace de coopération, en l'honneur de Maurice Flory* (Paris: Economica, 1994).

Henry, J.R., and Groc, C. (eds), *Politiques méditerranéennes entre espace étatique et espace civil* (Paris: Karthala, 2000).

Henry, J.-R., and Schäfe, I. (eds), *Mediterranean Policies from Above and from Below* (Baden-Baden: Nomos, 2009).

Lager, C., 'Euro-Arab Dialogue Forum, a New Instrument to Reinforce Euro-Arab relations' CERMAM (November 2005), http://www.cermam.org/en/logs/zoom/1st_euroarab_dialogue_forum_a/ (accessed 13 January 2010).

Levallois, L., 'Preface', in I. Urbain, *L'Algérie française: indigènes et immigrants* (Paris: Séguier, 2002 [1862]).

Perroux, F., *L'Europe sans rivages* (Paris: Presses Universitaires de France, 1954).

La politique de coopération avec les pays en voie de développement, Rapport de la Commission d'études, set up by the decree of 12 March 1963 (Paris: Ministère chargé de la réforme administrative, July 1963).

Sarkozy, N., 'Message à la communauté musulmane de la Seine-Saint-Denis' (Union des Associations Musulmanes de la Seine-Saint-Denis, 12 December 2008, http://www.uam93.com/ (accessed 12 February 2010).

Sebbar, L., *La Seine était Rouge* (Arles: Actes Sud, 2009).

Vauzelle, M., 'Ouverture du Forum des autorités locales et régionales de la Méditerranée' Forum des autorités locales et régionales de la Méditerranée (Marseille, 2008), http://www.commed-cglu.org/IMG/pdf/Actes_Forum_FR.pdf (accessed 13 January 2010).

Vedrine, H., *Rapport pour le Président de la République sur la France et la mondialisation* (Paris: Fayard, 2007).

Visier, C., *L'Etat et la coopération, la fin d'un monopole: l'action culturelle française au Maghreb* (Paris, L'Harmattan, 2004).

IVANO BRUNO

French Foreign Policy in the Mediterranean: Europeanization between Collective and National Approaches

The events that shook the Arab world in early 2011 have raised profound questions about the EU and its strategy in the region. For some, the Arab Spring has proved that EU initiatives in the Mediterranean were inadequately formulated and should be deemed a failure. Despite the criticism, the EU seems to have taken a strong stance towards Libya and Syria and has responded to the needs of Egypt and Tunisia with added financial allocations. As usual, reaching a coherent EU position has taken time and does not preclude further member states' power struggles, negotiations and compromises. The question that remains unanswered however is what the EU approach towards the 'new' Arab partners will be and, more importantly, to what extent member states will play a role in shaping and influencing EU Mediterranean policy in order to reflect their preferences and interests. This chapter will argue that in line with its traditional foreign policy orientation, France has attempted to increase its regional influence in the Mediterranean by articulating its preferences at the supranational level. In this context, this chapter will attempt to highlight the role and relevance of France and other actors key to the development of the Euro-Mediterranean Partnership and, specifically, the Union for the Mediterranean.

The Union for the Mediterranean (UfM) initially launched by President Sarkozy is the latest addition to the EU's Mediterranean policy framework. At the Marseille meeting in November 2008, EU member states endorsed the UfM as the EU's new multilateral dimension in the Mediterranean, thus 'upgrading' the existing Barcelona Process. The initiative was initially perceived as reflecting French priorities and interests in the region and generated negative reactions from leaders across the EU.[1] So how does a French initiative become Europeanized to an EU policy and more importantly, why? The case of the UfM provides a fascinating example demonstrating the complexities that lie at the heart of the process of policy formulation in the EU and the dynamics which underpin member states' interests and preferences and their projections at the EU level.

This chapter stems from an interest in the concept of Europeanization and the mechanisms that inform it. Focusing on the Mediterranean and starting from an EU perspective, this chapter will examine the processes of Europeanization of French foreign policy and the impact of French preferences and interests on EU initiatives in the Mediterranean. The main arguments advanced in the chapter are that today, just as in the past, French foreign policy in the Mediterranean manifests a Gaullist orientation underpinned by principles of grandeur and exceptionalism. The way in which various French leaders have gone about achieving their goals, however, has been different at different times in history. In the past, France has been able to upload successfully and project a set of policy preferences in the Mediterranean and at the EU level, thus reflecting a bottom-up understanding of the process of Europeanization. In the case of the UfM, the chapter will argue that although the traditional orientation of France's foreign policy has not shifted, the means to attain its goals have. Importantly, the case of the UfM provides an interesting example of Europeanization, one not related to the process of projecting and uploading a set of policy preferences at the EU level but, rather, one that is related to the political costs of not doing so.

This chapter will firstly outline the EU multilateral policy framework in the Mediterranean and the factors that contributed to the construction of an EU framework in the region. Secondly, it will focus on French policy preferences and interests in the Mediterranean. A central aspect of this section will be to identify traditional French foreign policy characteristics and trace their articulation at the EU level in terms of Europeanization.

---

1   R. Balfour, 'The Transformation of the Union for the Mediterranean', *Mediterranean Politics*, 14/1 (2009), 100.

Finally, the chapter will focus on the UfM and evaluate French motives and EU reactions to the initiative. It will evaluate the degree to which the UfM can be explained through the process of Europeanization and evaluate to what extent France policy orientation has influenced the EU agenda in the Mediterranean. Overall, in this chapter I will contend that the process of Europeanization can tell us much about French policy interests and preferences and their projection in the Mediterranean even though Europeanization itself does not always take place according to a vertical process of uploading-downloading. Moreover, I will also argue that French foreign policy has manifested a continuity in terms of the principles and tenets of Gaullism and grandeur despite an apparent shift under President Sarkozy.

The EU in the Mediterranean

The development of a seemingly coherent EU agenda in the Mediterranean and the institutionalization of relations with Mediterranean and North African (MENA) partners can be traced to the early 1970s. External developments have been crucial in determining the pace and nature of EU regional engagement, characterized by an initial period of intensification resulting in the Global Mediterranean Policy (GMP) in 1972 and a second period in the mid-1990s culminating in the Barcelona Process. This section of the chapter will thus focus on the factors that resulted in the development of an EU agenda in the Mediterranean.

During the 1970s the precarious geo-political situation in the MENA region and Arab peninsula seemed to affect European societies in an unprecedented manner. The threat of terrorism and Arab nationalism emanating from the Mediterranean seemed to directly reach Europe through the tragic events that took place at the Munich Olympics in 1972 and the shock of the first OPEC oil crisis embargo of 1973. The uncertain political context and troubled relations with the European Community (EC) at the time provided the rationale for a more coherent engagement with the region through the Euro-Arab Dialogue and the launch of the GMP, an initiative originating from French efforts under President Pompidou and backed by Italy and Germany.[2] A number of European countries at the time adopted a pragmatic policy towards Israel in order to gain economic benefits from the Arab bloc, primarily access to oil. This was Pompidou's priority and as such he was able to guarantee French preferential access to Algerian oil and maintain the primacy of French national interests in the region. From a European perspective, the broad objectives of the GMP were to enhance trade relations and economic development with MENA partners.[3] Formulated along the lines of the first Lomé Convention signed in 1975, the provisions of the Co-operation Agreements under the GMP included the facilitation of trade and aid assistance. The agreements offered partner states in the Mediterranean preferential treatment and free access to EU markets for some industrial and agricultural exports, albeit with exceptions for the most sensitive sectors for EU member states.[4] In France in particular, the agricultural lobby proved very influential in limiting concessions on imports from the Mediterranean.[5]

The end of the Cold War and the changing international system saw a renewed European engagement in the Mediterranean. Aside from the obvious commercial interests, the increasing levels of illegal migration towards Europe in the 1980s coupled with the rising phenomena of Islamic fundamentalism and terrorism contributed to the perception of the

2 H.A. Jawad, *Euro-Arab Relations: A Study in Collective Diplomacy* (Reading: Ithaca Press, 1992), 23–4; see also D. Bechev and K. Nicolaidis, 'The Union for the Mediterranean: A Genuine Breakthrough or More of the Same?', *The International Spectator*, 43/3 (2008), 14; S. Stavridis and N. Waites, 'The European Union and Mediterranean Member States', in T. Couloumbis, S. Stavridis, T. Veremis and N. Waites (eds), *The Foreign Policies of European Union's Mediterranean States and Applicant Countries in the 1990s* (London: Macmillan, 1999), 29.
3 C. Bretherton and J. Vogler, *The European Union as a Global Actor* (London: Routledge, 2006), 155.
4 J. Licari, *Economic Reforms in Egypt in a Changing Global Economy* (Paris: OECD, 1997), 23.
5 Jawad, *Euro-Arab Relations*, 23.

Mediterranean as an area of security concerns.[6] Political Islam has always been a constitutive part of the political scene of some MENA countries, but the success of radical Islam witnessed in the run-up to the turn of the century was unprecedented and troublesome for Europe. The violent strain of political Islam in the region emerged primarily through the radicalization of a young and disillusioned generation. In its most violent incarnation, Islamic fundamentalism rested on the experience of many Jihadists who had returned from conflict with the Soviet Union in Afghanistan and who were inspired by the Islamic revolution in Iran. Examples of prominent groups include al Gamma't al-Islamiyya in Egypt and the Islamic Salvation Front (FIS) (now an outlawed Islamist party) in Algeria. The FIS leaders Abassi Madani and Ali Belhadj were able to preach to the young and disenfranchized sections of society. Following the arrests and dispersion of many FIS members in the aftermath of the cancelled 1992 elections in Algeria, the Armed Islamic Group (GIA) emerged as a splinter of the FIS and quickly became a terrorist group. This period marked the beginning of a serious challenge to secular Muslim societies in the region, a challenge that originated from within and was driven by Islamic militants inspired by the Revolution in Iran.[7]

Europe was not immune to these developments and countries such as France were directly affected. Events such as the hijacking of an Air France flight in 1994 and the wave of attacks on the Paris metro that took place a year later, both carried out by the GIA,[8] strengthened the perception amongst European leaders and public opinion that the Mediterranean was increasingly becoming a source of insecurity and instability for Europe.

---

6   Stavridis and Waites, 'The European Union and Mediterranean Member States', 29. See also E. Mortimer, 'Europe and the Mediterranean: The Security Dimension', in P. Ludlow (ed.), *Europe and the Mediterranean* (London: Brassey's, 1994), 111–20.
7   Stavridis and Waites, 'The European Union and Mediterranean Member States', 31.
8   N. Bouteldja, 'Who Really Bombed Paris?', *The Guardian* (8 September 2005).

More importantly, these violent events seemed to reinforce the perception of sharp divisions between Western and Islamic traditions and societies.[9]

By the mid-1990s and the end of the Cold War, a number of factors combined to make the possibility of an EU multilateral and all-encompassing framework in the Mediterranean possible. Internationally, the changing geo-political context and the wave of democratization that followed[10] seemed to encourage the establishment of multilateral frameworks aimed at collective responses, and signalled a period of policy re-orientation for many Mediterranean partners. Internally, the Treaty of the European Union (TEU) provided an explicit impetus for the development of a 'partnerships' dimension in the Mediterranean which, coupled with the changing international structure and almost unprecedented progress in the Middle Eastern Peace Process (MEPP) at the time, presented the EU with the opportunity to play a more prominent role as a regional actor.[11] Following the Barcelona Conference in 1995, the Euro-Mediterranean Partnership (EMP) was launched, also known as the Barcelona Process. It can be argued that both the opportunity for the EU to play a greater role in the region in terms of economic and political power, as well as the perceived threats originating from MENA countries, combined to produce a new multilateral framework for co-operation with Mediterranean partners aimed at achieving a common area of peace, stability and prosperity. The Euro-Mediterranean Partnership (EMP) provided a new and comprehensive dimension to relations with Mediterranean partners articulated under the 'three basket' logic. Firstly, it reinforced the most important objective which was to strengthen the EU approach in the region by developing a Free Trade Area (FTA) based on a neo-liberal economic orthodoxy. The EMP complemented EU trade and aid policies by providing increased funding and financial assistance for those partners who wished to embark on trade

---

9   Stavridis and Waites, 'The European Union and Mediterranean Member States', 31.
10 S. Huntington, 'America in the World', *The Hedgehog Review* (2002), 7–18.
11 Bretherton and Vogler, *The European Union as a Global Actor*, 155.

liberalization and economic modernization.[12] Secondly, the EMP aimed to institutionalize security co-operation and political dialogue with partner states with the objective of encouraging domestic political liberalization, the lack of which was understood in Europe to be an important factor in the rise of Islamic radicalism and militancy. Finally, the partnership aimed to increase dialogue at the cultural and societal level between the two shores of the Mediterranean. Therefore, under the EMP a process of economic modernization and reforms coupled with a degree of political reform aimed at institutional and legislative convergence with EU rules and standards was broadly envisaged.[13] In doing so, the EMP reinforced the element of conditionality in traditional aid terms and introduced, theoretically at least, an element of political conditionality by linking aid to the respect of human rights and democracy as stated in the provisions of the various Association Agreements (AA) with Mediterranean partners.[14] In this context, the EMP seemingly enabled the EU to address its security concerns and economic interests in the region and boost its perceived role as a normative and civilian power.[15]

In the context of EU policy towards the Mediterranean, the EMP approach could be described as a comprehensive one, linking socio-economic development to stability and, ultimately, to European security. As we will see, the role and preferences of member states is central to the understanding and explanation of the emergence of the policy. As Blunden

---

12 Stavridis and Waites, 'The European Union and Mediterranean Member States', 33; see also Bretherton and Vogler, *The European Union as a Global Actor*, 156.
13 The policy appears to follow a comprehensive approach to security based on spill-over and functionalist logic of exporting EU rules and standards: see M. Pace, 'Norm Shifting from EMP to ENP: The EU as a Norm Entrepreneur in the South?', *Cambridge Review of International Affairs*, 20/4 (2007), 660, and S. Biscop, 'The European Security Strategy and the Neighbourhood Policy: A New Starting Point for the Euro-Mediterranean Security Partnership?', paper presented at the EUSA Ninth Biannual International Conference (March–April 2005), http://aei.pitt.edu/2984/02/Paper_EUSA_ESS-EMP.doc (accessed 20 August 2010).
14 Bretherton and Vogler, *The European Union as a Global Actor*, 156.
15 Pace, 'Norm Shifting from EMP to ENP', 660. See also Bretherton and Vogler, *The European Union as a Global Actor*, 156.

has argued, the EMP is a good example of a policy shaped by member states' interests. In fact, the policy was initially shaped by French strategic perspectives, policies and interests, allowing relations with Mediterranean partners to be extended at the EU level. Blunden adds that the EMP is seen as having allowed French interests and influence, both strategic and commercial, to be expanded in the region.[16] This argument will be evaluated in the next section, which will outline traditional French foreign policy characteristics before focusing on the identification of French policy preferences and interests in the Mediterranean area and their articulation at the EU level, i.e. the eventual Europeanization of French preferences and interests. The aim is to evaluate the extent to which these characteristics can be identified in the Mediterranean context and whether they can be seen as having contributed to the formulation of an EU policy framework in the terms outlined above.

France and the Mediterranean: Reinforcing Traditional Foreign Policy Roles?

France has traditionally retained a Gaullist orientation in its foreign policy and a vision that reflected aspirations of French exceptionalism, grandeur and international influence.[17] Since the creation of the European Coal and Steel Community in 1951 these characteristics have increasingly shaped and provided the rationale for French policy towards Europe aimed at projecting French power and grandeur 'within' and 'through' Europe. These

---

16 M. Blunden, 'France', in I. Manners and R.G. Whitman (eds), *The Foreign Policies of European Union Member States* (Manchester: Manchester University Press, 2000), 40.

17 P.G. Cerny, *The Politics of Grandeur: Ideological Aspects of de Gaulle's Foreign Policy* (Cambridge: Cambridge University Press, 1980); R.Y. Wong, *The Europeanization of French Foreign Policy: France and the EU in East Asia*. (Basingstoke: Palgrave Macmillan, 2006).

foreign policy characteristics have translated over time into a dominant political discourse that has been seen as 'trapping' French policy-makers within specific role conceptions related to power, politics and identity.[18] The end of the Cold War meant that French foreign policy aspirations were expected to be re-oriented and address the shift from the East–West to a North–South axis of tensions;[19] a shift seemingly characterized by cultural as opposed to ideological cleavages.[20]

However, by the end of the Cold War, French officials seemed more concerned with the preservation of French values and universalism in the face of the emerging challenge brought on by globalization/Americanization and the spread of an Anglo-Saxon value system, ideology and culture.[21] Due to its historical and colonial legacy, the Mediterranean region and Arab countries in particular held a special importance for France in retaining its international and regional leadership in an increasingly competitive geopolitical context.[22] It could be argued that Europe has consistently been exploited in order to maintain French leadership in the Mediterranean and influence in Brussels. As Jawad notes, as early as the inception of the GMP, various French administrations realized that the European track was a complementary if not a better means to realize their Mediterranean ambitions and interests.[23] Therefore, and in the context of declining French international prominence, political elites in Paris become increasingly convinced that the EU was the vehicle through which French power and prestige could be restored. As Blunden so insightfully notes: 'Europe is

18 H. Drake, 'France, Europe and the Limits of Exceptionalism', in T. Chafer and E. Godin (eds), *The End of the French Exception? Decline and Revival of the 'French Model'* (Basingstoke: Palgrave Macmillan, 2010), 193.
19 L. Meyrede, 'France's Foreign Policy in the Mediterranean' in Couloumbis et al. (eds), *The Foreign Policies of the European Union's Mediterranean States and Applicant Countries in 1990s*, 45; Blunden, 'France', 40; A.G. Hargreaves, 'Immigration, Ethnicity and Political Orientations in France', in T. Chafer and B. Jenkins (eds), *France: From the Cold War to the New World Order* (London: Macmillan, 1996), 207.
20 Huntington, 'America in the World'.
21 Blunden, 'France', 22.
22 Meyrede, 'France's Foreign Policy in the Mediterranean', 42.
23 Jawad, *Euro-Arab Relations*, 22.

to France what the United States is to Britain, the optimum multiplier of national power.'[24] It has been argued that France has consistently tried to gain advantage from a strong EU in international and regional affairs and as an 'alternative', at times even as a 'second option' or instrument to maximize French interests.[25] France has also emerged as a key player in balancing the EU centre of gravity between its eastern and southern dimensions and preventing a dominant German centre of gravity influencing EU external policy.[26] Therefore, although multilateral initiatives have been promoted by France, the overall strategic rationale has been to maintain international and regional influence through the EU and, at the same time, retain its own bilateral channels of influence. The result in the Mediterranean has been a French foreign policy approach operating on two levels: a national and a European one.[27] Some have defined this policy approach in dialectic terms, at times facilitating the EU approach whilst at others moments obstructing it.[28]

French Foreign Policy Preferences and Interests

Traditionally, France's foreign policy towards the Mediterranean has retained a strong bilateral emphasis, particularly where Paris enjoyed a shared history and cultural influence considered to be exercised through the diplomatic use of the concept of francophonie.[29] As we will see, the Mediterranean has been and remains central to the exceptional role and responsibilities France perceives for itself as a regional and international

---

24 Blunden, 'France', 22.
25 D. Schmid, 'France and the Euro-Mediterranean Partnership: The Dilemmas of a Power in Transition', in H.A. Fernandez and R. Young (eds), *The Euro-Mediterranean Partnership: Assessing the First Decade* (Madrid: FRIDE, 2005), 98.
26 P. Fysh, 'Gaullism and the New World Order', in Chafer and Jenkins (eds), *France: From Cold War to the New World Order*, 182; see also Blunden, 'France', 39.
27 Blunden, 'France', 23–39.
28 Schmid, 'France and the Euro-Mediterranean Partnership', 96.
29 Schmid, 'France and the Euro-Mediterranean Partnership', 96.

power. Two areas of keen French interest and historical ties have been the Maghreb and to a lesser extent Lebanon. Algeria and Morocco held particular importance for French political leadership in the Maghreb, exercised predominantly through economic means. In fact, both states' external trade flows were and still are primarily oriented towards France, while local economies benefit from a range of French supporting instruments in terms of development and debt relief. Moreover, France to this day is the main foreign investor, creditor and bilateral aid donor for both Morocco and Algeria. Commercial interests remain an important aspect of French foreign policy in the region and have been promoted though favouring economic reforms and structural adjustment designed to attract more French firms to operate in the region.[30] Aside from commercial interests, French policy preferences towards the Maghreb have been shaped by security concerns related to the stability of North African regimes and the emergence of political challenges associated with radical Islam. Mass migration and the large Maghrebi diaspora in France have meant that at least in the French media, migration has become highly politicized.[31] This perception was reinforced by the systematic demonization of immigration by the National Front (FN) and by the apparent spillover of the Algerian civil war in France, most notably through the terrorist attacks of 1995 on the Paris Metro and a year later on the Port-Royal RER station. Finally, the shocking events of 11 September 2001 consolidated the perception of the Mediterranean as a region vulnerable to radical Islamist forces and, thus, presenting a security concern to Europe.

In the shifting geo-political environment of the 1990s, the stability of North African regimes became a priority for France, a policy that eventually contributed to the undermining of its legitimacy and, thus, its ability to exert influence in the region. It could be argued that French relations with Maghrebi regimes have been characterized by common interests as

---

30   Meyrede, 'France's Foreign Policy in the Mediterranean', 42–7. See also Schmid, 'France and the Euro-Mediterranean Partnership', 95–7.
31   Hargreaves, 'Immigration, Ethnicity and Political Orientations in France', 210. See also Meyrede, 'France's Foreign Policy in the Mediterranean', 46.

much as they have been undermined by the legacy of colonialism. The aims of French foreign policy towards the region have been to achieve a high degree of collaboration and co-operation over the security agenda and guarantee regional stability and the status quo. This collaboration itself had the effect of legitimizing Arab regimes and justifying domestic repression of Islamic political opposition on security grounds. North African regimes have been aware of French priorities and formulated their policy accordingly by enhancing co-operation over migration control, access to energy and by facilitating the promotion of European business interests in the region. French officials gradually became aware that as far as Arab public opinion was concerned, France's legitimacy and thus its political weight were increasingly undermined by its colonial legacy and the pragmatic nature of its relations with domestic regimes.[32] This situation for France resulted in EU initiatives being increasingly perceived by Paris as an 'unparalleled vehicle for reasserting French influence in the region'.[33]

French policy preferences in the Mediterranean rested on commercial and security interests and on the willingness by the various leaders in Paris to hold on to its 'special role' in the region. Maintaining power in a post-colonial context translated into an emphasis on bilateral relations and the projection of France's interests along political and commercial tracks. For example, Mitterrand's socialist administration identified its policy objectives in the region as the stability of the Maghreb and the related issue of managing migration to France, the commercial importance of the Mediterranean for French firms and the primacy of North Africa with regards to energy access (in particular Algerian gas). Mitterrand, who held the post of Overseas Minister during the Fourth Republic and was Minister of the Interior and Minister of Justice for part of the Algerian War (when Algeria was considered an integral part of French territory), was perceived as being familiar with the region and having a deep knowledge of Maghrebi affairs.[34] For a socialist president, it could be argued

---

32   Meyrede, 'France's Foreign Policy in the Mediterranean', 46.
33   Blunden, 'France', 39.
34   Meyrede, 'France's Foreign Policy in the Mediterranean', 54–6.

that Mitterrand displayed a strong inclination to follow a Gaullist tradition in foreign policy. In fact, he attempted to promote a multilateral and independent European security and defence structure through the launching of the Eurocorps and the 'Western Mediterranean Forum', eventually extended to form the '5+5 Framework' in the Mediterranean. Tellingly, however, his Mediterranean vision was one predominately focused on the Maghreb where France would have a comparative advantage to play a dominant role and enhance its international status vis-à-vis German and British leadership. Essentially, he saw the promotion and strengthening of the EU-level agenda in the Mediterranean as the best method to achieve French interests in the region.[35]

Under the conservative government of President Chirac, a degree of continuity initially seemed to characterize the French approach towards the Maghreb. Bilateral and personal relations between French leaders and their North African counterparts remained instrumental in maintaining France's 'special' position. President Chirac also relied extensively on his personal relations with King Hassan II of Morocco, whose friendship is seen as having translated into intensified co-operation in the military field. Similarly, Ben Ali of Tunisia studied at the French military school of Saint-Cyr and maintained close relations with members of the French administration, while former Lebanese prime minister Rafic Hariri, who lived in Paris for a long time, was also a personal friend of President Chirac. Even though French Mediterranean policy was based on continuity with the previous administration, under the presidency of Chirac it was characterized by a stronger conviction that France's best interest lay within Europe. Under Chirac, in fact, Paris engaged with the Mediterranean in its entirety and, importantly, more explicitly through the EMP. Unlike his predecessor, Chirac seemed to expand its strategic vision beyond the Maghreb and towards the Mediterranean as a whole. This was to project France as a leader of a cluster of southern EU members including Spain, Italy and Portugal working to balance the EU in its external dimension

---

35   Meyrede, 'France's Foreign Policy in the Mediterranean', 55–6; see also Fysh, 'Gaullism and the New World Order', 182.

and counterbalance the German-sponsored eastern dimensions.[36] Under Chirac, France promoted its role as a champion of European interests in the Mediterranean and, together with its southern EU member states, these interests were projected as collective European ones, thus also of concern to northern European member states.[37]

The Articulation of French Interests and Policy Preference at the EU Level

Before focusing on the articulation of French policy preferences and interests in the Mediterranean, it is necessary to briefly define what is meant by the process of Europeanization. Europeanization can be conceptualized in terms of being both 'top-down', suggesting the transposition of EU-level norms and rules to member (and non-member) states, but also as being 'bottom-up', reflecting the upwards transfer of member states' policy preferences to the European level.[38] Thus Europeanization can be understood as the approximation of policy preference and interests which leads to: '[...] a negotiated convergence in terms of policy goals, preferences and even identity between the national and the supranational levels'.[39] This chapter focuses on a bottom-up conceptualization of Europeanization in which member states, in this case France, are seen as active, primary actors in the formulation of EU policies through the projection of national interests and preferences at the EU level. That said, it is important to remain aware of how other factors and actors, including Spain and Egypt, have respectively

---

36   R. Gillespie, 'Spanish Protagonismo and the Euro-Mediterranean Partnership Initiative', *Mediterranean Politics*, 2/1 (1997), 38–9.
37   Meyrede, 'France's Foreign Policy in the Mediterranean', 64–5.
38   K. Featherstone, 'Introduction: In the Name of Europe', in K. Featherstone and C.M. Radaelli (eds), *The Politics of Europeanisation* (Oxford: Oxford University Press, 2003), 1–27. See also K.E. Howell, *Up-loading, Downloading and European Integration: Assessing the Europeanization of UK Financial Services Regulation* (Ashcroft international Business School APU, 2002), 2–6.
39   Wong, *The Europeanization of French Foreign Policy*, 6.

influenced and impacted on French foreign policy in the Mediterranean and in doing so served to shape national preferences.

During the late 1980s and early 1990s, Mitterrand's vision to maintain French influence and promote a European-wide security engagement with the Maghreb countries found strong support with his Spanish counterpart Felipe González and even with the Algerian president Chadli Bendjedid. Following the failure of the Hispano-Italian proposal for the Conference for Security and Co-operation in the Mediterranean (CSCM) in 1991, Spanish officials seemed to concede to French leadership and turned their attention towards the Maghreb and the 'Five + Five framework', an area of clear Spanish interest.[40] Similar to France, Spain's interests in the Maghreb were characterized by the threat of terrorism, increasing migratory flows from North Africa and the smuggling of drugs from Morocco in particular.[41] The Spanish approach focused on co-operation rather than confrontation and was based on the consolidation of bilateral relations through a process of enhancing political stability by channelling economic resources in the region, 'as a preventive measure in favour of Spanish security'.[42] Even though at the time the dynamics of Mediterranean policy were also characterized by Franco-Hispanic rivalry, particularly with regards to influencing and shaping the EU cultural framework,[43] the

[40] The reasons for the lack of progress on the CSCM track are seen as stemming from the Algerian crisis in 1992 as well as the isolation of Libya by the international community. Moreover, the initiative also generated a reluctant response by the US unwilling to see the development of a framework for security co-operation extended to the sensitive Mashreq. M. Salomon, 'Spain: Scope Enlargement towards the Arab World and Maghreb', in F. Algieri and E. Regelsberger (eds), *Synergy at Work: Spain and Portugal in European Foreign Policy* (Bonn: Europa Union Verlag, 1996), 104–5.

[41] Gillespie, 'Spanish Protagonismo and the Euro-Mediterranean Partnership Initiative', 35–40.

[42] C.J. Echeverria, 'Spain and the Mediterranean', in Couloumbis et al. (eds), *The Foreign Policies of the European Union's Mediterranean States and Applicant Countries in 1990s*, 102–3.

[43] Schmid, 'France and the Euro-Mediterranean Partnership', 100.

nature of interaction has been one dominated by co-operation activities.[44] Therefore, French and Spanish security policy objectives in the region strongly converged and both actors combined to lobby for more financial support and economic engagement from the European Community towards the Maghreb. Both member states argued for intensified relations towards Morocco, Algeria and Tunisia and promoted sustained efforts towards integration with European markets and socio-economic modernization. The threats perceived as emanating from Islamic fundamentalism remained a priority and were seen as a key factor in explaining France's supportive stance towards the Algerian regime in the annulment of the elections in 1991/92 following the victory of the FIS. The rise of Algerian fundamentalism has been a key concern for French politicians, resulting in support by Paris for Algerian institutions and the heavy-handed tactics adopted to repress Islamists.[45]

Gillespie notes that the best way to describe the EMP is a 'Hispano-French and southern Mediterranean partnership' characterized by trade-offs with northern member states.[46] Although Spain could be seen as having initiated the idea of the partnership, the role of Chirac in this context (and later in the context of the nascent European Neighbourhood Policy) was key in balancing German influence over the EU's external policy and budgetary allocation. With regards to the EMP, French and Spanish officials were fundamental in promoting and reinforcing the Mediterranean initiative at EU level as well as actively supporting Mediterranean partners. Under French influence, and with the support of Spain, the Commission eventually proposed a re-orientation of EU policy towards the Mediterranean which marked a second period of intensified EU engagement.[47] Gillespie notes that the role of Madrid

---

44   Gillespie, 'Spanish Protagonismo and the Euro-Mediterranean Partnership Initiative', 38.
45   M. Evans, 'Franco-Algerian Relations in the Post-Cold War World', in Chafer and Jenkins (eds), *France: From the Cold War to the New World Order*, 228–9.
46   Gillespie, 'Spanish Protagonismo and the Euro-Mediterranean Partnership Initiative', 46.
47   Meyrede, 'France's Foreign Policy in the Mediterranean', 59.

should not be over emphasized but taken as a constitutive part of a collective Mediterranean–EU effort. Nevertheless, the Euro-Mediterranean Conference in 1995 and the beginning of the Barcelona Process were seen as a great success for Spanish foreign policy and the acknowledgment of Madrid's role in the region.[48] Importantly, the eventual success in achieving a compromise over the budgetary allocation was seemingly due to the influence and skill of Spanish president Felipe González, particularly with regards to his influential and positive relationship with his German colleague Helmut Kohl. Even though France can be considered the most influential actor in the Mediterranean, Spanish support meant that Paris could be more effective in promoting its agenda and, thus, promote French interests at the EU level and play a key role in forging a truly collective European approach in the region. Meyrede argues that the success of negotiating the EMP showed that France was still able to retain the initiative on the international stage and at EU level. It emphasized the role of Paris as a key negotiator with southern EU members as well as a key facilitator in potentially difficult relations between the EU and sensitive partners such as Algeria, Tunisia and Morocco.[49]

There is little doubt that both Mitterrand and Chirac held true to the traditional characteristics of French foreign policy in the Mediterranean. Although evaluating the internal dynamics that shaped the foreign policy approaches of both leaders goes beyond the remit of this chapter, it can be noted that a degree of continuity can be identified in terms of Mediterranean policy. That said, it must also be noted that a dialectical relationship exists in France's policy between bilateral and multilateral approaches in the region. Nevertheless, French foreign policy in the Mediterranean seems to be underpinned by the principles of French grandeur and exceptionalism in international relations which, ultimately, translate into a policy and policy discourse of power and self-sufficiency.

---

48 Gillespie, 'Spanish Protagonismo and the Euro-Mediterranean Partnership Initiative', 46. See also Echeverria, 'Spain and the Mediterranean', 109.
49 Meyrede, 'France's Foreign Policy in the Mediterranean', 69.

The Re-orientation of Egyptian Foreign Policy

The following section will focus on the role of Egypt in contributing to the EU approach and framework in the Mediterranean. Egypt provides an interesting perspective in this regard as it has a distinct historical legacy with regards to France, compared to the Maghreb. Moreover, the importance of its geo-political role in the region, particularly with regards to the Arab-Israeli issue, is also very relevant. As will be demonstrated below, Egypt has consistently proved a key actor in the development of multilateral EU initiatives in the Mediterranean and continues to be perceived as such.

The first Gulf War in Iraq and later the events of 11 September 2001 signalled a turning point for a re-orientation of the foreign policies of partners such as Egypt, eventually resulting in a rapprochement with western European powers and the EU itself. No doubt the discourse used by the George W. Bush administration at the time and formulated along the lines of 'either you are with us or against us' was also an important factor in this context. In fact, increased American involvement in the Mediterranean after September 11 was opposed by France. This resulted in French resistance to American participation in EMP structures and a lack of enthusiasm over NATO-oriented initiatives in the Mediterranean as well as objections to interference in domestic matters of Arab states.[50] The 'traditional' nature of foreign policy under Chirac was most evident in the President's position over the war in Iraq. This Gaullist orientation and attitude reinforced the already pragmatic approach of the French administration towards Mediterranean partners. As in the past, French concerns remained focused on the impact of rapid economic and political reforms and the internal pressures and instability that this process could bring to MENA partners. The objective was to avoid forcing Mediterranean partners into a position where they would have to take numerous unpopular decisions, thus producing social unrest and favouring Islamic political opposition.[51]

---

50   Schmid, 'France and the Euro-Mediterranean Partnership', 97.
51   Meyrede, 'France's Foreign Policy in the Mediterranean', 66.

Chérigui notes that following the first Gulf War, Egyptian foreign policy in the Mediterranean became more active with the aim of establishing a leadership role in the region for Cairo. A number of factors seem to have contributed to this re-orientation towards the Mediterranean. The Gulf War resulted in a stronger and explicit rapprochement by Egypt with the West and Europe, and with France and Germany in particular. Moreover, the return of the Arab League's headquarters in Cairo, together with the presence of Egyptian personalities at the helm of other international organizations such as the UN secretary general Boutros Boutros-Ghali, was seen as a favourable strategic environment for the re-launching of Egyptian interests in the region and the promotion of Cairo in terms of a mediating role in the Middle East.[52] More importantly, Egyptian leaders seemed intent on preventing their isolation from the Maghreb and from EU engagement as a result of the existing '5+5 framework'. The Egyptian position could be seen as one aimed at counterbalancing the initiative by attempting to extend the EU framework to the east of the Mediterranean in order to be fully integrated in the EU Mediterranean project. President Mubarak had in fact proposed the idea of a Mediterranean Forum to the Parliamentary Assembly of the Council of Europe on 11 November 1991. El Sayyed Selim argues that although the security dimension of the Forum was an important element for Cairo, the immediate priority was seen as establishing a space for common co-operation on an economic, technical and cultural basis.[53] Moreover, Chérigui adds that although France was initially resistant to the proposal as it would create conflict with its direct interests and leadership in the '5+5' initiative in the Maghreb, agreement over the establishment of the Mediterranean Forum ultimately signalled and consecrated the Franco-Egyptian and Euro-Egyptian rapprochement.[54] The EMP presented the second component of Egypt's Mediterranean policy. The Egyptian administration responded positively to the proposal of association in the EMP

52   H. Cherigui, *La Politique Mediterranéenne de la France: entre diplomatie collective et leadership* (Paris: L'Harmattan, 1997), 159–60.
53   M. El-Sayyed Selim, 'Egypt and the Euro-Mediterranean Partnership: Strategic Choice or Adaptive Mechanisms?', *Mediterranean Politics*, 2/1 (1997), 70–4.
54   Cherigui, *La Politique Mediterranéenne de la France*, 160–4.

even though a number of issues remained sticking points in the negotiations. Overall, however, the objectives of the EMP were deemed as complementary to Egyptian ones, particularly with regards to economic modernization and development. Egyptian foreign policy orientation in this context is seen as an important factor which contributed to 'extend' the EU Mediterranean vision beyond the Maghreb to the east and embrace the Mediterranean in its entirety. For France, promoting the Mediterranean Forum was also in line with its preference to Europeanize its policy in the Mediterranean and extend its area of influence beyond the Maghreb.[55]

It could be argued that French policy in the Mediterranean has retained an element of this bilateral-multilateral dialectic, at times undermining multilateral approaches in favour of national interests and French exceptionalism. This seems evident in the French insistence on maintaining its dominant role in frameworks such as the '5+5' approach focusing on a sub-regional dimensions in the Maghreb, as well as supporting the Mediterranean Forum with its focus in the Mashreq. Schmid argues that French leaders have struggled to adjust to the changing Mediterranean realities even though the fact that they have become more sensitive to multilateral trends can be interpreted as a sign of weakness.[56] That is to say that increasing French reliance on the EU framework in the Mediterranean also translates into the perception of a less dominant role for Paris in the region vis-à-vis its former area of influence. However, it seems that under various French administrations the principles of Gaullism emerge as characteristics of policies towards the Mediterranean and Europe. In terms of the process of Europeanization, the establishment of the EMP indicates that a degree of convergence and overlap can be identified between French and Spanish preferences and interests and that a genuine collective process of negotiation and bargaining combined to achieve collective agreement. One could thus deduce that together with the geo-political environment of the time and arguably the role of actors such as Spain and Egypt, a genuine process of policy preference approximation took place. France has been

---

55   El-Sayyed Selim, 'Egypt and the Euro-Mediterranean Partnership', 75–9.
56   Schmid, 'France and the Euro-Mediterranean Partnership', 101–2.

able to project its national interests at the EU level and was instrumental in the forging of the EMP, albeit with support from other members and favourable international conditions. However, is also true that France was able to exploit the Mediterranean space for its own priorities and strategic objectives of power and influence. As Schmid has noted, although France remains one of the key defenders of the EMP, 'French exceptionalism itself accounts for many of the paradoxes' of the partnership.[57]

The Union for the Mediterranean: Renewed French Dynamism?

President Sarkozy first mentioned the idea of a Mediterranean Union during his electoral campaign in 2007. Although the electoral context cannot be ignored,[58] the French leader based his argument in favour of the UfM by pointing out the shortcomings of the Barcelona Process and the need to re-evaluate European approaches and mechanisms in the Mediterranean. In this respect, the Marseille Ministerial Conference in 2008 was a turning point for Mediterranean partners as it demonstrated the difficulties of finding a common ground.[59] EU members had wished to see political reform with the assumption that it would lead to greater stability and security, while southern Mediterranean members such as Italy, France and Spain perceived such reforms as potentially undermining their political stability. Moreover, from the perspective of Mediterranean partners the EMP was seen as addressing EU security concerns at the expense of their own and, thus, they wished to limit the objectives of the partnership to development

---

57   Schmid, 'France and the Euro-Mediterranean Partnership', 102.
58   Bechev and Nicolaidis, 'The Union for the Mediterranean', 14.
59   R. Aliboni and F. Ammor, 'Under the Shadow of "Barcelona": From the EMP to the Union for the Mediterranean', *Euromesco Papers* 77 (January 2009), http://www.euromesco.net/images/paper77eng.pdf (accessed 20 October 2010).

and economic integration within the context of a broad diplomatic dialogue. At face value one could argue that the argument for the UfM rested on sound evaluations of the state of affairs in the region and the need for fresh ideas. The concept of a 'Union of the Mediterranean' (UfM) was first put forward by President Sarkozy on 23 October 2007, emphasizing a renewed French dynamism, the shared experience of the Mediterranean peoples and the need to turn a new page in North–South relations:

> Just after the Second World War [...], Europe set off on the path of peace and fraternity. France calls on all the Mediterranean peoples to do the same thing, with the same goal and the same method. [...] We will build the Mediterranean Union, as Europe's union was built, on the basis of a political determination stronger than the memory of suffering, on the basis of the conviction that the future counts for more than the past.[60]

The president went on to spell out in more detail what his vision for the Mediterranean consisted of. Apart from formulating the idea in different terms from existing EU frameworks, the UfM was to be based on the same pragmatic rationale upon which the inception of the European Community rested. The UfM was thus to be a project-oriented, intergovernmental union focusing on specific sectors for co-operation.

> The Mediterranean Union must be pragmatic: the union will vary according to circumstances [*à géometrie variable*], with members joining together on projects as and when they wish. Just as Europe began with coal and steel and atomic energy, the Mediterranean Union will start with sustainable development, energy, transport and water. [...] The Mediterranean Union will first of all be a project-based union. But with a single goal: to make the Mediterranean the world's largest testing ground for co-development where development is decided on together and managed together, where freedom of movement of people is built together and managed together, and where security is organized together and guaranteed together.[61]

---

60 BBC, 'Nicolas Sarkozy's Victory Speech Excerpts', http://news.bbc.co.uk/1/hi/world/europe/6631125.stm (accessed 10 August 2009).
61 BBC, 'Nicolas Sarkozy's Victory Speech Excerpts'.

The message from the French President was clear in distinguishing the UfM from already existing EU policy-frameworks where the lack of success could be addressed by empowering southern partners in that process. Sarkozy's proposal envisaged France as a leading force and driver of a new, more effective agenda in the Mediterranean based on summitry and Paris' pivotal role in the region.

The Europeanization of the Union for the Mediterranean

One of the most controversial issues in the UfM, however, was that Sarkozy's project initially intended to be restricted to the coastal states of the Mediterranean basin and was presented as an explicit alternative to the existing EMP. Reactions to the initiative were critical from a number of states. Turkey immediately rejected the idea, perceiving it as a French attempt to present Ankara with an alternative to accession.[62] A number of commentators have argued that France has never hidden the fact that it opposed Turkish accession giving the Turks enough reasons to doubt the real French motives behind the initiative.[63] Italy and Spain also expressed concerns that the initiative could damage relations with the wider EU. In fact, at the Rome meeting in December 2007, the Spanish and Italian prime ministers ensured that the existing EU policies in the Mediterranean would not be compromised.[64] Germany provided the strongest opposition to the initiative. Chancellor Merkel initially expressed particular concern over the fact that the initiative should be restricted to the coastal states, arguing that this format would only create internal EU divisions and tensions by drawing northern and southern member states closer to

---

62 M. Emerson, 'Making Sense of Sarkozy's Union for the Mediterranean', *CEPS Paper No. 155*, www.ceps.eu (accessed 7 February 2009).
63 A. Le Gloannec, 'Sarkozy's Foreign Policy: Substance or Style?' *The International Spectator*, 43/1 (2008), 15–21. See also Emerson, 'Making Sense of Sarkozy's Union for the Mediterranean'.
64 Balfour, 'The Transformation of the Union for the Mediterranean', 100.

their respective geographical proximities.⁶⁵ Although northern European states have been traditionally sceptical and unwilling to share the burden of EU policies in the south Mediterranean, the events of 9/11 and later the terrorists attacks in Madrid and London contributed to a change of perception in this regard. Germany has been a strong supporter of the Barcelona Process since its inception even though German foreign policy circles of late have regarded the EMP as the 'playground of a few southern European states' and an inadequate forum in which to address political and security concerns, mainly hampered by the different positions on the Arab–Israeli issue. Over the last few years, Germany has tried to balance its position between having a say in Euro-Mediterranean affairs and guaranteeing appropriate resources directed at EU engagement in the eastern dimension. This bargaining process has been relatively straightforward to date with an accepted division of labour between northern and southern EU members as defenders of their interests in their respective neighbourhoods. The initiative of President Sarkozy was seen in German circles as jeopardizing this balance and the long-standing Franco-German alliance at the heart of the EU. In the eyes of Chancellor Merkel, such a 'union' with access to EU budget and resources would certainly create internal tensions and potentially crack the heart of the EU. The stance taken by Germany and the issue related to intra-EU bargaining for resources resulted in a 'domino effect' refusal by non-Mediterranean member states to contribute financially to the Union.⁶⁶

It could be argued that German opposition was not so much the catalyst for the re-structuring of Sarkozy's plan as much as its inapplicability was. In fact, the architect of the initiative, neo-Gaullist presidential adviser and speech writer Henri Guaino, initially foresaw an intergovernmental framework made up of sovereign Mediterranean states and with minimum involvement of the Commission. The challenges associated

---

65   Emerson, 'Making Sense of Sarkozy's Union for the Mediterranean'.
66   T. Schumacher, 'From Paris with Love? Euro-Mediterranean Dynamics in Light of French Ambitions', *The GCC-EU Research Bulletin*, 10 (2008), 8–11. T. Schumacher, 'Germany: A Player in the Mediterranean', *IEMed/CIDOB Yearbook 2009: Med2009 – Mediterranean Yearbook* (Barcelona: CIDOB Press: 2009).

with this vision and which included getting collective EU agreements on the various concessions on movement of people, technical co-operation and harmonization with the *aquis communautaires* meant that eventually the UfM had to be scaled down, the final outcome bearing only a vague resemblance to the origin idea.[67] Eventually, the Franco-German summit in Berlin in March 2008 paved the way for the Europeanization of the project and the UfM was agreed with a degree of compromise and by integrating it into existing EU structures in the Mediterranean. The inclusion of non-Mediterranean member states reflected concerns over budgetary allocation,[68] while the focus on project-oriented initiatives was retained and reinforced by the inclusion of the Commission who privately voiced its concern over the compatibility of French and EU policies.[69] The result is that the UfM is now considered an 'upgrade' as opposed to an alternative to the EMP, benefiting from existing EMP funding instruments and reflecting ENP priorities thus, fully integrated with EU structures and mechanisms.[70] But can we talk in terms of the Europeanization of the original French initiative? It seems that France's rationale lies in its interests in addressing key national geopolitical priorities such as illegal migration, security, energy, trade and investment, all compatible with EU interests and strategic approach in the Mediterranean under the EMP structures. There is little evidence that the UfM came about because of specific new shared interests or even a degree of overlap and negotiated policy convergence amongst two or more member states. On the contrary, it seems that the UfM was originally an idea of President Sarzoky which was uploaded to the EU level in order to maintain the internal balance of the EU, particularly with regards to budgetary matters, and to retain a degree of coherence in the external EU approach in the Mediterranean. Allowing the French initiative to go ahead as envisaged by Sarkozy would have resulted in an overlap of already complicated institutional structures

---

67   Bechev and Nicolaidis, 'The Union for the Mediterranean?', 13–15.
68   R. Brizzi, 'Francia: Sei mesi per l'Europa', *Il Mulino* (no date), 708–20 (715), www.ilmulino.it (accessed 14 February 2008).
69   Balfour, 'The Transformation of the Union for the Mediterranean', 100.
70   Bechev and Nicolaidis, 'The Union for the Mediterranean', 15–16.

in the Mediterranean and the drawing of power away from member states and the Commission. Thus an original incompatibility existed between the idea of Sarzoky and the EU agenda and structure in the Mediterranean, making it difficult to define the agreement of the UfM as the result of a process of Europeanization. Although the unilateral effort by the French President points to a degree of continuity with Gaullist foreign policy principles of leadership and specific role conceptions, it is not clear whether the French initiative was simply an impulsive and ill-defined proposal to be understood in the context of domestic politics or a genuine attempt at reinvigorating French foreign policy.

French Motivations for the Union for the Mediterranean

As the initiative was proposed during Sarkozy's electoral campaign, many commentators have deduced that the idea was merely dictated by domestic political needs and as an attempt to attract the largest possible number of voters of Maghrebi and non-Maghrebi origins. In fact, some of the arguments presented by Sarkozy during the electoral campaign and in the context of the Mediterranean seemed to be aimed at pre-empting the position of the National Front on migration and asylum, issues highly resonant with the French electorate. He thus advertised the establishment of a new minister for immigration, integration, national identity and co-development in order to curb illegal migration and asylum.[71] Although it remains difficult to ascertain to what extent the elections provided the rationale for the initiative, it could be argued that in an electoral context, the idea and concept of the Mediterranean is a 'safe' if not 'winning' concept in French domestic political discourse. France's historical role and legacy in the Mediterranean have helped to load the concept with a positive perception in the public

---

71 T.C. Tasche, 'The Project of a Union for the Mediterranean: Pursuing French Objectives through the Instrumentalisation of the Mare Nostrum', *L'Europe en formation*, 365 (Summer 2010), 53–70, http://www.cife.eu/UserFiles/File/EEF/356/EEF356-4TT.pdf (accessed 24 October 2010).

sphere, one which is directly related to French identity and which finds resonance across the board, when compared, for instance, to discourses of Atlanticism or German-axis related policies.

Providing a different explanation, Le Gloannec argues that the UfM was an attempt at a renewed French dynamism in foreign policy. He notes in fact that: '[...] on most foreign policy accounts, the new President seems to have taken his distance from Jacques Chirac',[72] manifesting a degree of ambiguity between a realist and idealist approach. This is evident in Sarkozy's approach towards NATO, Israel, Syria and, specifically, in the Mediterranean. The wish to strengthen European defence was one of the first indications that Sarkozy thrived on a leading European role. Advocating a fairer economic global environment, where all actors play by the same rules and the same restrictions, was also seen as a sign of French dynamism in European and international affairs.[73] The position over Israel, the newfound love for the US, the strong stance with regards to Iran and the attempt to re-engage within NATO all point to a more innovative and active French foreign policy.[74] However, it seems that the foreign policy envisaged by the French administration manifests ambiguities such as the attempted rapprochement with Putin's Russia and the Libyan president Muammar Gaddafi, while at the same time wanting to be a beacon for human rights in the world. In this context, the role of the Quai d'Orsay under socialist foreign minister Bernard Kouchner makes for an interesting analysis. In the aftermath of Sarkozy's electoral victory, the appointment of a leading socialist figure at the helm of the foreign ministry was presented as a significant coup and seemingly justified claims of a 'new' foreign policy direction envisaged by the president. Kouchner is not a diplomat; on the contrary he worked as an activist and is famously known as the founder of Médecins Sans Frontières. According to Kouchner's rhetoric, the Quai d'Orsay was to follow a humanitarian and value-based diplomacy where

---

72   Le Gloannec, 'Sarkozy's Foreign Policy: Substance or Style?', 15.
73   D.G. Dimitrakopoulos, A. Menon, and A.G. Passas, 'France and the EU under Sarkozy: Between European Ambitions and National Objectives?', *Modern and Contemporary France*, 17/4 (2009), 455–61.
74   Le Gloannec, 'Sarkozy's Foreign Policy: Substance or Style?', 15.

and when possible. This orientation signalled a clear shift and distance from the pragmatism of Chirac and the traditional foreign policy preferences in Paris. These expectations, however, failed to materialize. French foreign policy under Sarkozy has in fact been driven by an interest-based and personalized style of diplomacy, the result of which has been the marginalization of the Quai d'Orsay.[75] The concentration of foreign policy decision-making into the hands of Elysée officials – and notably, for the Middle East and Africa portfolio, into the hands of then-Elysée general secretary Claude Guéant – has been well documented.[76] A reversal of this situation has only become a possibility in 2011, with the events of the Arab Spring and the appointment of Alain Juppé to the role of foreign minister. Up until this point, all of the above tends to indicate that the president has systematically bypassed the Quai d'Orsay on key foreign policy decisions by concentrating decision-making at the Élysée and on personalized diplomatic relations. Under the French presidency of the EU in the second half of 2008, Sarkozy was able to promote and successfully take credit for the UfM while also securing support from EU and non-EU partners, in particular that of Egypt and President Mubarak. Although it seems at times that Sarkozy's foreign policy and his vision of Europe are unlikely to shift from a traditional Gaullist orientation, the means by which Gaullist principles seem to be achieved have manifested a change in approach. However, the question remains; why pursue a project that in its agreed version seems like a blueprint of the 'Barcelona Process'?

Aliboni et al. note that France has alluded at a lack of EU motivation in implementing the Mediterranean agenda.[77] From an EU perspective,

---

[75] P. Bernard, 'Jean-Christophe Rufin au "Monde": Le Quai d'Orsay est un ministère sinistré', *Le Monde* (6 July 2010), http://www.lemonde.fr/afrique/article/2010/07/06/jean-christophe-rufin-au-monde-le-quai-d-orsay-est-un-ministere-sinistre_1383868_3212.html (accessed 20 October 2010).

[76] M. Kamdar, 'The End of the Affair', *Foreign Policy* (October 2010), http://www.foreignpolicy.com/articles/2010/10/19/the_end_of_the_affair (accessed 29 October 2010).

[77] R. Aliboni et al., 'Putting the Mediterranean Union in Perspective', *Euromesco*, Paper 68 (Barcelona: Euromesco, 2008), http://www.euromesco.net/images/paper68eng.pdf (accessed 20 October 2010).

Sarkozy clearly alluded to a failure to implement the concept of 'shared ownership' in the Mediterranean by focusing simply on the trade and economic dimensions of the partnership. In fact, the concept of 'shared ownership' was a central element in promoting the UfM to southern Mediterranean partners. Even though southern partners may have divergent preferences with regards to the bilateral or multilateral co-operation with the EU, the initial reaction of key states such as Morocco and Egypt, who was offered the first co-presidency, have been positive (Egypt reacted positively to the initiative as it reflected its preference for a multilateral track in the Mediterranean and enhanced its regional role by being granted the co-presidency).[78]

However, it seems that the initiative evolved out of three distinct motives identified by the French leader: the marginalization of the Mediterranean from the world economy, the shortcomings manifested to date by the EU's Mediterranean policy and the perceived erosion of the role of France in the Mediterranean. It has also been noted that the initiative was aimed at allowing new diplomatic channels to be opened both in the Mediterranean and Middle East.[79] In this respect, the so-called 'Club Med' seems to manifest a willingness to restore French diplomatic grandeur in the region by engaging in summitry with regional leaders and support a more active French foreign policy vis-à-vis its EU partners. The rapprochement with Syria and the attempt to mediate over the Palestinian question all point to a willingness by France to play a more central and leading role in the region at a time when the US administration seemed to have left an opening for potential mediators.[80] This argument is consistent with the French historical and political role in the region which sees Paris still exerting a degree of political influence linked to its colonial past.

---

78   Bechev and Nicolaidis, 'The Union for the Mediterranean', 14–16.
79   Alibioni et al., 'Putting the Mediterranean Union in Perspective'.
80   S. Erlanger and K. Bennhold, 'Bold Mediterranean Summit Opens Diplomatic Door', *The New York Times* (13 July 2008), http://www.nytimes.com/2008/07/13/world/europe/13iht-summit.4.14457192.html?_r=1 (13 February 2009).

Conclusion

This chapter started with considerations regarding the nature of EU policy in the Mediterranean in the context of the ongoing Arab Spring and the extent of member states' influence over that policy. Issues of interest have been related to the question of Europeanization and how and why member states' policy preferences are up-loaded to the EU level. Looking at the experience of the EMP, we could argue that French interests and influence in the Mediterranean have been increasingly attained through multilateral frameworks and, specifically, through the EU. France has been successful in projecting its national priorities and preferences at the EU level and has retained a high degree of influence both within the EU's Mediterranean dimension and within the MENA region. No doubt France's colonial and cultural legacy in the Mediterranean, and in the Maghreb in particular, have been a determining factor in the formulation of the EU agenda and policy framework in the region. However, if French activism was successful in re-directing the EU's attention to the Mediterranean, this was also due to a dynamic Spanish leadership that shared regional interest and concerns with Paris. Similarly to France, Spain has lobbied and successfully projected its national interests at the EU level. Moreover, the role of non-EU members such as Egypt provides an interesting perspective on the construction of an EU Mediterranean space and on debates related to the process of Europeanization. Thus, we can argue that the EU agenda in the Mediterranean to date has been highly influenced by French priorities and interests and eventually shaped by a 'bottom-up' process of Europeanization, understood in terms of the approximation and convergence of policy preferences at the EU level.

In the context of the UfM, however, the initiative of President Sarkozy cannot be considered in similar terms and as the result of a process of Europeanization. In fact, the initial proposal for the UfM potentially undermined the EU agenda in the Mediterranean in its entirety. When considering the tenets that underpinned the logic of the EMP, the UfM seems very weak in its political substance and promotion of civil society, a characteristic which could result in further politically legitimizing regimes which systematically undermine human rights and democratic practices and, thus, seriously dent EU credibility. Importantly, the principle of co-ownership which was presented by Sarkozy as the added value of the UfM and endorsed at the highest political level appears to be not complementary with the EU supranational approach in the region. On the contrary, it is more likely that EU–Mediterranean summitry and the inter-governmental framework underpinning increased co-ownership will undermine any process of political reform as originally envisaged in the Barcelona Process, and the ability of the EU to exercise conditionality. Therefore, and in the context of Europeanization, it is difficult to consider the UfM as reflecting a genuine process of policy approximation, projection of shared interests and overlap of policy preferences. The UfM might indeed address policy issues and areas of concern to member states, but unlike the dynamics which underpinned the establishment of the Barcelona Process, Sarkozy's initiative is a purely unilateral exercise which has seemingly been Europeanized due to the high political and financial costs associated with any alternative of not doing so.

Since the beginning of 2011 the political scenario in the southern Mediterranean has changed dramatically, making the study and understanding of the concept of Europeanization all the more relevant. In this chapter I have attempted to highlight to what extent influential EU members such as France have been able to shape and influence EU Mediterranean policy to reflect national and international interests. The first section of the chapter confirmed that France has followed a pragmatic and self-interested approach in the Mediterranean, one which has been mainly driven by the notion of French influence, exceptionalism and historical-role conception in the region. Although French security and commercial interests have been the priorities of French engagement with Mediterranean, regional influence and power can be said to have characterized relations at the EU level. As we have seen above, President Sarkozy's foreign policy approach seems to indicate the shift away from the traditional tenets of Gaullism illustrated by a number of examples, be it his stance with regards to the US and NATO or over the impact of globalization. As the case of the UfM shows, Sarkozy's overall goals and policy strategy are not that different from his predecessors but, rather, it is the method and style that differ, not the substance. Like his predecessors, Sarkozy has remained true to the principles of Gaullism and French grandeur in the Mediterranean. Whether the shift in approach indicates that he is the first French leader to be aware of the difficulties for French foreign policy's tradition to adapt to new international and regional realities remains to be seen.

Bibliography

Azzam, M., 'Islam: Political Implications for Europe and the Middle East', in P. Ludlow (ed.), *Europe and the Mediterranean* (London: Brassey's, 1994), 89–104.

Balfour, R., 'The Transformation of the Union for the Mediterranean', *Mediterranean Politics*, 14/1 (2009), 99–105.

BBC, 'Nicolas Sarkozy's Victory Speech Excerpts', http://news.bbc.co.uk/1/hi/world/europe/6631125.stm (accessed 10 August 2009).

Bechev, D., and Nicolaidis, K., 'The Union for the Mediterranean: A Genuine Breakthrough or More of the Same?', *The International Spectator*, 43/3 (2008), 13–20.

Bernard, P., 'Jean-Christophe Rufin au "Monde": Le Quai d'Orsay est un ministère sinistré', *Le Monde* (6 July 2010), http://www.lemonde.fr/afrique/article/2010/07/06/jean-christophe-rufin-au-monde-le-quai-d-orsay-est-un-ministere-sinistre_1383868_3212.html (accessed 20 October 2010).

Biscop, S., 'The European Security Strategy and the Neighbourhood Policy: A New Starting Point for the Euro-Mediterranean Security Partnership?', paper presented at the EUSA Ninth Biannual International Conference (March–April 2005), http://aei.pitt.edu/2984/02/Paper_EUSA_ESS-EMP.doc (accessed 20 August 2010).

Blunden, M., 'France', in I. Manners and R.G. Whitman (eds), *The Foreign Policies of European Union Member States* (Manchester: Manchester University Press, 2000), 19–43.

Bouteldja, N., 'Who Really Bombed Paris?', *The Guardian* (8 September 2005).

Bowen, N., 'France, Europe, and the Mediterranean in a Sarkozy Presidency', *Mediterranean Quarterly*, 18/4 (2007), 1–16.

Bretherton, C., and Vogler, J., *The European Union as a Global Actor* (London: Routledge, 2006).
Brizzi, R., 'Francia: Sei mesi per l'Europa', *Il Mulino* (no date), 708–20, www.ilmulino. it (accessed 14 February 2008).
Cerny, P.G., *The Politics of Grandeur: Ideological Aspects of de Gaulle's Foreign Policy* (Cambridge: Cambridge University Press, 1980).
Cherigui, H., *La Politique Mediterranéenne de la France: entre diplomatie collective et leadership*. (Paris: L'Harmattan, 1997).
Copsey, N., *Member State Policy Preferences on the Integration of Ukraine and the Other Eastern Neighbours* (Swedish International Development Agency, 2008).
Del Sarto, R.A. and Schumacher, T., 'From EMP to ENP: What's at Stake in the European Neighbourhood Policy in the Mediterranean', *European Foreign Affairs Review*, 10 (2005), 17–38.
Dessouki, A.E., 'Regional Leadership: Balancing off Costs and Dividends in the Foreign Policy of Egypt', in A.E. Dessouki and B. Korany (eds), *The Foreign Policies of Arab States: The Challenge of Globalization* (Cairo: The American University in Cairo Press, 2008), 167–93.
Dimitrakopoulos, D.G., Menon, A., and Passas, A.G., 'France and the EU under Sarkozy: Between European Ambitions and National Objectives?', *Modern and Contemporary France*, 17/4 (2009), 451–65.
Drake, H., 'France, Europe and the Limits of Exceptionalism', in T. Chafer and E. Godin (eds), *The End of the French Exception? Decline and Revival of the 'French Model'* (Basingstoke: Palgrave Macmillan, 2010), 187–202.
Echeverria, C.J., 'Spain and the Mediterranean', in T. Couloumbis, S. Stavridis and T. Veremis (eds), *The Foreign Policies of the European Union's Mediterranean States and Applicant Countries in the 1990s* (London: Macmillan, 1999).
El-Sayyed Selim, M., 'Egypt and the Euro-Mediterranean Partnership: Strategic Choice or Adaptive Mechanisms?', *Mediterranean Politics*, 2/1 (1997), 64–90.
Emerson, M., Noutcheva, G., and Popescu, N., *The European Neighbourhood Policy Two Years On: Time Indeed for an 'ENP Plus'* (Brussels: CEPS, 2007).
Emerson, M., and Tocci, N., 'A Little Clarification, Please, on the Union for the Mediterranean', *CEPS Commentary*, www.ceps.eu (accessed 13 February 2009).
Emerson, M., 'Making Sense of Sarkozy's Union for the Mediterranean', *CEPS Paper No. 155*, www.ceps.eu (accessed 7 February 2009).
Evans, M., 'Franco-Algerian Relations in the Post-Cold War World', in T. Chafer and B. Jenkins (eds), *France: From the Cold War to the New World Order* (London: Macmillan, 1996), 219–32.

Erlanger, S., and Bennhold, K., 'Bold Mediterranean Summit Opens Diplomatic Door', *The New York Times* (13 July 2008), http://www.nytimes.com/2008/07/13/world/europe/13iht-ummit.4.14457192.html?_r=1 (accessed 13 February 2009).

Featherstone, K., 'Introduction: In the Name of Europe', in K. Featherstone and C.M. Radaelli (eds), *The Politics of Europeanisation* (Oxford: Oxford University Press, 2003), 1–27.

French Embassy in the United Kingdom, 'State visit to Morocco: Speech by M. Nicolas Sarkozy, President of the Republic' (2007), http://www.ambafrance-uk.org/President-Sarkozy-on-Mediterranean,9743.html (accessed 10 August 2009).

Furness, M., and Bodenstein, T., 'The Edges of External Governance: Europe and its Mediterranean Neighbours', paper presented at the UACES 37th Annual Conference (Birmingham, September 2007).

Fysh, P., 'Gaullism and the New World Order', in T. Chafer and B. Jenkins (eds), *France: From Cold War to the New World Order* (London: Macmillan, 1996), 181–92.

Gillespie, R., 'Spanish Protagonismo and the Euro-Mediterranean Partnership Initiative', *Mediterranean Politics*, 2/1 (1997), 33–48.

Hargreaves, A.G., 'Immigration, Ethnicity and Political Orientations in France', in T. Chafer and B. Jenkins (eds), *France: From the Cold War to the New World Order* (London: Macmillan, 1996), 207–18.

Howell, K.E., *Up-loading, Downloading and European Integration: Assessing the Europeanization of UK Financial Services Regulation* (Ashcroft international Business School APU, 2002).

Huntington, S., 'America in the World', *The Hedgehog Review* (2002), 7–18.

Jawad, H.A., *Euro-Arab Relations: A Study in Collective Diplomacy* (Reading: Ithaca Press, 1992).

Johansson-Nogues, E., '"A 'Ring of friends"? The Implications of the European Neighbourhood Policy for the Mediterranean', *Mediterranean Politics*, 9/12 (2004), 240–7.

Kamdar, M., 'The End of the Affair', *Foreign Policy* (October 2010), http://www.foreignpolicy.com/articles/2010/10/19/the_end_of_the_affair (accessed 29 October 2010).

Kelley, J., 'New Wine in Old Wineskins: Promoting Political Reform Through the New European Neighbourhood Policy', *Journal of Common Market Studies*, 44/1 (2006), 29–55.

Kennedy, P., 'Spain', in I. Manners and R.G. Whitman (eds), *The Foreign Policies of European Union Member States* (Manchester: Manchester University Press, 2000), 105–27.

Le Gloannec, A., 'Sarkozy's Foreign Policy: Substance or Style?', *The International Spectator*, 43/1 (2008), 15–21.

Licari, J., *Economic Reforms in Egypt in a Changing Global Economy* (Paris: OECD, 1997).
Meyrede, L., 'France's Foreign Policy in the Mediterranean' in T. Couloumbis, S. Stavridis, T. Veremis and N. Waites (eds), *The Foreign Policies of the European Union's Mediterranean States and Applicant Countries in 1990s* (London: Macmillan, 1999), 40–72.
Mortimer, E., 'Europe and the Mediterranean: The Security Dimension', in P. Ludlow (ed.), *Europe and the Mediterranean* (London: Brassey's, 1994), 105–26.
Pace, M., 'Norm shifting from EMP to ENP: The EU as a Norm Entrepreneur in the South?', *Cambridge Review of International Affairs*, 20/4 (2007), 659–75.
Parfitt, T., 'Europe's Mediterranean Designs: An Analysis of the Euromed Relationship with Special Reference to Egypt', *Third World Quarterly*, 18/5(1997), 865–81.
Pfeifer, K., 'How Tunisia, Morocco, Jordan and even Egypt became IMF "Success Stories" in the 1990s', *Middle East Report*, 210 (1999), 23–7.
Salomon, M., 'Spain: Scope Enlargement towards the Arab World and Maghreb', in F. Algieri and E. Regelsberger (eds), *Synergy at Work: Spain and Portugal in European Foreign Policy* (Bonn: Europa Union Verlag, 1996), 90–110.
Schmid, D., 'France and the Euro-Mediterranean Partnership: The Dilemmas of a Power in Transition', in H.A. Fernandez and R. Young (eds), *The Euro-Mediterranean Partnership: Assessing the First Decade* (Madrid: FRIDE, 2005), 92–102.
Smith, K.E., 'The Outsiders: The European Neighbourhood Policy', *International Affairs*, 81/4 (2005), 757–73.
Stavridis, S., and Waites, N., 'The European Union and Mediterranean Member States', in T. Couloumbis, S. Stavridis, T. Veremis and N. Waites (eds), *The Foreign Policies of European Union's Mediterranean States and Applicant Countries in the 1990s* (London: Macmillan, 1999), 22–39.
Tasche, T.C., 'The Project of a Union for the Mediterranean: Pursuing French Objectives through the Instrumentalisation of the Mare Nostrum', *L'Europe en formation*, 365 (Summer 2010), 53–70, http://www.cife.eu/UserFiles/File/EEF/356/EEF356-4TT.pdf (accessed 24 October 2010).
Wong, R.Y., *The Europeanization of French Foreign Policy: France and the EU in East Asia* (Basingstoke: Palgrave Macmillan, 2006).

DAVID STYAN

The Revival of Franco-Libyan Relations

Introduction: Trade with Tripoli, France's Mediterranean Vision and the *volte-face* of 2011

This chapter analyses the re-establishment of diplomatic and economic ties between France and Libya during the period 2003–10. Its focus is thus on events *before* the upheavals in the Arab world and the Western intervention in Libya of 2011. Set against the overlapping ties which were carefully woven between Paris and Tripoli over the previous seven years, France's leading role in the March 2011 military campaign to overthrow Gaddafi appears highly paradoxical. While it is not the purpose of this chapter to examine such paradoxes, in the light of the events of 2011 it is worth prefacing this chapter by noting some key aspects of, and questions raised by, France's leading military role and the abrupt change in stance of France, Britain, Italy and the US towards Gaddafi and his government.

In Libya, three weeks of violent clashes followed demonstrations which began in Benghazi on 15–17 February, inspired in part by events in neighbouring Tunisia, from where President Ben Ali had fled on 14 January, and Egypt, where President Mubarak had been forced out of office on 11 February. Mubarak had been a key supporter of Sarkozy's Union for the Mediterranean. French diplomacy and statements on the uprisings in Tunisia and Egypt had vacillated. Policy uncertainty coupled with a growing sense of complacency and complicity culminated in the dismissal of the foreign minister, Michèle Alliot-Marie, on 27 February. Hesitant and confused French diplomacy in February had also thrown a critical light on the years of rapprochement between Paris and Tripoli which are the focus of this chapter.[1]

As violence in Libya escalated, by 1 March President Sarkozy and his new foreign and defence ministers were considering a no-fly zone and military options towards an increasingly belligerent Gaddafi government.[2] On 10 March France was the first power to grant full diplomatic recognition to the opposition Libyan National Council; France then lobbied hard for what became UN resolution 1973. Adopted by the Security Council on 17 March, this authorized 'all necessary measures' to protect Libyan civilians, with France's Rafale and Mirage-2000 jets reported to be the first western combat aircraft over Libya just a few hours later. With France playing a key role in the military campaign, the assumption of NATO command of all allied military actions over Libya on 27 March also marked a significant practical step in the reintegration of France within NATO.

Several questions arise from the juxtaposition of the years of rapprochement between Libya and France (and by extension Britain, Italy and the other powers with whom, as the following text emphasizes, France was competing intensely for influence and contracts in Libya prior to 2011), and the subsequent hostility to Gaddafi. Firstly, to what degree, if at all, was the abruptness of President Sarkozy's attacks on Gaddafi and his government linked to the apparent hast and nature of the relations re-established between France and Libya between 2007 and 2010? Secondly, to what extent were the policy decisions taken against the Gaddafi government in spring 2011 a reaction not primarily to events in Libya itself, but rather a belated attempt to reassert French policy-making in the rapidly changing diplomatic context of early 2011 in the Mediterranean, in which President Sarkozy and the Fillon government appeared to have lost their bearings? I will briefly return to these questions in the conclusion, once

---

1   Not least because Alliot-Marie's partner, Patrick Ollier, was both minister for parliamentary relations and president of the Franco-Libyan friendship parliamentary group.
2   N. Nougayrède, 'Paris n'exclut pas une interdiction de survol de la Libye', *Le Monde* (3 March 2011), 8.

the rapprochement between France and Libya prior to 2011 has been analysed.

Prior to the events of 2011, it appeared simultaneously odd and obvious to focus specifically on the revival of ties between Paris and Tripoli in this volume examining France's wider ties with the Mediterranean. Odd because France patently has fewer historical ties with Libya than with almost any country around the Mediterranean; having none of the cumbersome colonial baggage, ties of trade or migration associated with her former colonies in the Maghreb or the Levant. The choice of Libya appears all the more anomalous given the Libyan leader was the only head of state *not* to have signed President Sarkozy's flagship regional initiative, the Union for the Mediterranean (UfM) at its launch in July 2008.

Yet a focus on Libya is simultaneously an obvious choice for students of French foreign policy under Nicolas Sarkozy. In the first six months of his presidency, France's links with Libya were rarely out of the news. In terms of both theatre and controversy, they overshadowed France's bilateral ties with any of its more established partners in the Mediterranean. Within weeks of coming to power he had twice despatched his then wife Cécilia to untie the final strand of the complex HIV/Bulgarian nurses dossier, which had blocked the EU's 'normalization' of ties with Libya since 2004.[3] With his wife having apparently played a key role in ending the nurses' eight-year ordeal on 24 July 2007, just hours later President Sarkozy himself arrived in Tripoli. Relishing his new presidential role on an Arab and African stage, he immediately proceeded to sign new arms and nuclear contracts. This sparked outcry and unease in France and elsewhere, prompting calls for a full parliamentary investigation into the links between the nurses' release and French contracts. As the initial, hyperactive, media-obsessed version of what Olivier Duhamel dubbed 'Starkozysme' got into its stride, controversy continued to surround the Libyan dossier for the next six months.[4] In December 2007 Sarkozy became the first EU head of state to welcome

---

3 Libya sentenced five Bulgarians and one Palestinian to death on allegations they had infected over 400 children with HIV/AIDS in a Benghazi hospital in 1998.
4 O. Duhamel and M. Field, *Le Starkozysme* (Paris: Seuil, 2008).

Colonel Gaddafi on an official visit. In the process he comprehensively sidelined both his foreign minister, Bernard Kouchner, and Rama Yade, under-secretary of state for human rights. Yade fulminated as Gaddafi planted his tent for a week in Paris in December, famously proclaiming that France was 'not a doormat on which dictators, terrorists or otherwise, can wipe away the blood of their infamy'. In the process Yade – France's first contemporary black African minister and the youngest member of the cabinet – earned a (no doubt suitably delicate) touch across the knuckles from her boss.[5]

This chapter examines the resumption of Franco-Libyan diplomatic and commercial ties prior to 2011 via a series of questions. Firstly it sketches the evolution of bilateral ties between France and Libya in 2007, asking how and why ties were restored in what, at least to French media and opposition politicians, appeared such a hasty and deeply controversial manner. Did the burgeoning Libyan relationship tell us something about Sarkozy's style of diplomacy, including as it did the unorthodox deployment of his second (soon to be divorced) wife as a fixer and the promotion of an apparently impetuous and cut-throat commercial policy with a state that just recently had accepted responsibility for one of the bloodiest terrorist attacks ever committed against a French target (the September 1989 bombing of Flight 722)? How were widespread moral qualms over renewed ties with Libya so easily overridden by the lure of nuclear and military sales? To what extent did another high policy priority of the new president, countering illegal immigration into the EU from Libya, weigh in the equation?

Far from Franco-Libyan relations being exceptional, the chapter argues that Sarkozy's embrace of Gaddafi corresponded to two deeper, essentially economic policy trends, the first being the continuity of long-standing French commercial strategies towards oil-rich states under the Fifth Republic. Policy towards Libya has far more in common with that

---

5   The full quote: 'la France ne doit pas recevoir ce baiser de la mort' and '[Gaddafi] doit comprendre que notre pays n'est pas un paillasson sur lequel un dirigeant, terroriste ou non, peut venir s'essuyer les pieds du sang de ses forfeits', 'Rama Yade convoquée à l'Elysée après ses critiques contre Kadhaf', *Libération* (10 December 2007).

towards Iraq, Saudi, UAE or other oil-rich states than the more dense and convoluted historical, political and cultural ties with the relatively populous and poorer states of Algeria, Morocco or Tunisia.[6] Secondly, reinforcing this point, that French policy was simply doing what all other OECD states, headed by the UK and Italy, were pursuing in trying to re-enter Libyan markets as sanctions were progressively lifted from 2003 onwards. France was particularly well-positioned to benefit from this, both because of earlier military ties, and success in developing and exporting the technology, including nuclear, that a rehabilitated Libya sought.

It was this, rather than the fact that France advanced a 'Mediterranean geo-strategic vision', newly re-packaged by the incoming president, which explains the evolution of policy, including as it did several threads from other French foreign policy objectives, including both controlling illegal migration into the EU and reforming France's military presence in Africa. The chapter thus also examines the paradox that, twenty years after Libyan and French troops fought each other in northern Chad, the two countries found themselves sharing similar regional agendas in the Sahel, notably towards civil wars in both Chad and Sudan's Darfur.

The final section highlights how the controversy surrounding the apparent links between 'Cécilia's' liberation of the nurses and the signing of Franco-Libyan contracts in July 2007 prompted the most substantive Parliamentary Commission of Enquiry into foreign policy ever undertaken in the Fifth Republic. The enquiry's findings serve the Franco-Libyan dossier in two ways. Firstly the 500-page report provides unprecedented detail into the evolution of Franco-Libyan relations between 2003 and 2007. Secondly the enquiry provides a trial run for an element within the broader constitutional changes subsequently proposed by President Sarkozy in May 2008 – that of making French foreign policy more accountable to the National Assembly.

---

[6] My book, *France and Iraq: Oil, Arms and French Policy Making in the Middle East* (London: I.B. Tauris, 2006), examines the evolution of policy towards oil-rich states.

Background: The Fifth Republic's Relations with Libya

As Colonel Gaddafi celebrated forty years in power in 2009, it should be recalled that France's commercial ties with Libya are just as old. Thus, recently renewed ties are very much a *revival* of a lucrative past relationship, interrupted first by war over Chad, then Libyan terrorism in the Sahel in the 1980s.

As part of de Gaulle's reinvention of French Arab policy in the mid-1960s, in April 1968 a Libyan delegation followed Egyptian and other Arab missions in seeking support and arms from Paris. When Gaddafi seized power on 1 September 1969, Pompidou swiftly recognized his government, which in turn promptly confirmed the purchase of 110 Mirage V planes, which had been under negotiation with King Idris's officials since 1968. At the time the deal – which was then the largest single overseas sale of French weaponry ever made – was kept secret. Details were leaked the following year, as Pompidou struggled to contain the anger of pro-Israeli lobbies in France and the US over France's ostensibly 'pro-Arab' policy post-1967. Controversy was compounded by Libya's use of Egyptian pilots and the subsequent 'loan' of planes to Egypt in 1973.[7] Dassault's delivery and servicing of the Mirage Vs over the coming years facilitated further military supplies. These were incorporated into a broader bilateral commercial accord in 1974. More quixotically, there are also perennial rumours, widely circulated among older French army officers, that Gaddafi had a French father.[8]

---

7   Styan, *France and Iraq*, 65, 86. See also G., Georgy, *Le Berger des Syrtes* (Paris: Flammarion, 1996), a memoir by France's ambassador to Libya in the 1970s.
8   A. Guidicelli, 'Gadaffi, the Corsican connection', *Bakchich*, http://www.bakchich.info/Gaddafi-the-Corsican-connection,02718.html (accessed 30 July 2008). While these need not detain us here, it is a historical curiosity that Gaddafi appears to have had links with influential French Gaullists well *before* coming to power. Jacques Benoist-Méchin, whose numerous works on the Arab world influenced Gaullist policy, apparently met Gaddafi clandestinely in early 1967. See Styan, *France and Iraq*, 121.

From 1979 onwards relations soured over Libya's claims to the Aouzou Strip which runs the length of its southern border with Chad. This progressively prompted Libyan incursions and Gaddafi's backing for rebels opposed to successive Chadian governments in Ndjamena. Chad's external security was guaranteed under a defence pact with France, in keeping with most African colonies which gained independence from Paris in 1960. In 1983 Libya invaded, prompting France's Operation Mantra in August 1983. This was replaced by Operation Épervier in February 1986. France then facilitated Chadian president Hissene Habré's eventual victory over Libyan troops in 1987. Over two decades later 1200 French troops are still stationed in Chad. Sovereignty over the Aouzou strip was awarded to the Chadian state by the International Court of Justice in 1994. Yet in September 1989 a French UTA aircraft DC10 ('Flight 722') exploded en route from Ndjamena to Paris, killing 170. Thus, shortly after the Lockerbie and Berlin attacks, France found herself in the same judicial-financial stand-off with Libya over terrorism as other Western powers. The UTA 722 trial in Paris of six men *in absentia* in 1999, including a brother-in-law of Gaddafi, the deputy head of Libyan intelligence, was followed in August 2003 with an acknowledgement of Libyan involvement to the UN, and financial compensation in January 2004.

France Embraces Gaddafi: July to December 2007

The key elements which propelled Franco-Libyan ties into the headlines during July 2007 were evoked in the introduction. The purpose of this section is to summarize selected aspects of closer bilateral ties between July and December 2007, and the French media and political reactions they provoked. As is explained in the subsequent section, despite the furore and hype, much of the reporting of Sarkozy's improbable *automne libyen* rests on a somewhat illusory misreading of both the motivations and mechanisms via which Sarkozy, on coming to power, simply consolidated several years of earlier, painstaking preparation by agencies of the French state for a resumption of commercial ties with Libya.

In essence the events of mid-2007 revolved around four distinct pieces of the Franco-Libyan jigsaw, itself a sub-puzzle within a larger EU–Libyan equation. Schematically these can be identified as: 1) the EU's need to secure the liberation of detained medical personnel; 2) Sarkozy's deployment of his wife as *émissaire spéciale* as part of these negotiations; 3) Sarkozy's visit to Tripoli the day after their liberation and the immediate signature of accords relating to the sale of military and nuclear technology to Libya; 4) the ensuing Parliamentary Enquiry into the chronology and linkages between these events.

The Nurses' Liberation

The six Bulgarian nurses and a Palestinian doctor incarcerated in Libya since 1999 due to a HIV-contaminated blood scandal were released on 24 July 2007. Following their imprisonment, torture and sentencing to death, negotiations over their release had become bound-up with the payment of compensation to families of the 500 families of the children infected with the HIV virus in Benghazi Children's Hospital. The nurses were released following the second visit of Nicolas Sarkozy's then wife, Cécilia, to Libya in July. She was accompanied by Claude Guéant, general secretary to the president, and the European commissioner for external relations, Benita Ferrero-Waldner. The Austrian Ferrero-Waldner had spearheaded attempts to release the nurses since her appointment in November 2004.

Cécilia Sarkozy's involvement prompted a barrage of criticism in France. While her husband praised her '*travail tout à fait remarquable*', other ministers stressed that she had no official mandate. *Le Figaro* noted that: 'Irrespective of the criticisms [...] by placing his wife at the forefront of complex international negotiations, the head of state has opened

The Revival of Franco-Libyan Relations

a new form of personalized, family-based diplomacy.'[9] In fact, as *Le Monde* noted, Cécilia was merely accompanying two high-placed French and EU officials. Yet while the use of private emissaries is hardly unique in the annals of Franco-Arab relations, there was considerable semantic confusion around her role (the terms 'wife and mother', 'presidential emissary' and 'the symbol of European humanitarian concerns' were all variously used by officials).[10]

Sarkozy's success in securing the release of the nurses appears to have been above all a question of timing. Negotiations had reached a critical stage, and although Tony Blair had assiduously courted Libya, and it had reportedly been British diplomats who had first elaborated the key face-saving mechanism (the Benghazi International Fund, from which compensation was paid, facilitating the dénouement of the nurses' saga), Blair left office shortly after Sarkozy came to power on 16 May. As Marc Pierini, the EU diplomat who headed the BIF from its creation in 2006, stated pithily during his hearing before the official French enquiry:

> In terms of personal relations, Colonel Gaddafi did know Tony Blair well, but he had left office on 27 June. As for Mr Prodi, he wasn't in a position to offer nuclear reactors for sale, while Germany's policy was only to sell weapons to NATO states, except in very limited circumstances, which weren't applicable to Libya. All these considerations meant that, of the EU states, only France could really address Libya's needs.[11]

The focus of the chapter is to highlight the underlying, longer-term economic logic and mechanisms of French policy. While we won't dwell on the (highly mediatized) role of Cécilia Sarkozy, it is nevertheless noteworthy that Pierre Moscovici's view, stated in the introduction to the enquiry's report, is explicit regarding her role in the liberation of the nurses:

9   S. Bussard, 'Cécilia Sarkozy, la diplomatie du troisième type', *Le Temps* (25 July 2009).
10 '*Happy end* en Libye' (English in the original), *Le Monde* (25 July 2009).
11 'Commission d'enquête sur les conditions de liberation des infirmières er du médecin bulgares détenus en Libye et sur les récents accords franco-libyens' (Paris: Assemblée nationale, 22 January 2008), Rapport No. 622, 96. All translations by the author.

All the testimonies received by the commission demonstrated that Mrs Sarkozy, the President's 'personal envoy', played a decisive role in the liberation of the Bulgarian nurses which was agreed directly in a conversation between Colonel Gaddafi and her, the details of which no one else is privy to.[12]

The parliamentary enquiry provides extensive evidence as to the length and complex chain of events leading up to the liberation. While much of this was beyond the influence of France, there is no doubting Sarkozy's interest in the matter; significantly, he even mentioned the nurses in an emotionally charged aside during his presidential acceptance speech on the evening of his election.[13]

Nuclear and Military Contracts: Areva ... and Dassault?

At the root of the scandal, controversy and resulting enquiry was the fact that Sarkozy signed a series of bilateral commercial accords with Libya the day after the nurses were released. In early August the Libyans publicized the fact that two contracts had also been finalized with French industrialists and armament companies, linking them to the French president's brief visit. Arriving in Libya on 25 July, Sarkozy was accompanied by foreign minister Bernard Kouchner and his secrétaire d'État à la coopération, Jean-Marie Bockel. Somewhat ironically, given her subsequent criticism of Gaddafi in December, the secrétaire d'État aux affaires étrangères chargée des droits de l'Homme, Rama Yade, as well Claude Guéant, appears to have made several *aller-retours* to Tripoli with Cécilia, the nurses and Commissioner Ferrero-Waldner. Although the minutiae need not concern us, seven separate accords were signed. These covered: a general partnership, defence and arms supplies; civil nuclear technology (with the largely state-owned Areva conglomerate); as well as four separate agreements on co-operation in the

---

12 'Commission d'enquête' (2008), 8.
13 In fact it was not in the written version, but an ad-lib (inadvertently referring to 'infirmières libyennes') which features in the filmed version of the 'Salle Gaveau' speech, 6 May 2007.

domains of research, education and health. Noteworthy is the fact that the general accord contains clauses covering policy co-operation over migration and security issues, while also stressing the importance of *'Mediterranean solidarity'*.

All the accords were published in the enquiry report issued in early 2008. It is notable that the Defence Accord, the source of considerable press speculation, was published in the *Journal officiel* on 10 October 2007, thus honouring a foreign policy pledge to promote greater transparency by making public France's bilateral defence accords.

The contracts immediately provoked controversy in France, with the Parti socialiste first secretary François Hollande declaring the lack of transparency over the deal 'troubling' after hearing Bernard Kouchner deny any linkage between the nurses' liberation and arms deals. This despite Gaddafi's son claiming that the 25 July accord covered joint military exercises and Libya's purchase of Milan anti-tank missiles.[14] Notwithstanding the controversy and confusion surrounding bilateral ties during July–September 2007, the enquiry did clearly establish the chronology and context of the sales, which had indeed been in preparation since 2004, as outlined below. Indeed it appears to have been the Libyan authorities who sought publicity for both the arms and nuclear deals with France, the enquiry stating:

> It appears that, just as had occurred for the arms contract, Libya deliberately sought to publicize its nuclear accord with France, in order to underscore to the international community that Libya had again become a legitimate and respectable state.[15]

By May 2009 it appeared that the military deal initialized in July 2007 envisaged the eventual supply not only of the Milan missiles, but also thirty-five helicopters, and potentially fourteen Rafale planes. Thus while the contracts actually signed in July 2007 were relatively small financially, their intention was to lay the foundations for larger contracts for Dassault,

---

14 Agence Française de Presse (AFP), 'Accord France-Libye', 1 August 2007.
15 'Commission d'enquête' (2008), 92.

the manufacturer of both Libya's aging *Mirages*, and the *Rafale*, which had yet to find confirmed overseas buyers.[16]

The Parliamentary Enquiry

In the week following the release of the nurses and signature of contracts there was considerable public and political disquiet over the deal. In early August the socialist group (SRC) in the National Assembly called for a full parliamentary enquiry into the links between the nurses' release and the arms contracts. This was supported by the Assembly's UMP's president Bernard Accoyer, and backed fully on 3 August by the offices of both the president and the prime minister, François Fillon, which stated that an enquiry:

> [...] will both provide total transparency on the questions [raised by parliament relating to the nurses' release and contracts] and simultaneously address the desire for full transparency [over its conduct of foreign policy] as requested by the state authorities.[17]

The establishment of the *Commission d'Enquête sur les conditions de libération des infirmières et du médecin bulgares détenus en Libye et sur les récents accords franco-libyens* was approved unanimously on 11 October 2007 when the National Assembly reconvened. Presided by Pierre Moscovici (PS) with Axel Poniatowski (UMP) as *rapporteur*, the enquiry held a comprehensive series of hearings with most key actors.[18] The final 500-page report was published on 22 January 2008 and provides one of the most detailed dissections of French foreign policy decision-making ever published. In some respects the filmed proceedings are even more revealing; the demeanours, intonations, evasions and wry smiles from the likes of Bernard Kouchner and Claude Guéant often speaking louder than the

---

16   Agence Française de Presse (AFP), 'Défense: Sarkozy soucieux d'activer un partenariat stratégique avec Tripoli' (18 May 2009).
17   Reuters, 'Sarkozy et Fillon pour une commission d'enquête' (3 August 2007).
18   The notable exceptions were Cécilia Sarkozy and the president himself.

dry prose of the report.[19] The testimonies of civil servants normally in the shadows, including Alain Bugat, administrateur général du Commissariat à l'énergie atomique (CEA), and François Lureau, Délégué général pour l'armement (DGA), provide unprecedented public insight into policy making.[20] The enquiry in many respects represented an attempt *avant la lettre* to promote President's Sarkozy's declared desire to render foreign policy making more accountable to parliamentary scrutiny. As the president of the enquiry stated:

> Parliament has decided, for the first time in the Fifth Republic, to exercise its powers to be informed of major diplomatic initiatives; in doing so they are putting into practice one of the commitments the President of the Republic has proposed to modernize our institutions.[21]

This refers to the substantial constitutional changes proposed by Sarkozy in 2007 and eventually voted through on 22 July 2008. These provide for greater parliamentary scrutiny of foreign policy, hitherto minimal in the Fifth Republic.[22]

Commercial Competition between OECD States to 'Normalize' Commercial Ties with Triopli

One should be wary of the notion of French 'exceptionalism' regarding foreign policy; French agencies – both diplomatic and commercial – act both in competition with, and mostly in a similar manner to, their OECD rivals.

19 All of these sessions were videoed and are available via the Assemblée's website See http://www.assemblee-nationale.fr/13/dossiers/infirmieres_bulgares_accords_franco_libyens.asp (accessed 14 August 2009).
20 See Styan, *France and Iraq*, for my earlier attempts to reconstruct such mechanisms for the Fifth Republic's relations with Iraq.
21 'Commission d'enquête' (2008), 10.
22 Ibid.

The chronology of Libya's rehabilitation as a 'legitimate' and lucrative commercial partner for OECD states is well established: the UN lifted sanctions on 12 September 2003; Libya renounced Weapons of Mass Destruction (WMD) on 19 December 2003, a decision in which Tony Blair and British diplomats claimed a key part.[23] In March 2004 Libya signed the Vienna protocol of the Non-Proliferation Treaty and joined the International Atomic Energy Agency. The USA lifted sanctions the following month, while initially retaining Tripoli on its 'terror list'. Blair visited Libya on 25 March 2004, Shell announcing significant contracts the following day. The first French commercial ministerial delegation had visited Libya two weeks earlier. The EU then lifted its embargo on arms sales to Libya in October 2004, opening the way for further contacts and contracts. In November 2005 Boeing announced the sale of two 737–800 planes; in December the Italian company Agusta won a major contract for helicopters in the face of stiff competition from French-backed Eurocopter. In May 2007 Tony Blair again visited, with British Petroleum subsequently announcing its 'single biggest exploration commitment' in the huge onshore Ghadames and offshore Sirt basins.[24]

Italy appeared particularly well placed to benefit from Libya's opening. In October 2004 Berlusconi and Gaddafi opened the pipeline constructed by Italian oil groups ENI and Agip linking Mellitah on Libya's west coast to Gela in Sicily. New commerce and colonial ties were intertwined; in 2004 Gaddafi had agreed that 20,000 Italians expelled in 1970 would be allowed to visit their old properties in Libya, while in August 2008, during Berlusconi's visit to Benghazi, Italy agreed to pay five billion dollars in reparations for colonial rule, reportedly linked to a Libyan undertaking to curb illegal migration. The agreement was ratified during a further trip to Tripoli by the Italian premier in March 2009 and the restoration of ties culminated in the June 2009 visit by the Libyan leader to Rome. Maintaining the colonial thread, he arrived replete with a photo of Libya's nationalist hero Omar Mukhtar (taken just before the rebel was hanged by

---

23 J. Kampfner, *Blair's Wars* (London: Free Press, 2003).
24 BP press release, 29 May 2007.

The Revival of Franco-Libyan Relations

Italy in 1931) pinned to his chest. Large numbers of Italian firms, including Finmeccanica, have signed significant contracts in Libya since Italy's rapprochement.

Italian oil giants ENI and Agip have long-standing investments in Libya. More surprisingly, when in early 2005 the first post-embargo oil contracts were awarded, they included awards to US companies. US policy towards Libya had shifted under pressure from oil interests; notably the August 2008 agreement exempting Libya from legal penalties, opening the way for an exchange of ambassadors in January 2009, the US ambassador subsequently talking enthusiastically about military co-operation and the sale of weaponry, while praising the newly appointed foreign minister Moussa Koussa.[25]

Just as other OECD states had sought, with varying degrees of success, to renew commercial ties with Libya after 2004, France followed. Thus the contracts of July 2007 were, in practice, the fruit of several years' work. Following agreement over financial indemnities for the victims of the 1989 UTA attack on 9 January 2004, meetings were held between respective ministers of foreign affairs. France's minister of external trade, François Loos, visited Tripoli shortly afterwards. In April 2004 Paris hosted the first formal official visit to Europe of Libya's premier, Choukri Ghanem, who met with President Jacques Chirac. In April 2004 Chirac made the first ever visit of a French head of state to Libya, with minister of defence Michèle Alliot-Marie subsequently signing a letter of intent on arms supplies in 2005.

These political and diplomatic contacts provided the context for the re-launch of economic ties, with around thirty French companies investing in Libya by 2007. Between 2004 and 2007 these included: Total (2005 exploration contract); EADS (Airbus sales initialled in December 2007); Thales (competing with Italians for radar installation); Areva (December 2007, 300 million euro contract for electricity generation and distribution,

---

25   H. Saleh, 'Thaw Spurs Prospect of Arms Sales to Libya', *Financial Times* (11 March 2009).

plus discussions on civil application of nuclear technology, notably for desalination); as well as Alcatel-Lucent (telecommunications).[26]

In terms of arms, France – who, as explained earlier, had furnished Libya with Mirages and other weaponry in the 1970s – aimed to renovate existing arms as well as sign new contracts. This was the purpose of the framework defence agreement signed by Michèle Alliot-Marie, minister of defence, on 5 February 2005. This in turn led to a sharp increase in applications for export licences (required for exhibitions and demonstrations as well as eventual sales) to France's Commission interministérielle pour l'étude des exportations de matériel de guerre (CIEEMG) from 2005 onwards.

While Libya sought to purchase a wide range of military equipment, French exporters faced stiff competition, both from OECD states, particularly the UK and Italy, as well as Russia. Ironically, one of the two 'French' contracts to be actually signed was with MBDA, a defence group involving joint French, Italian and British capital. The only other firm contract was signed by Dassault in November 2006 and concerned the renovations ('retrofit') of Libya's ageing Mirage F1s.[27]

On appointment as defence minister by Sarkozy in 2007, Hervé Morin concluded that France was lagging behind its main competitors in terms of military sales to Libya. Noting that Germany, Australia, Austria, China, Italy, Poland, Russia and the UK all appeared to have stolen a march on France, he concluded that: 'Honestly, we had been more cautious than our competitors before agreeing contracts with the Libyan authorities.'[28]

The acceleration of economic ties from 2007 onwards was accompanied by the re-launch of the Franco-Libyan Chamber of Commerce. Originally created in 1971, following a template adopted for many other key export markets, the CCFL acted as a lobbying, facilitation and access group for French investors in Libya. It published a regular newsletter and worked closely with the French embassy export promotion agencies in the country.[29]

26 'Commission d'enquête' (2008), 62.
27 Ibid., 66.
28 Ibid., 68.
29 Chambre de Commerce Franco-Libyenne, 'Lettre du CCFL' (September 2009), 5. The CCFL was headed by Michel Casals since its renaissance in 2007. Its *Lettre* and

Franco-Libyan Relations and the Wider Mediterranean

Bilateral relations are rarely one-dimensional; while the revival of commercial ties appears to have been uppermost in French policymakers' minds, three other aspects of Franco-Libyan ties are significant: controlling migration flows from Africa, enhancing pan-Mediterranean regional co-operation, and France's regional policies towards the Sahel and central Africa.

Migration

Particularly since 2005, French attitudes towards Libya, like those of Italy and the EU in general, have been shaped by concerns over illegal migration into the EU from Libyan shores. Nicolas Sarkozy first visited Libya when still interior minister, meeting Gaddafi on 6 October 2005. During the visit he signed an agreement relating to co-operation over illegal migration with his Libyan counterpart. Two months later, on 21–22 December, Brice Hortefeux, then Secrétaire d'Etat aux collectivités territoriales, signed a letter relating to security with his Libyan counterpart. Hortefeux, a close confidant of Sarkozy, was appointed to the newly baptized 'Ministry for Immigration, Integration, National Identity and Development Solidarity' in the first Fillon cabinet in 2007. The controversially renamed ministry neatly, albeit divisively, encapsulated how the French right perceive the link between migration, identity and 'co-développement', the latter term often a euphemism for the repatriation of migrants. Hortefeux accompanied Sarkozy to Libya in July 2007. Article 5 of the general accord signed during the visit concerned migration; it committed France and Libya to

---

website most acutely reflect the acute competition between French, German and Italian businesses for key investments. A separate indication of rapprochement came with the creation in 2006 of a more academic and cultural Association France-Libye, headed by archaeologist André Laronde, 'Who's Who: André Laronde', *Maghreb Confidential* (11 February 2006).

'reinforce their co-operation against illegal migration, notably via support for home countries'.

Sarkozy made immigration controls one of his two headline themes of the 2008 French presidency of the EU. Migration controls also featured in his plans to reform France's African policy.[30] Controls included the reinforcement of border controls, and strengthening the EU's embryonic Frontex force – joint-naval missions designed to secure the Union's borders against illegal migration. Having partially stemmed the flow of migrants into the Canaries and Spanish enclaves in northern Morocco, after 2006 the EU and Frontex paid greater attention to the Libyan coasts, from which thousands of African migrants paid traffickers in the hope of reaching Lampedusa, Malta or Italy.[31] In August 2009 European audiences were again confronted with the spectacle of dying migrants huddled in boats while coastguards of Malta, Italy and the EU's Frontex squabbled over the legal responsibilities and humanitarian consequences.[32]

As the country, after Malta, most directly receiving African migrants transiting via Libya, Italy played a key role in the immigration dossier. During the landmark opening of the Libya–Sicilian pipeline in October 2004, weeks before the EU had confirmed the lifting of the arms embargo, Gaddafi was reported as having warmly thanked Italy for their role in ensuring the embargo was lifted, press reports noting that:

> Rome pressed its EU partners to lift the embargo to allow Libya to beef up its border controls with patrol boats, all-terrain vehicles and surveillance equipment to combat illegal immigration to Europe via Italy.[33]

---

30   D. Styan, 'Sarkozy's African Policies: Framing Reforms and "Rupture"', paper to Symposium on Franco-British Cooperation in Africa, Portsmouth (26 June 2009).
31   D. Styan, 'The Security of Africans beyond borders: migration, remittances and London's transnational entrepreneurs', *International Affairs*, 83/6 (2007), notably 1177–8.
32   Extensive coverage of these issues can be found at http://migrantsatsea.wordpress.com and http://fortresseurope.blogspot.com (both accessed 24 August 2009).
33   Alexander's Gas and Oil Connections (11 November 2004), http://www.gasandoil.com/goc/news/nte44584.htm (accessed 24 August 2009).

Largely unnoticed in northern Europe, on 5 May 2009 the Italian government signed a bilateral agreement between Italy and Libya which enabled it to turn back would-be refugees from Libya. This came into force immediately, with the Italian navy intercepting and repatriating migrants.[34] In late July 2009 the EU announced a package of eighty million euros 'to assist Libya with dealing with illegal migration' (twenty million euros for centres to receive asylum seekers, sixty million euros to help manage illegal immigration, presumably on the southern border with Niger and Chad). Libya later asked for a far larger financial package, prompting the then EC Justice Commissioner Jacques Barrot (a member of President Sarkozy's UMP) to state that '[Libya is] demanding impossible things [...] We proposed 20 million euros, but they are asking for 200 to 300 million.'[35]

Libya and the Union for the Mediterranean

The complex interconnections and contradictions between diverse dossiers concerning France's ties with the Mediterranean are well illustrated by Libya's anomalous role in Sarkozy's flagship project, L'Union pour la Méditerranée (Union for the Mediterranean). At the launch of the UfM in July 2008, forty-three countries were represented, including every single Mediterranean state *except* Libya. Claude Guéant had visited Tripoli in late June; the Spanish Foreign Minister Miguel Angel Moratinos followed to plead for Libyan participation in the UfM just prior to the launch. All in

---

34 The policy is beyond the scope of this chapter, but it is noteworthy that in August 2009 Italy's minister of the interior claimed a 92 per cent reduction in the arrivals of irregular migrants since 5 May, asserting that 1,345 migrant arrived in Italy compared with 14,220 over the same period in 2008.

35 'Libya Seeks €200–€300 Million from EU in Exchange for Assistance with Migrants', *Migrants at Sea*, http://migrantsatsea.wordpress.com/2009/07/27/libya-seeks-e200-e300-million-from-eu-in-exchange-for-assistance-with-migrants-news/ (accessed 14 August 2007).

vain, Gaddafi's opposition appears to have rested on two points; firstly, echoing unease over the UfM in several EU capitals, he noted the proliferation of existing agencies and multilateral Mediterranean initiatives – most notably the Barcelona Process – explicitly doubting the usefulness of the multiplicity of overlapping regional initiatives.

Secondly, Libya objected to the geo-political inconsistency of the UfM. While the EU insisted that *all* EU members participate (meaning the UfM includes northern EU states with no Mediterranean coast), to the south only states bordering the Mediterranean were admitted (plus Jordan and Mauritania). Thus pan-Europeanism trumped Libya's attachment to pan-Arab and pan-African unity. Gaddafi reaffirmed his attachment to the '5+5' dialogue launched in Rome in 1990.[36] He also noted that the UfM risked being seen as a Western attempt to further divide Arab and African regimes, playing into the hands of Islamists, stating in Tripoli that:

> I believe that this project of the Union for the Mediterranean will simply spur illegal immigration and terrorism; Islamists will justify jihadist attacks on the grounds that [the UfM] represents European colonization.[37]

Libya, Chad and French African policy

The bilateral linkages on oil, arms and immigration discussed above cannot be isolated from broader regional concerns linking Paris and Tripoli. President Sarkozy's trip to Libya in July 2007 occurred against a background

---

[36] The 5+5 initiative regroups Portugal, Spain, France, Italy and Greece (plus Malta) with North African states. Libya chaired the group in 2008; in May 2009 members agreed to joint military initiatives. AFP, 'Réunion sur la sécurité et la formation en Méditerranéan' (17 May 2009). Much of the controversy focuses on the UfM's overlap with the 'Barcelona Process', launched in 1995, which encompasses thirty-eight countries. See K. Kausch and R. Youngs, 'Europe and North Africa; the end of the Euro-Med vision', *International Affairs* 85/5 (September 2009).

[37] Chambre de Commerce Franco-Libyenne, 'Lettre du CCFL' (September 2009), 5.

The Revival of Franco-Libyan Relations

of heightened diplomatic attention paid to the troubled Sudanese province of Darfur. Both French NGOs and Kouchner, then – lest it now be forgotten – still a darling of the Parti socialiste, had ensured that Darfur played a prominent symbolic role in the 2007 presidential campaign. Sarkozy's politically canny appointment of Kouchner as foreign minister propelled Darfur further up the agenda of both France and the EU, with a series of conferences being held, both in Paris and Libya (which hosted an African Union/UN conference in mid-July 2007) on the issue. Such interventions culminated in the authorization in September 2007 by the UN of the despatch of a 3,500-strong EUFOR military force to Eastern Chad, home to 250,000 refugees from neighbouring Darfur.[38]

Libya played a crucial role in the complex set of alliances and proxy forces straddling the Chad–Sudan border and Darfur. Between 2005 and 2010 a de facto state of war existed between the Chadian and Sudanese presidents, Idriss Déby and Omar el-Bashir, with rebels supported by Sudan narrowly failing to overthrow Déby in February 2008 and cross-border raids recurring repeatedly since late 2005.[39] Having been at war over Chad in the 1980s, Tripoli and Paris now found themselves uneasy allies of Idriss Déby. Thus during the February 2008 rebel assault in the Chadian capital, it was only ammunition supplied by Libya, whose transport was facilitated by French forces holding Ndjamena airport, that allowed Déby to withstand the rebels. In May 2009 Sarkozy explicitly called upon Libya to help find a solution to Chad's civil war following another series of rebel assaults. Libyan agreement and assistance is essential if Sarkozy or his successors are to achieve their long-term aim of withdrawing French troops from Chad and Central African Republic, thus significantly reducing the

---

38   On the role of Africa and Darfur in the presidential campaign, see R. Banégas, R. Marchal and J. Meimon, *Le fin du pacte colonia? Politique Africaine* (Paris: Karthala, 2007). For a definitive overview of France's highly convoluted policies on Chad and Darfur, see R. Marchal, 'Understanding French Policy toward Chad/Sudan? A Difficult Task', http://blogs.ssrc.org/darfur/2009/06/04/understanding-french-policy-toward-chadsudan-a-difficult-task-1/ (accessed 8 June 2009).

39   Civil war in Chad was reignited in 2005 by Déby's constitutional reforms, designed largely to allow his continued control of Chad's oil revenues.

cost of France's military presence in Africa. The 2008 EU experiment of EUFOR Tchad-RCA was in part a step towards this, with French troops being switched from their bilateral *Épervier* mission in Chad to the multilateral force to provide security in Eastern Chad for Darfurian refugees and Chadians displaced by civil war.

President Gaddafi also played an influential role throughout the Sahel and West Africa, and thus was of importance to broader French policy in the Sahel. Co-operation with Libya over African issues are viewed in Paris as assisting President Sarkozy's declared ambitions to effect a 'rupture' in Franco-African relations. One element of policy change is the need to drastically reduce the numbers, and cost, of French troops in Africa, a central plank of the June 2008 Defence White Paper.[40] The closure of military bases in Africa is in turn tied to a broader shift in French priorities towards the Gulf and oil-rich states. In May 2009 President Sarkozy inaugurated the new French base in Abu Dhabi, the United Arab Emirates capital, eclipsing France's military position in Djibouti. While France currently retains a reduced force in Djibouti, part of their facilities have now been leased to the US, Washington's first permanent garrison in Sub-Saharan Africa. This raises a final area of military policy convergence between not just France and Libya, but Tripoli and Western powers more generally – that of countering terrorism and Islamism. As noted earlier, the US now has limited military and counter-terrorism training ties with Libya.[41] Thus bilateral ties between Paris and Tripoli are embedded in a wider series of regional relationships, stemming largely from France's convoluted postcolonial military and political ties with Chad, the Central African Republic and Sahelian states, rather than simply France's Mediterranean partners within the UfM.

---

40   Livre Blanc, *Défense et sécurité nationale* (chair J.-C. Mallet, 17 June 2008), http://www.ladocumentationfrancaise.fr/rapports-publics/084000341/index.shtml (accessed 12 August 2008). See also Styan, 'Sarkozy's African Policies'.
41   The so-called Trans Saharan Counterterrorism Initiative, follow-on to the Pan Sahel Initiative (PSI) that began in 2002 training troops in four Saharan states: Mali, Mauritania, Niger and Chad.

Conclusion: Revived Franco-Libyan Relations versus Multilateral Mediterranean Policy?

What is striking about this analysis of the various strands which together comprised France's revived policy towards Libya is how little of the substantive fabric of *bilateral* ties stemmed from broader French policy ideas and initiatives towards the 'Mediterranean'. France clearly had active and multifaceted policies towards Tripoli; yet multilateral initiatives which rest on the notion of the Mediterranean as a geo-political concept appear marginal to policy, featured only rhetorically, if at all. Thus, analysing the evolving policy content of relations with Libya serves to nuance the 'Mediterranean dimension' of French policy, particularly when one notes the cacophony of overlapping and contradictory institutional initiatives (the Barcelona Process, the '5+5' formula, or President Sarkozy's UfM).

As such, European states' relations with Libya to 2010 were a partial microcosm of broader ties between northern and southern Mediterranean states. On the one hand there is intense commercial and diplomatic competition between EU states, particularly for military and infrastructure contracts in the Libyan case. Yet there is also a need for co-operation with Libya over pan-Mediterranean policy issues – both between individual EU states and within the EC as an institution. The chapter focused on migration policies, but the same may be true of environmental and energy policies. Thus, whether 'the Mediterranean' as a geo-political space is viewed in Paris as being strategic or marginal, or indeed if 'Mediterranean policy' can be analysed as being more than the sum of relations with individual states, depends on the specific policy focus.

Aside from these broad considerations, the chapter has sketched partial answers to the questions flagged in the introduction. Clearly the energetic, highly personalized nature of President Sarkozy's diplomacy towards Libya in mid-2007 epitomized the French president's desire to stamp his authority on foreign policy. Yet the text suggests there was little 'uniquely French', or indeed specific to Sarkozy's own style or goals, about either the objectives or mechanisms of French policy towards Libya. President Sarkozy merely consolidated commercial rapprochement between France and Libya, underway since 2004. France was simply following other OECD states in trying to re-enter Libyan markets as sanctions were progressively lifted from 2003 onwards. Indeed, several factors suggest that, despite being well-positioned having supplied Libya with arms in previous decades, France lagged somewhat behind British and Italian commercial rapprochement with Tripoli in the period 2004–7.

The chapter also highlighted a significant unintended outcome of revived Franco-Libyan ties, namely the controversy sparked by Sarkozy involving his second wife in what were essentially longstanding European Union negotiations to free the Bulgarian nurses, just hours before he signed commercial contracts. The controversy prompted the most substantive Parliamentary Commission of Enquiry into foreign policy ever undertaken in the Fifth Republic. This provides a wealth of detail on how priorities and decisions were arrived at in different branches of the French state. For scholars of France, it also provides a practical example of what may become increasingly the norm in terms of parliamentary scrutiny of foreign policy, making French foreign policy more accountable to the National Assembly.

The introduction posed two speculative questions in the light of the French rupture with Gaddafi in the rapidly changing regional and diplomatic context of March 2011. Clearly France was not alone in abruptly breaking ties with Gaddafi, despite the years of patient lobbying and diplomacy to re-establish commercial ties with Tripoli, as narrated above. Despite the commercial rivalry, notably between Britain, Italy and France to renew economic links with Tripoli, all three countries broke relations once Gaddafi's legitimacy was undermined, both by the rebellion in the east, and the Libyan regime's bellicose threats and violence against civilians. Thus it is difficult to see a direct link between the specificity of France's rapprochement with Libya under Sarkozy since 2007 which has been examined here, and the subsequent rupture of ties.

However, France did move first in recognizing the Libyan National Council, reflecting its hawkish attitude in pushing for the isolation of Gaddafi evident from late February onwards. This was both a relatively high-risk and a highly personalized strategy. Firstly the specific decision to recognize the LNC was taken without any consultation, not simply with any allied states but even with colleagues close to the president. It was the 'media-philosopher' Bernard-Henri Lévy who was instrumental in prompting Sarkozy to take this step.[42] France's own foreign minister, Alain Juppé, was completely unaware of the decision prior to learning of it via the media while at an EU summit in Brussels, convened on 10 March to discuss policy towards Libya. These early, and sharply personalized, moves to decisively distance France from Gaddafi may be seen in part as a reaction by Sarkozy to the barrage of criticism, both in 2007 for having allowed Gaddafi to visit Paris but also for his lack of clarity in articulating a specifically French response to the changes evident in the Arab world in the first months of 2011.

This links to the second question raised in the introduction: whether France's abrupt policy change over Libya reflected not so much a shift in bilateral relations with Libya, but rather a belated attempt to reassert French policymaking in the rapidly changing diplomatic context of early 2011 in the Mediterranean. The speed with which the decisions both to back military action against Gaddafi, and to unilaterally support the Benghazi-based rebel LNC, can be plausibly seen as President Sarkozy drawing a firm line under the hesitations and confusions which marked French policy towards the initial revolts in Tunisia and Egypt. This was important not simply for France's regional standing but also for Sarkozy's support within France in general and within his ruling UMP in particular, where his hesitancy in February in sacking foreign minister Michèle Alliot-Marie, tainted by ties to the ousted Tunisian president, had drawn extensive criticism.

Bibliography

Banégas, R., Marchal, R., and Meimon, J., *Le fin du pacte colonial? Politique Africaine* (Paris: Karthala, 2007).
Chambre de Commerce Franco-Libyenne, 'Lettre du CCFL' (2009), http://www.chambre-de-commerce-franco-libyenne.org/ (accessed 24 October 2009).

42 'Le jour où "BHL" est devenu le porte-parole de l'Elysée', *Le Monde* (12 March 2011), 5.

'Commission des affaires étrangères sur le projet de loi constitutionelle de modernisation des institutions de la Ve République' (14 May 2008), http://www.assemblee-nationale.fr/13/rapports/r0890.asp (accessed 24 August 2009).

'Commission d'enquête sur les conditions de liberation des infirmières er du médecin bulgares détenus en Libye et sur les récents accords franco-libyens' (Paris: Assemblée nationale, 22 January 2008), Rapport No. 622, http://www.assemblee-nationale.fr/13/rap-enq/r0622.asp (accessed 14 August 2009).

Duhamel, O., and Field, M., *Le Starkozysme* (Paris: Seuil, 2008).

Georgy, G., *Le Berger des Syrtes* (Paris: Flammarion, 1996).

Glaser, G., and Smith, S., *Sarko en Afrique* (Paris: Plon, 2008).

Guidicelli, A., 'Gadaffi, the Corsican connection', *Bakchich* (19 February 2008), http://www.bakchich.info/Gaddafi-the-Corsican-connection,02718.html (accessed 30 July 2008).

Kampfner, J., *Blair's Wars* (London: Free Press, 2003).

Kausch, K., and Youngs, R., 'Europe and North Africa: The End of the Euro-Med Vision', *International Affairs* 85/5 (September 2009).

La Chaîne parlementaire assemblée nationale (LCPAN), 'Forces Françaises à l'étranger: rester, mais où ?' (28 January 2009), http://www.lcpan.fr/emission/72662/video (accessed 1 February 2009).

Livre Blanc, *Défense et sécurité nationale* (chair J.-C. Mallet, 17 June 2008), http://www.ladocumentationfrancaise.fr/rapports-publics/084000341/index.shtml (accessed 12 August 2008).

Marchal, R., 'Understanding French Policy toward Chad/Sudan? A Difficult Task' (2009), http://blogs.ssrc.org/darfur/2009/06/04/understanding-french-policy-toward-chadsudan-a-difficult-task-1/ (accessed 8 June 2009).

Sarkozy, N., 'Discours', 17th Ambassadors' Conference (26 August 2009), http://www.elysee.fr/documents/index.php?lang=fr&mode=view&cat_id=7&press_id=2851 (accessed 10 April 2010).

Styan, D., *France and Iraq: Oil, Arms and French Policy Making in the Middle East* (London: I.B. Tauris, 2006).

Styan, D., 'The Security of Africans beyond Borders: Migration, Remittances and London's Transnational Entrepreneurs', *International Affairs*, 83/6 (2007), 1171–1191.

Styan, D., 'Sarkozy's African Policies: Framing Reforms and "Rupture"', paper to Symposium on Franco-British Cooperation in Africa, Portsmouth (26 June 2009).

PART 2

Cultural Dialogue and Cross-Fertilization

GÉRALD ARBOIT

France and Cultural Diplomacy in the Mediterranean

On 4–5 November 2008, while the meeting of the foreign affairs ministers of the member countries of the 'Barcelona Process: Union for the Mediterranean' was drawing to a close, the Etats-généraux culturels méditerranéens (Mediterranean Cultural Forum or MCF) were getting under way in the same location. For the first time since the launch of the Euro-Mediterranean partnership, the cultural constituent appeared to be linked to the political dialogue between the Mediterranean's northern and southern shores. Better, the apparent continuity between the two events led one to believe that a more pro-active approach in the development of cultural cooperation was to be seriously endorsed by partners on each side of the Mediterranean. The most important aspect to this was that it took up and developed further a French initiative launched by President Jacques Chirac in Barcelona on 28 November 2005 to celebrate the tenth anniversary of the Euro-Mediterranean partnership. Reasserting from the start of his first mandate (1995–2002) his intentions to follow an Arab policy, with strong Gaullist overtones,[1] Chirac knew the extent to which cultural action is based on 'une certaine idée de la France' and is a powerful tool to demonstrate and enhance France's power on the international stage. President Chirac was determined to give clear doctrinal direction to France's foreign cultural action and he intended to use the Mediterranean as his privileged arena of influence to put into practice the ideas underpinning his cultural diplomacy. The first meeting was held in Paris in September 2006 under the title 'Dialogue des peuples et des cultures'. This particular initiative, which was planned to last two years, came to an end in Alexandria in January 2008.

---

1 M. Vaïsse, *La puissance et l'influence: La France dans le monde depuis 1958* (Paris: Fayard, 2008), 351–5 and 363–70. See also J.R. Henry's chapter in this volume.

Launching his project for Mediterranean union within the first minutes of his mandate (6 May 2007), Nicolas Sarkozy decided to continue with his predecessor's cultural ambition; but the two initiatives were to remain distinct from each other.

Culture has usually been seen as an element of soft power. In France's case, it has been as much an instrument of co-operation as of competition, even rivalry, between States. On 6 August 1982, the Conférence mondiale sur les politiques culturelles, organized by the United Nations Educational, Scientific and Cultural Organization (UNESCO), had taken the initiative in defining culture:

> as the set of traits – distinctive, spiritual and material, intellectual and emotional – which characterize a society or social group. In addition to the arts and letters, it encompasses ways of life, fundamental human rights, value systems, traditions and beliefs.[2]

The appearance of culture in the domain of international relations corresponded to a new awareness on the part of States that culture really matters. In the case of France, Chirac and then Sarkozy's involvement were, in fact, part of a long tradition in the field of cultural diplomacy. Culture has been 'a major, original element' of France's foreign action, and the Mediterranean has been its preferred space for this action.[3] France therefore has had at its command a body of cultural doctrine for this area into which the Etatsgénéraux culturels fitted. But as is often the case, France has moved forward on its own without any real co-ordination with its potential partners. In fact, it has been hard to reconcile France's conception of culture with the political problems which have shaken the Mediterranean since 2001, without forgetting, of course, a series of Franco-French debates which reveal a lack of consensus on what France's cultural diplomacy ought to be.

---

2   'Declaration on Mexico by the United Nations Educational, Scientific and Cultural Organization', http://portal.unesco.org/culture/fr/files/12762/11295422481mexico_fr.pdf/ mexico_fr.pdf (accessed 25 March 2011).

3   R. Young, *Marketing Marianne: French Propaganda in America 1900–1940* (Piscataway, NJ: Rutgers University Press, 2004).

France's *doctrine culturelle* for the Mediterranean

France could no longer remain indifferent to the establishment of a global society that is increasingly dependent, for its prosperity and *rayonnement*, on knowledge transfer and cultural management. French leaders of all political hues are often tempted to develop a dual discourse about the fate of France, a discourse which has acquired a strong resonance since the terrorist attacks on New York and Washington on 11 September 2001. Either France is condemned to a slow but certain national decline, and there are enough *déclinologistes* on the market to point out what is wrong and what should be reformed;[4] or France is presented as an eternal nation who is able to re-invent herself and whose national values are certain to provide a sense of direction to all in a global, confused and disorganized world. France's perception of herself – which transcends all political divides – is primarily as a universal power, showing the rest of the world the road to follow.[5] Some powerful, normative decisions shaping France' foreign policy ensued from this unique, universal mission: one can even argue that this fixed idea explains 'France's refusal to adapt to changing international conditions'.[6] In contrast to the United States, often defined by their ability to display all the attributes of military might and hard power, France's desire to project her cultural might has been particularly prevalent in periods of domestic soul searching and perceived weakness on the international stage; the Third Republic had made good use of it during its opportunist period (1879–98), and then in the period following the First World War (1919–29), whereas Vichy – as well as the leaders of the Resistance in the post-war period – turned it into a keystone of French foreign policy.

---

4   For a good illustration of déclinologie, see I. Ramonet, 'Liberté, équalité, sécutité', *Le Monde Diplomatique* (April 2006).
5   S. Hoffman, 'Paradoxes of the French Political Community', in Stanley Hoffmann et al., *In Search of France* (Cambridge, MA: Harvard University Press, 1963).
6   R.C. Macridis, 'French Foreign Policy', in *Foreign Policy in World Politics* (Englewood Cliffs, NJ: Prentice-Hall, 1962), 63.

Within this context, the launch of great ventures, in the Mediterranean as elsewhere, even at home, is first of all rooted in a sense of insecurity – and not in the desire to turn grand ideals into reality or even in a thirst for reforms. In spite of its proclaimed unchangeable universalist objectives, French cultural diplomacy is remarkably flexible in its objectives and often appears to provide a form of justification which, in fact, adapts to whatever has to be justified. How to explain, otherwise, the obvious contradiction between Chirac's reassertion that France is *'une république une et indivisible'*, reluctant to embrace cultural pluralism at home and the promotion of cultural diversity on the international level?[7] President Chirac knew how to make use of the diplomatic clichés and in the defence of his position put forward grand concepts, such as sovereignty and equality of nations in a global world and the importance that must be granted to the elegance of the French language, the language of negotiation and agreement *par excellence*. In this respect, Nicolas Sarkozy uses the same language: France's cultural power, rather than any military or economic hegemony, and her moral superiority, rather than the crass domination of a globalized Hollywood culture, are presented in abstract and symbolic terms whereas her material capabilities and are rarely mentioned and practical objectives are often eluded. For instance, the *atelier* (working group) 'Fractures culturelles, mémoires, histoires, préservation des patrimoines' of the the Etats généraux culturels méditerranéens, has been seen by political partners on each side of the sea as the most interesting of all. However, between the first meeting in Paris (2005) and the last one in Alexandria (2008), no one was able to define exactly what the *atelier* was about. Less than 30 per cent of its participants turned up at the Seville meeting in June 2007, whilst the historians, a minority within the group, were divided in their approach. The first meeting in Paris aimed to focus mainly on the Israeli–Palestinian conflict and proposed a new revision of school textbooks. The second

---

7 C. Bédarida and B. Gurrey, 'Jacques Chirac célèbre la diversité culturelle', *Le Monde* (3 February 2003). See also how French intense lobbying to secure the Convention on the protection and promotion of diversity cultural expressions by UNESCO: http://www.unesco.org/new/en/culture/themes/cultural-diversity/2005-convention (accessed 6 June 2011).

meeting limited itself to producing a catalogue of arguments, based on philosophical concepts, which debunked 'false memory'. The third meeting returned to the initial idea of evaluating textbooks. The most interesting aspect was that neither the history teaching department of the Council of Europe nor the Georg Eckert Institute for International Textbook Research had been invited to these meetings. As always in France, this exaltation of French cultural diplomacy, which no one would describe as nationalist, is more concerned with its own moral and political standing than with real power, and what is essential is keeping within this.

Since the end of the Cold War, it seems that the West has lost some of its bearings in the Mediterranean and this must be taken into account to make sense of French cultural diplomacy. The first Gulf War was a serious incubator of the misunderstandings which were to come.[8] President Chirac, from the start of his mandate, asserted that the Mediterranean had to become a space for further political cooperation and an ambitious project of the European Union. 'Having destroyed a wall in the East, Europe must from now on build a bridge to the south.'[9] This speech, given on 8 April 1996 at the University of Cairo, marked France's willingness to engage with the consequences of her own history in the Mediterranean, of her geographical (and thus political, economic and cultural) proximity with other Mediterranean countries and societies and of the presence on her territory of substantial communities of immigrants, both from the northern and southern shores, who have recently arrived on her territory or who are now deeply integrated into her social fabric. For President Chirac, the framework for this action was the Euro-Mediterranean Conference in Barcelona which had opened in November 1995. In Barcelona, Chirac invited the partners to participate in an *'atelier culturel méditerranéen'*, which would be organized in Paris in 2006 and from the southern shore

8   F. Hoveyda, *Que veulent les Arabes?* (Paris: First, 1991). This book was published again, unsurprisingly, in March 2004 by Page après Page, with a preface by G. Millière, a neo-conservative essayist.

9   J. Chirac, 'Discours à l'Université du Caire' (1996). Presidential discourses are available on the site of the Elysée Palace: http://www.elysee.fr/elysee/elysee.fr/francais_archives/interventions/discours_et_declarations (accessed 28 June 2009).

it appears that Egypt would become a major actor in the development of this cultural dialogue. Ten years later, on 20 April 2006, whilst inaugurating the French university in Cairo, President Chirac praised his counterpart, Hosni Mubarak, who demonstrated 'the same confidence in the future, the same view of the world, and the same vision of the dialogue necessary between cultures'. He recalled above all how much 'from one shore to the other of the Mediterranean, [he and Mubarak] had not stopped putting it into practice, this dialogue, which may appear to be natural, but which isn't necessarily so'. Chirac concluded his discourse with Gaullist overtone, defining this cultural dialogue as *'une ardente obligation'* (an imperative duty).[10] Thus, cultural diplomacy is very much part of French foreign policy and some have argued that, in this respect, France is a rather exceptional country, being one of the rare and one of the first states to have developed this sort of strategy.[11]

From France's point of view, cultural diplomacy became an important method of maintaining relations when the political dialogue was interrupted. France had previously used this method in the Arab world on several occasions, after the Suez crisis (1956), in the aftermath of the Algerian war (1954–62) and since the reorientation of French foreign policy towards the Mediterranean (1995). What is original in Chirac's approach is the fact that he intended to use European construction and its mode of governance as a model for his Mediterranean diplomacy. Indeed, the president made explicit references to the objectives prescribed in the White Paper on European governance to define his Mediterranean strategy: 'the strengthening of the dialogue with civil society, particularly businesses and NGOs' which is a core element of European governance should also become a template for

---

10 Ibid.
11 M. Vaïsse, *La puissance et l'influence: La France dans le monde depuis 1958* (Paris: Fayard, 2008), 523–52. See also: L. Dollot, *Les relations culturelles internationales* (Paris: Presses universitaires de France, 1968); S. Balous, *L'action culturelle de la France dans le monde* (Paris: Presses universitaires de France, 1970); F. Roche and B. Pigniau, *Histoires de diplomatie culturelle des origines à 1995* (Paris: La documentation française, 1995); J.-F. de Raymond, *L'action culturelle extérieure de la France* (Paris: La Documentation française, 2000).

relations across the Mediterranean.¹² Cultural diplomacy had long been considered the poor relative of diplomatic action by the ministry of foreign affairs¹³ and if the Mediterranean has for a long time been one of its priority areas, particularly North Africa, its relative importance had somewhat diminished since its 1970s heyday.¹⁴ The reforms undertaken by the ministry of foreign affairs from 1996 onwards aimed, according to those who were promoting them, 'to reinforce the coherence of international co-operation, to allow it to have its proper place within France's foreign action and to promote the active participation of actors from civil society in its elaboration and implementation'.¹⁵ To support these objectives a new Direction générale de la coopération internationale et du développement (DGCID) was created to replace the defunct Ministère de la coopération (abolished in March 2008). This sought to rationalize and use more strategically resources, in a context of budgetary reductions.

This acknowledgement of the growing importance of civil society on the international stage required projects on culture, usually managed by professional diplomats, to be redesigned in order to secure the participation of local associations and groups leading to the creation of new partnerships. This was particularly true in the Mediterranean basin where there is a particularly dense network of associations and NGOs, such as in Morocco or Italy, or where immigrants' communities on the northern shores of the sea are well structured: this favours the development of partnerships between and across the two shores. The development of such partnerships went beyond the promotion of relationships between migrant communities separated by the sea. First, associations and NGOs from the northern shores also entered the fray and linked up with southern

12   J. Chirac, 'Intervention du président de la République au sommet Euro-Méditerranéen de Barcelone' (2005), http://www.elysee.fr/elysee/elysee.fr (accessed 28 June 2009).
13   A. Bry, *La Cendrillon du Quai d'Orsay, avril 1945–décembre 1998* (Paris: Grou-Radenez, 1999).
14   P. Cabanel (ed.), *Une France en Méditerranée: Ecoles, langue et culture françaises, XIXᵉ–XXᵉ siècles* (Paris: Creaphis, 2006).
15   Bry, *La Cendrillon du Quai d'Orsay*, 59.

groups, ensuring substantial, decentralized co-operation surrounding the management of Mediterranean issues; second, actors from civil societies managed to expand upon the modest place which was originally accorded to them in the Euro-Mediterranean partnership, and legal practitioners and women became particularly active. The crisis of the Barcelona Process even helped intensify their action in order to compensate for the failures of inter-governmentalism.

The long-standing cooperation between the Quai d'Orsay and the Alliances françaises[16] had doubtlessly prepared the Ministry of Foreign Affairs office for this sort of development, even if the development of partnerships between the Quai d'Orsay and civil society associations was not always particularly easy. On the one hand, the *'énarques'*[17] having replaced the *'agrégés',*[18] the DGCID quickly became *un monstre ingouvernable*. Likewise the Haut conseil de la coopération internationale, established on 10 February 1999 as a consultative authority to the prime minister, revealed itself to be a tool for the management of political tensions, due to the cohabitation between a conservative president and a socialist prime minister, rather than an effective institution, promoting and coordinating international cooperation.[19] On the other hand, by getting involved in such

---

16   Created 21 July 1883 by the diplomat Paul Cambon, a French resident in Tunisia, the Alliance française was a private initiative which was set up to allow France to bounce back following the French defeat in 1870 by reinforcing its cultural influence abroad. Apolitical and non-religious, it is the model for all subsequent cultural bodies such as the Deutsche Akademie (which became the Goethe-Institut in 1955) which first appeared in 1925; the British Council, founded in 1934 (its first office outside the British Isles was in Alexandria four years later); and even the Instituto Cervantes, even though this has been dependent on the Spanish ministry for foreign affairs since its creation in 1990.

17   Former students of the École nationale d'administration (ENA), an institution for the initial training of top civil servants from France and abroad.

18   Participants in a competitive examination for the recruitment of high-school teachers in subjects which are the object of teaching in high-school establishments (in general, high schools and preparatory classes for these participants).

19   France experienced her third period of cohabitation from 2 June 1997 until 21 April 2002 between President Chirac elected by a popular majority, on 17 May 1995, and

partnerships, NGOs and other representative groups of civil society kept their distance from state institutions, being somewhat wary of the consequences of entertaining too strong connections with 'those in power'. As a result, partnerships were often very loose and alliances between groups tended to be ephemeral, with the exception of alliances between universities[20] often described as an 'epistemic community',[21] which, since the onset of the Barcelona Process, has managed to maintain a degree of cohesion to the point that it is now able to formulate original and often alternative proposals rather than being simply critical of the entire process. It is moreover difficult to define the real motivation of these new actors who were as numerous as they were diverse: their motivations could be religious or political, they may support an alter-globalist agenda or a humanistic one, or they may seek to pursue personal or professional interests. If, admittedly, these new actors have a keen interest in cultural issues, they often seem to enter into logic of confrontation with state actors and prefer to develop arguments pertaining to human rights issues rather than explore further the cultural themes put forwards by the *ateliers* of the Mediterranean Cultural Forum.

---

    a left-wing parliamentary majority which resulted from the legislative elections of 1 June 1997, led by Lionel Jospin as prime minister.

20 F. Siino, 'Une politique méditerranéenne des chercheurs?', in J.H. Robert and G. Groc (eds), *Politiques méditerranéennes entre logiques étatiques et espace civil* (Paris: Karthala, 2003), 287–301.

21 Such a community, it has been argued, is based on 'temperate equality' (H. Longino, *The Fate of Knowledge*, Princeton: Princeton University Press, 2002), 'limited solidarity' (E. Lazega, *The Collegial Phenomenon: The Social Mechanisms of Cooperation Among Peers in a Corporate Law Partnership*, Oxford: Oxford University Press, 2001) and types of exchange based on critical discussion (A.I. Goldman 'Foundations of Social Epistemics', in A.I. Goldman (ed.), *Liaisons: Philosophy meets the Cognitive and Social Sciences*, Cambridge, MA: MIT Press, 1992, 179–207). However, a lack of cognitive certainty makes the activities of the agents in the domain of collective expertise highly dependent on the dynamics of social co-ordination. See B.Conein, 'Communautés épistémiques et réseaux cognitifs: coopération et cognition distribuée', *Revue d'Economie Politique*, 113 (2004), 141–59.

The limits to this form of political action became apparent in the way state actors defined priorities: these were neither immediately relevant, and not always decipherable by civil society actors. In France in particular, when civil society put forward radical cultural initiatives, representatives of the state were often reluctant to accept what they perceived to be a major departure from a rather traditional, if not elitist, conception of culture. In 2008, the Ministry for Foreign Affairs undertook another major restructuring of its internal organization, partly motivated by budgetary constraints, partly, it was argued, to provide effective strategic responses to global challenges. The Haut conseil de la coopération internationale was phased out in March 2008 and some of the DGCID's missions were given to a new Direction générale de la mondialisation. Education and training, as well as economic and social development were passed to the Agence française de développement, the central organization for aid, and to other organizations carrying out logistical functions or specialized technical assistance. In this context; the creation in 2007 of the Association des Ateliers culturels méditerranéens – presided over by Jacques Andréani, a retired ambassador who remained an active French diplomat – played a part in re-orientating public action towards civil society. But some observers, such as Dominique Wolton, were hard to convince:

> In a time of globalization, France has to be at the forefront of a great political offensive – no longer in terms of power but of influence; this influence is to be found in her capacity to act in the fields of culture, art, science and communication. And yet at the same time when globalization is indirectly paying tribute to us and allowing us to have a return on our investment – that is to say, to reap the benefits of what our cultural network abroad has achieved over a century – then we decide to tighten our belts![22]

On 10 June 2009, Senator Jacques Legendre from the cultural affairs select committee and Senator Josselin de Rohan from the foreign affairs, defence and armed forces select committee produced an unequivocal report on the

---

22   O. Quirot, 'Dominique Wolton: "La France brade son réseau culturel à l'étranger"', *Le Nouvel Observateur* (11 December 2008).

foreign cultural action that had been underway for a year.²³ For two decades national representatives had not stopped worrying about the decline in French cultural influence in the world. In the summer of 2009, comments expressing scepticism about the cultural dimension of the Union for the Mediterranean appeared in the French media.²⁴ In turn, they noted the lack of audacity and vision 'on both shores of the Mediterranean at a crucial moment when new talents and new projects were about to come to the fore'.²⁵

From Cultural Dialogue to Cultural Policy? The French Contribution

Les Ateliers culturels mediterranéens launched with great pomp at the Elysée in September 2006 were an opportunity for Jacques Chirac to refute again the Huntingtonian idea that there is a clash of civilizations. Seeking to promote a dialogue between peoples and cultures (*dialogue entre les peuples et les cultures*), the Ateliers were in fact a series of three meetings with representatives of civil society organizations active in the 'Europe–Mediterranean–Gulf' network. They were held in Paris, Seville (July 2007) and Alexandria (January 2008). The last meeting was to finalize the creation of an epistemic network, and devise an ambitious plan of action from the conclusions presented by its six working groups.²⁶ To achieve that, a level of continuity between each meeting was necessary. Alas, the opposite turned out to be the case. Only 26 per cent of the participants of the

23   Le rayonnement culturel international: une ambition pour la diplomatie française (Sénat, 2009), http://www.senat.fr [accessed 28 June 2009].
24   R. Michel and J.-F. Boyer, 'L'union pour la Méditerranée et la télévision', *Le Figaro* (13 August 2009).
25   Ibid.
26   1) 'Cultural divides, memories, histories, heritage preservation'; 2) 'Images, writing, cultural and artistic creation and circulation'; 3) 'Religion and society'; 4) 'Social and cultural modernization'; 5) 'Education'; 6) 'Shared values, common values'.

first meeting attended the second one. These meetings also revealed tensions between French and Spanish organizers (from the Fundación Tres Culturas[27]), the two European partners accusing one another of minimizing their respective contribution to the Ateliers and squabbling about the best way to organize them.

The development of the Mediterranean cultural working groups was based on the too often neglected idea that culture was a 'factor of economic and social development', 'an essential factor in promoting cultural diversity' and fostering the development of conviviality and long-lasting friendships across borders. At no time, however, was the existence of a Mediterranean culture debated. Of course, without fail, the role of agriculture, the importance of the alphabet, the presence of three great monotheist religions and three great empires (Greek, Roman and Ottoman) were mentioned as to explain how brilliant and diverse civilizations emerged and merged around the edges of this *al-bahr al-abyad al-mutawassat* ('white sea in the middle'). The question remains as to whether the Mediterranean is still a culturally relevant region now that we live in a time of 'shock about the lack of cultures', to quote Hubert Védrine.[28] Since September 2001, every attempt to define a Mediterranean culture and to discuss the way it must be approached has been met with scepticism, and generated more anxiety than hope. All sorts of criticisms were levelled at the definition of the Mediterranean as a coherent and relevant cultural arena. In particular, the recent development of post-modern issues (on *métissage*, hybridity and cultural globalization) have contributed to the discrediting of the very idea of a Mediterranean cultural sphere. Nevertheless, and regardless of the thorny debates about the nature – or even the existence – of a Mediterranean culture, most actors involved in the Ateliers agreed that to take the Mediterranean as a frame of reference meant on the one hand, that debates did not fall into the vain exaltation of the small cultural differences that can be found around the

---

27   The Fundación Tres Culturas, created in 1999, is based in Seville and aims to promote progress and tolerance among the three main cultures of the Mediterranean: European, Arab and Jewish cultures. See http://www.tresculturas.org (accessed 1 January 2011).

28   H. Védrine, 'Les Etats-Unis: hyper-puissance ou empire?', *Cités*, 20/4 (2004), 147.

France and Cultural Diplomacy in the Mediterranean 145

Sea – a 'narcissism about small differences'[29] – and on the other hand, that the excessive fascination that Europeans feel for a parochial conception of their own culture was neutralized.

Table 1: Plan of Action at Alexandria, 21 January 2008

1. *History and memory* – new collective historical frames of reference – better knowledge of memories – policy of memories and reconciliation
2. *Images and writing* – help with production and diffusion of audio-visual material – support with translation – support with artistic exchanges and creation – support with freedom of information and media independence
3. *Religious events and society* – creation of a denominational consultative council with a role in co-ordination and contact – set up a system of international training – comparative study of religions – create cultural centres of information about religions
4. *Social modernization* – create a structure for debates and proposals – create a network for high-level exchange and dialogue – set up an information network
5. *Education and research* – optimize use of existing European programmes – creation of a programme of co-operation and mobility
6. *Common values, shared values* – 'Charte sur le dialogue des cultures' – set up a system for monitoring the findings and proposals of the Atelier culturel

---

29 P. Burke, 'Cultural Exchange as Appropriation and Adaptation', in D. Albera, A. Block and C. Bromberger (eds), *L'Anthropologie de la méditerranée* (Paris: Maisonneuve & Larose/Maison méditerranéenne des sciences de l'homme, 2002), 555–6.

This is why the plan of action adopted at Alexandria looked like *un inventaire à la Prévert*.[30] Strangely enough, this plan of action does not make any reference – indeed, seems to ignore – the work undertaken in the same field by other organizations, such as such as the Council of Europe and UNESCO in the north, and the ALECSO (Arab League Educational, Cultural and Scientific Organization), the ISESCO (Islamic Educational, Scientific and Cultural Organization) and the IRCICA (Research Centre for Islamic History, Art and Culture) in the south. The absence from the Ateliers of the Georg Eckert Institute for International Textbook Research, whose expertise has been universally recognized, was particularly puzzling. This German research centre had not only already been involved in several international programmes[31] with UNESCO as well as with national governments to update textbooks – particularly history textbooks[32] – but since September 2001 it has established itself as the cornerstone in all initiatives aiming to modify the perceptions that are brought about through school teaching in Arab-Muslim and Western countries.

One of Chirac's responses to the terrorist attacks perpetrated in September 2001 was to reinforce the cultural dialogue between Islam and the Western world. The major challenge here was to find a way to insert this cultural dialogue into the EU cultural and foreign policy agenda. In this respect, following the creation of the Anna Lindh Foundation (ALF) in Alexandria on 20 April 2005,[33] the Council of European ministers for culture had adopted in 2007 a 'European agenda for culture', which recognized

---

30 Translator's note: a reference to a poem by Prévert, 'Inventaire', in which he lists items that do not appear to have any common theme.

31 R. Riemenschneider, 'L'Institut Georg Eckert, un lieu de mémoire des échanges internationaux', D. Groux and N. Tutuiaux-Guillon (eds), *Echanges internationaux et la comparaison en éducation. Pratiques et enjeux* (Paris: L'Harmattan, 2000), 242–6.

32 F. Pingel, *Guide UNESCO pour l'analyse et la révision des manuels scolaires* (Paris: UNESCO, 1999).

33 The Anna Lindh Foundation's purpose is 'to bring people together from across the Mediterranean to improve mutual respect between cultures'. Since its launch in 2005, the foundation has launched and supported action across fields impacting on mutual perceptions among people of different cultures and beliefs, as well as developing a region-wide network of over 3,000 civil society organizations from around forty-

the social, political and economic dimensions of culture.[34] Introduced as an arena for strategic action for achieving major community objectives, culture seemed to integrate the European Union's foreign relations. In 2008, the year of Euro-Mediterranean dialogue between cultures[35] had even been the occasion for the ALF to launch a collective action involving the whole of 'its' civil society (NGOs, cultural groupings, universities, local authorities and public institutions) – '1001 actions for dialogue'. Some thirty countries in the Mediterranean area held public events and initiatives to encourage dialogue. Although the focus was firmly on artistic questions in general and contemporary creation in particular – which obviously limited the scope of the exercise – civil society could, on this occasion, gauge the difficulty of receiving adequate funding from the ALF and realized how, at times, the foundation's funding mechanism failed to adapt to their needs. For a long time, the European Union's relative lack of interest in culture had led non-state actors in the Mediterranean area to favour the Ford Foundation's simpler support mechanisms.[36] Very present in the southern countries, this North American institution had always developed a more pragmatic and effective approach: Fernand Braudel had not been wrong when he asked for its support in establishing the Maison des Sciences de l'Homme in Paris (1962); neither had Boutros Labaki when he put forward the extent of the crisis of the university system in the south (2009).[37]

---

three countries. See the Anna Lindh Foundation website: http://www.euromedalex.org/ (accessed 1 January 2011).

34 The European Agenda for Culture, European Commission, 2007, http://ec.europa.eu/culture/our-policy-development/doc399_en.htm (accessed 1 January 2011).

35 Décision n° 1983/2006/CE European Parliament and European Council, 18 December 2006.

36 Created in 1936 by Edsel and Henry Ford, the Ford Foundation is an independent, non-profit, non-governmental organization. Its goals are to: (a) strengthen democratic values; (b) reduce poverty and injustice; (c) promote international cooperation; (d) advance human achievement. See http://www.fordfoundation.org (accessed 1 January 2011).

37 B. Labaki, 'Présentation', in B. Labaki (ed.), *Enseignement supérieur et marché du travail dans le monde arabe* (Beirut: Presses de l'Ifpo, 2009), 9–11.

Taking advantage of the French presidency of the European Union (July–December 2008), President Sarkozy sought to place the Mediterranean cultural dialogue firmly within the Euro-Mediterranean partnership. The launch of the Union for the Mediterranean (UfM) in Paris in July 2008 neglected the cultural dimension of the European partnership.[38] The Mediterranean Cultural Forum (MCF)[39] (or Etats géneraux culturels mediterranéens) organized in Marseille[40] between 4 and 6 November 2008 were the main event planned by the French Presidency of the European Union in the framework of 2008 – the European year of intercultural dialogue.[41]

38   The UfM pledged to make progress on six practical fields: (1) De-pollution of the Mediterranean; (2) Maritime and Land Highways; (3) Civil Protection; (4) Alternative Energies: Mediterranean Solar Plan; (5) Higher Education and Research, the Euro-Mediterranean University; (6) the Mediterranean Business Development Initiative.
39   The French ministry for foreign affairs defined the Mediterranean Cultural Forum as 'a think tank made of non-governmental actors involved in Euro-Mediterranean culture and representatives of public and private institutions with an interest in cultural matters'. See http://euro-mediterranee.blogspot.com/2008/11/etats-gnraux-culturels-mditerranen.html (accessed 1 January 2011).
40   In order to endorse its application for European Capital of Culture 2013, Marseille wanted to include in its bid the organization of a meeting of the type held by the Ateliers culturels. Its bid was indeed successful. On the other hand, the French ministry for foreign affairs also hoped that Marseille could become the headquarters of the newly created UfM secretariat. This failed as the secretariat remained based in Barcelona. See: *Dossier de sélection 2008*, http://www.marseille-provence2013.fr (accessed 1 January 2011).
41   At the level of the European Union, 2008 had been proclaimed the year of intercultural dialogue and this led to a flourishing of initiatives and declarations. The Euromed Conference of ministers of culture in Athens (Slovenian presidency) on 29 and 30 May 2008 was a platform for intensifying the pursuit of dialogue between the cultures in the Euro-Mediterranean area. It declared that more than ever – particularly in the area of culture – it was necessary to encourage dialogue between cultures and to reflect on the values that are commonly shared across cultures. See: Agreed Conclusions of the third Euro-Mediterranean Conference of Ministers of Culture Athens, 29–30 May 2008, http://www.eeas.europa.eu/euromed/docs/culture_concl_0508_en.pdf (accessed 1 January 2011). Other Mediterranean programmes equally experienced some progress in this year, whether it was at the level of UNESCO, its Arab equivalent ALECSO, the World Bank, the leadership of the European Commission, European or Mediterranean states such as Egypt, co-president of the UfM. In addition to this there

The aim was to provide additional input with new ideas, proposals and projects from non-government organizations and to bring their conclusions to decision makers, in this instance the meeting of foreign ministers from the Euromed: Union for the Mediterranean (UfM) meeting organized in Marseille at President Sarkozy's request. Unlike the Atelier culturels, whose members came from southern Europe, the Mediterranean and the Gulf, the Mediterranean Cultural Forum modelled itself on Euromed, that is, it included *all* EU member states, not only its southern members. More, the MCF brought into the process of cultural dialogue countries and observers who until then had not been members, such as countries from the Balkans and the Gulf or from the Arab League. Likewise, if it maintained a tradition of open debate, as is often the case in meetings of civil society organizations, its overall objectives brought the Forum closer to the processes of intergovernmental proceedings. In fact, the spirit of the Forum was much similar to the meeting of the foreign affairs ministers and the continuity between the two events was assured by the configuration of the meetings, the type of participants and the issues discussed. Thus, enthusiastic declarations were made: the MCF marked 'a new stage of Barcelona' (Taieb Fassi Fihri, Moroccan minister for foreign affairs), placing culture at the heart of its agenda; the MCF contributed to the 'overhaul of Barcelona' (Henri Guaino, special adviser to Nicolas Sarkozy), which started in Paris on 13 July 2008. Most of all, the general secretariat of the UfM was supposed to provide feedback to ministers, ministerial conferences and to Euromed's summits on projects emanating from the MCF, and would search for partners to finance the approved projects.

---

were initiatives favouring 'dialogue between Europe and its neighbouring countries from the south of the Mediterranean with a view to strengthening regional stability through better mutual awareness and comprehension' from the Council of Europe as much through the activity of its Direction générale de l'Education, de la Culture et du Patrimoine, de la Jeunesse et du Sport (Strasbourg) as its Centre Nord–Sud (Lisbon). Thus, concerning this very sensitive question about history – subject to all sorts of manipulation and widely used to their own advantage by politicians from all sides – three meetings (Copenhagen, Lisbon and Tunis) were held in the weeks preceding the meeting in Marseille, often gathering together the same organizations (UNESCO, ALECSO, the IRCICA and the Council of Europe) and the same experts, but above all dealing with the same concerns.

Table 2: The Ten Mediterranean Cultural Projects, 8 December 2008

1.	*Memories, histories, heritage* – didactic material for teachers presenting Mediterranean history as a collection of positive interactions, based on a multi-perspective approach and inter-cultural dialogue
2.	*Images, audiovisual, cinema* – Mediterranean audiovisual project on the production of images common to the Mediterranean – Mediterranean media watchdog
3.	*Writing, translation, libraries* – 'Translate in the Mediterranean'
4.	*Artists, artistic creation, artistic and cultural mobility* – cultural practices watchdog (creation and mobility)
5.	*Religions and societies* – platform for 'lay and religious actors' (analysis of the relationship between religions and societies, religious pluralism, the teaching of religion) – Erasmus project on theology
6.	*Modernization of societies and law* – Mediterranean network for working law (dialogue about rights, the law and human rights, legal co-operation)
7.	*Modernization of societies and law* – network of the 'Mediterranean women' networks (the status of women)
8.	*Education and universities* – creation of a Mediterranean Office for Youth
9.	*Education and universities* – 'Mediterranean education', a series of proposals for basic education – professional training – a 'Mediterranean Bologna' – a genuine Mediterranean Erasmus – inter-cultural teaching in education systems
10.	*Sustainability of the Ateliers culturels méditerranéens in Marseille* – meeting place – place for exchange – place where the actors in Mediterranean societies can work on important cultural and societal questions

A Culture of 'Façade'

Sarkozy's voluntarism was supposed to mark a turning point in the Euro-Mediterranean relationship. Until then, culture had never been taken into account beyond its simple educational or artistic dimension. The countries in the south were less than enthusiastic, arguing that the Anna Lindh Foundation had been created to bring together the two shores of the Mediterranean on a social and cultural level; but one of the MCF's ambitions was precisely to prove that projects could emerge and be supported outside the ALF. At the level of European institutions, it was feared that the development of a cultural agenda within the Euro-Mediterranean partnership would encroach on the directorate general for the education and culture portfolio. To drum up support for the UfM, Sarkozy's initial project had to be watered down (as explained elsewhere in this volume) and as such, the UfM's practical objectives are wholly neutral or technical: a search for a greater consensus around the nature and objectives of the UfM meant that more contentious issues were not discussed during the Paris summit in July 2008. Moreover, the power and scope of the general secretariat of the UfM, adopted on 3 November 2008 in Marseille, do not cover cultural ones and the fact that the Secretariat struggled to set itself up in the first semester of 2009 did not help either. As for the UfM Cultural Council, created by decree on 8 December 2008[42] and inaugurated the following 19 May at the Hôtel Matignon in Paris (the French prime minister's residence) all its energy seems to be resolutely directed towards securing for Marseille the title of European Capital of Culture 2013, as the city had been hugely involved in the promotion of France's Mediterranean policy, generating, encouraging and co-ordinating a whole range of public and private initiatives. The MCF itself did not seem to be able to make any substantial or practical proposals. If the Ateliers culturels had led to a plan of action, this was not the case for the MCF. Its eight

---

42 Decree no. 2008–1277 December 2008 creating a cultural council for the Union for the Mediterranean: See: http://www.legifrance.gouv.fr (accessed 1 January 2011).

working groups[43] had to generate 'Ten Mediterranean cultural projects' or '[projects] for the development of Mediterranean culture', in which the reference to the large-scale campaign '1001 Actions' by the Anna Lindh Foundation is clear: in this case, the ALF may well appear to be the place where feasible and practical cultural initiatives are to be developed. Thus, there was concern that all that was left for the Forum was declarations of intent that had no future due to a lack of funding. And indeed the lack of funding is an enormous functional stumbling block.

There would be no funds from the European Union before the new financial period for 2013–20 and the budget negotiations starting in 2010 were to be conducted in a difficult economic and financial context. In other words, this is where the MCF's biggest challenge lies: it is likely to become an empty shell due to a lack of money. Stripped progressively of its most important assets by the Cultural Council ('Project 3'[44] and 'Project 7' have been both absorbed by the Cultural Council, mostly to the benefit of Marseille and its bid to become European Capital of Culture in 2013),[45] having served as a bill poster for the UfM ('Projects 6, 8 and 9'), nothing but an empty shell remains of the structure of the MCF. Electoral squabbles brought the completion of 'Project 10' crashing down. This inevitably influenced the fate of the three remaining projects. The working group on 'Memories, histories and heritage' ('Project 1') managed to convene

43   The eight working groups were: (1) 'Memories, histories, heritage'; (2) 'Images, audiovisual, cinema'; (3) 'Writing, translation, libraries'; (4) 'Artists, artistic creation, artistic and cultural mobility'; (5) 'Religions and societies'; (6) 'Modernization of societies'; (7) 'Education and universities'; (8) 'Identities, cultures, values and visions of the Mediterranean'.
44   Project 3, for instance, was essentially about translation – an essential condition of dialogue between cultures, and recognized as a priority in May 2008 during the meeting of Euromed culture ministers in Athens. The project was absorbed by the Conseil Culturel and redirected to the benefit of Marseille: it became part of the bid 'Marseille 2013: European capital of Culture'.
45   *Premiére année du Conseil culturel de l'Union pour la Méditerranée: realisations 2009 et perspectives 2010*, http://www.conseilculturel-upm.gouv.fr/ (accessed 1 January 2011).

in May 2009, despite the fact that Israeli bombing of the Gaza Strip four months previously had created a climate which made it difficult to engage with such crucial issues: but, cruelly, there were no funds to advance the project. 'Project 5' on 'religions and societies' was also difficult to run given the tensions in the Middle East, even if initially it had found in the Conseil général des Bouches-du Rhône and the Caisse de dépôts et consignation strong financial backers. Yet, both as a result of the speculative financial crisis and of the Israeli military operations in Gaza, these long-term public investors preferred to withdraw their support. When projects re-enforced well established Euro-Mediterranean initiatives, they fared better. 'Project 2' (Images, audiovisual, cinema) contributed to reinforce the dynamics of Euro-Mediterranean audiovisual co-operation initiated by European Commission (the Euromed Audiovisual programme). The project that came off best was 'Project 9' (Education and Universities) because it entered clearly into the process launched by the UfM, being one of the priority issues adopted during the Paris Summit in July 2008 and confirmed by the Declaration of Marseille in November 2008.

The second stumbling block facing the MCF is international competition. As was already the case with the Ateliers culturels méditerranéens, the objectives may have been ambitious, but the financial means were limited.[46] Other international organizations, with similar objectives, may be able to draw on more important financial support. For instance, the United Nations' Alliance of civilizations[47] also aims to 'improve understanding and cooperative relations among nations and peoples across cultures and religions. It also helps to counter the forces that fuel polarization and extremism'. It seeks to 'initiate partnerships and to set up productive collaboration between diverse communities, but also to strengthen confidence

46 'Chirac's involvement in the Ateliers [had] its limits: but this too-ambitious policy [lacked] means.' M. Vaïsse, *La puissance ou l'influence? La France dans le monde depuis 1958* (Paris: Fayard, 2008), 410.
47 The Alliance of Civilizations (AoC) is an initiative proposed in 2005 by José Zapatero, then Spanish prime minister, at the 59th General Assembly of the United Nations (UN) and co-sponsored by the Turkish prime minister, Recep Erdoğan.

and reconciliation amongst cultures'.[48] The second summit of the Alliance of Civilizations in Istanbul in 2009 marginalized the MCF's contribution. Not only was the idea of a network of the Atelier culturel taken up again, but the issue of education received particular attention, notably the teaching of history as a way of preventing conflict and promoting inter-cultural understanding. It led to an initiative jointly proposed by the Council of Europe and the Research Centre for Islamic History Art and Culture (IRCICA)[49] to create pedagogical materials on relations between countries bordering the Mediterranean and between other regions of the world: this initiative entered into direct competition with the MCF's 'Project 1' one rather than as a supplement to it. Indeed, the Council of Europe was pursuing here a policy about the teaching of history, a policy which had been initiated at the time of its own creation in 1949. Its expertise in this field no longer needed to be proven to the point where the European Commission was happy to adapt for the EU the majority of the Council's recommendations. The Council had been the first and only intergovernmental authority to define, on 31 October 2001, fixed clear principles which should inform the methodologies used for the teaching history in a democratic, pluralist Europe.[50] Moreover, The Council of Europe's policy of regional co-operation, under way since the 1990s in Eastern Europe and

---

48 United Nations: Alliance of Civilizations: http://www.unaoc.org (accessed 1 January 2011).

49 Created following a Turkish proposal in 1976 during the Seventh Islamic Conference of Foreign Ministers IRCICA started its activities in 1980 as the first subsidiary organ of the Organisation of the Islamic Conference (OIC) concerned with culture. See http://www.ircica.org (accessed 1 January 2011).

50 Recommendation (2001) 15 regarding the teaching of history in Europe in the twenty-first century (adopted by the Committee of Ministers of the Council of Europe, 31 October 2001, during the 771st meeting of the representatives of ministers). See http://www.coe.int/t/dg4/education/historyteaching (accessed 1 January 2011). See also A. Cardwell, 'Les travaux du Conseil de l'Europe sur l'enseignement de l'Histoire', in M. Mathien, *La Médiatisation de l'Histoire. Ses risques et ses espoirs* (Brussels: Bruylant, 2005), 387–90, 394.

Russia, gave further credibility to its initiatives:[51] indeed, it had led to a series of complementary teaching materials (the Caucasus, 2003; the Black Sea, 2004;[52] Bosnia Herzogovina, 2008[53]), that the Council quickly transposed outside this zone (Japan, 2003; Cyprus, 2005–6[54]). This cultural activism was prolonged by formal relations (with ALF, the ALECSO) and informal ones (the IRCICA) which had been established as a result of the Council of Europe's policy of cultural co-operation with southern Mediterranean countries since 28 January 2003.[55] Competition between international institutions can have damaging effects. France, a founding member of the Council of Europe, had supported the Council policy, but started to criticize it for its lack of impact. At the end of 2007, arguing that financial difficulties may have hindered the work of the Council's North–South Centre (NSC),[56] France declared that 'that in the current climate of competition between institutions, the North–South Centre had not sufficiently made its mark or developed in recent years' for France to be able to continue working with it.[57] This did not in any way prevent the

---

51   Council of Europe/Ministry of Education and Sciences of the Russian Federation, *History Education in Europe. Ten Years of Co-operation between the Russian Federation and the Council of Europe* (Strasbourg: Conseil de l'Europe, 2006).

52   Council of Europe, *The Black Sea: A History of Interactions* (Strasbourg: Conseil de l'Europe/Gyldendal Undervisning, 2004).

53   L. Black, *Manual for History Teachers in Bosnia and Herzegovina* (Strasbourg: Conseil de l'Europe/OSCE/ Gouvernement du Canada, 2008).

54   Council of Europe, *The Use of Sources in Teaching and Learning History: The Council of Europe's Activities in Cyprus* (in English, Greek, Turkish), 1–2 (Strasbourg: Conseil de l'Europe, 2005–6).

55   Council of Europe: Recommendation (2003)1590, Cultural co-operation between Europe and countries in the south of the Mediterranean: http://assembly.coe.int (accessed 1 January 2011).

56   The Council of Europe's North–South Centre, also known as the 'European Centre for Global Interdependence and Solidarity' was created in 1989 to promote dialogue and cooperation between Europe, the south of the Mediterranean and Africa and to build a global citizenship based on human rights and citizens' responsibilities. See: http://www.coe.int/t/dg4/nscentre/default_fr.asp (accessed 1 January 2011).

57   Council of Europe/ North–South Centre (2007), Report from the 48th meeting of the member states' representatives committee. 30 October 2007: http://www.

French government, two years later, from seeking finance from the NSC for her MCF projects ('Project 1' and 'Project 5'). The sustainability of the projects themselves was put in a difficult position and the NSC withdrew its support, despite hosting a session for 'Project 1'. Financial, but certainly political considerations were at stake.

This background is a good example – if one is needed – of the fact that the meeting in Marseille was desperately short of political will. Until 2007, the main question was about knowing whether France's policy between the Arab states and Israel, and between the three Maghreb states (Morocco, Algeria and Tunisia), could be balanced. It was, clearly, difficult to find equilibrium. Jacques Chirac could pride himself on his permanent concern about avoiding a 'clash of civilizations' between Islam and the West. With Nicolas Sarkozy, the particulars were different. His Mediterranean initiative was part of the framework of the French presidency of the European Union and thus fell within its efforts to search for a consensus between different groups. Furthermore, the double meeting in Marseille also fell within the traditional Franco-Spanish competition in the field of Euro-Mediterranean policy. For the competition between Marseille and Barcelona had ended, on the eve of the opening of the MCF, in a crushing defeat for the former, with the granting of the general secretariat of the UfM – the very one which was to support the financing of the MCF's projects – to the latter. This success for Spain equally demonstrated the limits of the French initiative. Even if France convinced the Czech Republic to pursue and developed the Euro-Mediterranenean initiatives launched under the French presidency of the EU, Prague was not able to maintain the initial French momentum, not only because of a lack of funds, but also because mounting clashes between Gaza and Israel meant that the Czech presidency ended in paralysis for the Euro-Mediterranean process. 2009 was a lost year for the French-sponsored UfM as well as for the MCF, but not for Spain. Madrid, Paris's main European rival in the Euro-Mediterranean process, took over the rotating presidency of the European Union on 1 January 2010. Taking

---

coe.int/t/dg4/nscentre/statutory_bodies/ReportMS48_fr.pdf (accessed 1 January 2011).

advantage, then, of the opening of the European budget debate, Spain fully intended to relaunch the Barcelona Process in pushing forward her vast, but quite distinctive, cultural ambitions:

> A wide-ranging [cultural] programme will also be set up by the Casa de América (House of America), the Casa Sefarad-Israel (Sefarad-Israel House), the Casa Árabe (Arab House), the Casa África (the House of Africa), the Casa Asia (the House of Asia) and the new Casa del Mediterráneo (House of the Mediterranean), which will place Spain at a crossroads and will make it a meeting point just as it has been so many times throughout the past and just as it would like to remain in the future.[58]

This cultural programme, under the generic title of 'Culture, continents, civilizations: an alliance for the 21st century', is deeply rooted in Spain's own historical experience, intended to instil in Europe a vision of the Mediterranean space mixing 'the Arab world, which is always so close, [the] Mediterranean culture which […] is specific [to Europe] and the inspiring identity of […] the Sephardic world'.[59] In any case, this programme could not have been based on the MCF. Through its *global* ambition, it was close to the missions of the Council of Europe's North–South Centre, as well as those of the Alliance of Civilizations; in any case, these two institutions had already 'developed and deepened [their] co-operation'.[60] André Azoulay symbolizes the will to think the Mediterranean in a larger, more global context: powerful adviser to the king of Morocco, he is above all a member of the Comité des sages of the Alliance of Civilizations, whilst at the same time being vice president of the Fundación Tres Culturas (Seville) and president of Anna Lindh Foundation, which, thanks to a strong network has now carved for itself an important place with the Euro-Mediterranean process in order to 'rebuild the human and cultural links at the heart' of the

---

58   European Union: Objectives of the cultural programme during the Spanish presidency of the EU from 1 January 2010, http://www.eu2010.es (accessed 1 January 2011).
59   Ibid.
60   Council of Europe: North–South centre (2009), Editorial by D. Huber, executive director, 'Une année mémorable pour le Centre Nord–Sud' (16 December 2009), http://www.coe.int/t/dg4/nscentre/intro09_fr.pdf (accessed 1 January 2011).

Mediterranean region.[61] In other words France was no longer in control of the Euro-Mediterranean cultural process.

'There is a 2,000-year-old pact between France's greatness and others' freedom', the ageing General de Gaulle said to his loyal supporter André Malraux.[62] During the opening ceremony at the headquarters of the Fundación Tres Culturas, Régis Debray had scoffed at these 'brilliant professionals who use dialogue for the sake of dialogue' and other 'civilian theologians of dialogue who are the new opium of States'.[63] By launching themselves into a policy of power without giving themselves the means to do so, Jacques Chirac and Nicolas Sarkozy were forgetting that cultural action was not the same as foreign policy. Indeed, the refusal of the 'clash of civilizations' is a message which does seem to have proved really weak in a context of internationalization and cultural relativism, including at the highest levels of the state.

Taking this cultural as well as geopolitical background into consideration, what are the long-term prospects for the MCF? Neither Marseille, which got what it wanted in becoming the 2013 European capital of culture, nor France, which does not have the means to achieve its ambitions, have any interest in it. As for civil society, it lacks the organization and resources to have any influence on the MCF. The MCF working sessions planned for 2009 did not take place before the end of June 2010 and only two working groups are still working ('Project 1' and 'Project 5'). Project 5 (Religion and society) only survived because the Institut Catholique de la Méditerranée (ICM) in Marseille host its activities and had allocated a professorial chair to the subject.[64] But there is no indication that the development of didactic material on shared Mediterranean history will ever be published. Did we

---

61 'La Fondation Anna Lindh: un acteur pour le dialogue euro-méditerranéen', http://www.euromedalex.org (accessed 1 January 2011).
62 A. Malraux, *Les chênes qu'on abat* (Paris: Gallimard, 1971), 88.
63 R. Debray, *Un mythe contemporain: le dialogue des civilisations* (Paris: CNRS, 2007).
64 For instance, the ICM co-organized (with the MCF) a conference entitled 'La fécondité d'une confrontation entre théologie et sciences religieuses' in November 2010.

witness in Marseille on 4 and 5 November 2008 a rehearsal of the fate in store for the UfM? Would the 'new "*souffle*" of air' become a 'new *soufflé*', according to the neat phrase used by Roberto Aliboni, director of studies at the *Istituto Affari Internazionali* in Rome? For the time being, there is concern that cultural diplomacy is not a priority in the Mediterranean. It remains to be seen, as Henri Froment-Meurice was already wondering some years ago, whether the Ministry for Foreign Affairs is the institution most capable of 'managing this enormous machine'.[65]

Bibliography

Balous, S., *L'action culturelle de la France dans le monde* (Paris: Presses universitaires de France, 1970).
Bédarida, C., and Gurrey, B., 'Jacques Chirac célèbre la diversité culturelle', *Le Monde* (3 February 2003).
Black, L., *Manual for History Teachers in Bosnia and Herzegovina* (Strasbourg: Conseil de l'Europe/OSCE/ Gouvernement du Canada, 2008).
Bry, A., *La Cendrillon du Quai d'Orsay, avril 1945–décembre 1998* (Paris: Grou-Radenez, 1999).
Burke, P., 'Cultural Exchange as Appropriation and Adaptation', in D. Albera, A. Block and C. Bromberger (eds), *L'Anthropologie de la méditerranée* (Paris: Maisonneuve & Larose/Maison méditerranéenne des sciences de l'homme, 2002), 555–6.
Cabanel, P. (ed.), *Une France en Méditerranée: Ecoles, langue et culture françaises, XIXe–XXe siècles* (Paris: Creaphis, 2006).
Cardwell, A., 'Les travaux du Conseil de l'Europe sur l'enseignement de l'Histoire', in M. Mathien (ed.), *La Médiatisation de l'Histoire: Ses risques et ses espoirs* (Brussels: Bruylant, 2005), 386–401.
Chirac, J., 'Intervention du président de la République au sommet Euro-Méditerranéen de Barcelone' (2005), http://www.elysee.fr/elysee/elysee.fr (accessed 28 June 2009).

---

65 H. Froment-Meurice *Vu du Quai 1945–1983* (Paris: Fayard, 1998), 286–7.

Chirac, J., 'Discours à l'Université du Caire' (1996), http://www.elysee.fr/elysee/elysee.fr/francais_archives/interventions/discours_et_declarations (accessed 28 June 2009).
Conein, B., 'Communautés épistémiques et réseaux cognitifs: coopération et cognition distribuée', *Revue d'Economie Politique*, 113 (2004), 141–59.
Council of Europe, *The Black Sea: A History of Interactions* (Strasbourg: Conseil de l'Europe/Gyldendal Undervisning, 2004).
Council of Europe, *The Use of Sources in Teaching and Learning History: The Council of Europe's Activities in Cyprus* (Strasbourg: Conseil de l'Europe, 2005–6).
Council of Europe, 'Recommendation regarding the teaching of history in Europe in the 21st century' (2001) (adopted by the Committee of Ministers of the Council of Europe, 31 October 2001, during the 771st meeting of the representatives of ministers), http://www.coe.int/t/dg4/education/historyteaching (accessed 1 January 2011).
Council of Europe, 'Recommendation: Cultural Co-operation between Europe and Countries in the South of the Mediterranean' (2003), http://assembly.coe.int (accessed 1 January 2011).
Council of Europe/North–South Centre, 'Report from the 48th meeting of the member states' (30 October 2007), http://www.coe.int/t/dg4/nscentre/statutory_bodies/ReportMS48_fr.pdf (accessed 1 January 2011).
Council of Europe/Ministry of Education and Sciences of the Russian Federation, *History Education in Europe: Ten Years of Co-operation between the Russian Federation and the Council of Europe* (Strasbourg: Conseil de l'Europe, 2006).
Debray, R., *Un mythe contemporain: le dialogue des civilisations* (Paris: CNRS, 2007).
'Declaration on Mexico by the United Nations Educational, Scientific and Cultural Organization', http://portal.unesco.org/culture/fr/files/12762/11295422481 mexico_fr.pdf/ mexico_fr.pdf (accessed 25 March 2011).
Dollot, L., *Les relations culturelles internationales* (Paris: Presses universitaires de France, 1968).
Froment-Meurice, H., *Vu du Quai 1945–1983* (Paris: Fayard, 1998).
Goldman, A.I., 'Foundations of Social Epistemics', in A.I. Goldman (ed.), *Liaisons: Philosophy Meets the Cognitive and Social Sciences* (Cambridge, MA: MIT Press, 1992), 179–207.
Hoffman, S., 'Paradoxes of the French Political Community', in Stanley Hoffmann et al., *In Search of France* (Cambridge, MA: Harvard University Press, 1963).
Hoveyda, F., *Que veulent les Arabes?* (Paris: First, 1991).
Labaki, B., 'Présentation', in B. Labaki (ed.), *Enseignement supérieur et marché du travail dans le monde arabe* (Beyrouth: Presses de l'Ifpo, 2009), 9–11.

Lazega, E., *The Collegial Phenomenon: The Social Mechanisms of Cooperation Among Peers in a Corporate Law Partnership* (Oxford: Oxford University Press, 2001).

Longino, H., *The Fate of Knowledge* (Princeton, NJ: Princeton University Press, 2002).

Macridis, R.-C., 'French Foreign Policy', in *Foreign Policy in World Politics* (Englewood Cliffs, NJ: Prentice-Hall, 1962).

Malraux, A., *Les chênes qu'on abat* (Paris: Gallimard, 1971).

Michel, R., and Boyer, J.F., 'L'union pour la Méditerranée et la télévision', *Le Figaro* (13 August 2009).

Pingel, F., *Guide UNESCO pour l'analyse et la révision des manuels scolaires* (Paris: UNESCO, 1999).

Quirot, O., 'Dominique Wolton: "La France brade son réseau culturel à l'étranger"', *Le Nouvel Observateur* (11 December 2008).

Ramonet, I., 'Liberté, équalité, sécutité', *Le Monde Diplomatique* (April 2006).

Raymond, J.-F. de, *L'action culturelle extérieure de la France* (Paris: La Documentation française, 2000).

Riemenschneider, R., 'L'Institut Georg Eckert, un lieu de mémoire des échanges internationaux', in D. Groux and N. Tutuiaux-Guillon (eds), *Echanges internationaux et la comparaison en éducation: Pratiques et enjeux* (Paris: L'Harmattan, 2000), 242–56.

Roche, F., and Pigniau, B., *Histoires de diplomatie culturelle des origines à 1995* (Paris: La documentation française, 1995).

Siino, F., 'Une politique méditerranéenne des chercheurs?', in J.H. Robert and G. Groc (eds), *Politiques méditerranéennes entre logiques étatiques et espace civil* (Paris: Karthala, 2003), 287–301.

Vaïsse, M., *La puissance et l'influence: La France dans le monde depuis 1958* (Paris: Fayard, 2008).

Védrine, H., 'Les Etats-Unis: hyper-puissance ou empire?', *Cités*, 4/20 (2004).

Young, R., *Marketing Marianne: French Propaganda in America 1900–1940* (Piscataway, NJ: Rutgers University Press, 2004).

PATRICIA CAILLÉ

On the Confusion between Films about Women and Films by Women in the Construction of a Regional Cinema: The Maghreb vs. the Mediterranean

Moroccan, Algerian and Tunisian cinemas taken together constitute the 'Cinemas of the Maghreb'. Even though this term has contributed to producing significant knowledge about film in an area that remains underexplored,[1] the self-evident nature of the category 'Cinemas of the Maghreb' is deceptive. Hamid Nacify notes that regional cinemas have 'generally identified shared features of films from contiguous geographic regions without developing a coherent theory of regional formation'.[2] This chapter seeks to identify what this regional denomination covers and explores the ways in which films by women filmmakers from the Maghreb or of Maghrebi descent contribute to this regional construction. 'Cinemas of the Maghreb' is not the only term which covers this production, as other regional entities are cited at least as often,[3] with Moroccan, Algerian and Tunisian films being integrated into North African, Arab, African, Amazigh or Mediterranean cinemas. It is very common to see the national dimension resurfacing in the content of texts devoted to a region and they often

---

1   Notable works include: M. Berrah, 'Cinémas du Maghreb', *Cinémaction*, 14 (1981); R. Bensmaïa, *Experimental Nations: The Invention of the Maghreb* (Princeton: Princeton University Press, 2003); M. Serceau (ed.), 'Les Cinémas du Maghreb', *Cinémaction*, 111 (2004); R. Armes, *Cinémas du Maghreb: Images postcoloniales* (Paris: L'Harmattan, 2006); and D. Brahimi, *50 ans de cinéma maghrébin* (Paris: Minerve, 2009).
2   H. Nacify, 'For a Theory of Regional Cinemas: Middle Eastern, North-African and Central Asian Cinemas', *Early Popular Visual Culture*, 6.2 (2008), 97–102.
3   P. Caillé, '"Cinemas of the Maghreb": Reflections on the transnational and polycentric dimensions of regional cinemas', *Studies in French Cinema* (2012).

end up blocking the region out altogether.[4] Furthermore, a brief survey of the production of knowledge about Moroccan, Algerian and Tunisian films within the three countries shows that both articles and books focus on national cinemas rather than on regional Maghrebi cinemas.[5]

The term 'Cinemas of the Maghreb' is generally used to put together three national film industries that are considered to share a common landscape, a common Arab-Muslim culture, and a common history of colonization by the French.[6] Each of these film industries has contributed to the construction of national identity, and, as a result, these film histories are associated with independence and cover approximately the same time period. The three national cinemas are all economically and culturally dominated within a transnational film market. The similarities do not go much further and indeed what appear to be similarities may in fact mask discrepancies. For example, French colonization took different forms in each of the countries in question. The construction of national identity through films developed along diverse paths and consequently these three national film industries are significantly different: they have very distinct histories, heydays, film cultures and economies. Today, Morocco produces between fifteen and twenty films a year, Algerian film production came to a halt during the period of terrorism in the 1990s and has been very erratic since,

---

4   A. Khalil (ed.), 'Through the Lens of Diaspora: North African Cinema in a Global Context', *The Journal of North African Studies* 12.3 (2007), 273–5.

5   A. Gabous, *Les Femmes et le cinéma en Tunisie* (Tunis: Cérès Editions/Crédif, 1998); H. Khélil, *Abécédaire du cinéma tunisien* (Tunis: Hédi Khélil, 2007); H. Khélil, *Le Parcours et la Trace, Témoignages et documents sur le cinéma tunisien* (Tunis: Éditions Médiacom, 2002); S. Chamkhi, *Cinéma tunisien nouveau: Parcours autres* (Tunis: Sud Editions, 2002); S. Chamkhi, *Le Cinéma tunisien à la lumière de la Modernité: 1996–2006* (Tunis: Centre de Publication Universitaire, 2009); Ismaël, *Cinéma en Tunisie, Kaléidoscope d'une saison 2007/2008* (Tunis: Arts Distribution, 2008); K. Dwyer, *Beyond Casablanca: M.A. Tazi and the Adventure of Moroccan Cinema* (Bloomington: Indiana University Press, 2004); M.D. Jaïdi, *Le cinéma colonial: Histoire du cinéma au Maroc* (Rabat: Aljamal, 2001).

6   In relation to film, the Maghreb is very seldom extended to five countries; Moulay Driss Jaïdi raises this issue in 'Le Cinéma maghrébin: la quête vers un idéal', *Wachma* (2008), 3–4, 19–28.

while film production in Tunisia runs to about two or three films a year. A survey of film production in the 1970s would have produced a completely different picture with Algerian cinema leading the way. A regional category is constructed in different locations, and 'Cinemas of the Maghreb' as a category has a different significance within the Maghreb and outside. Even though cooperation and co-productions between the three countries of the Maghreb exist, it is a category that does not refer to a common market for films, shared personnel or audiences. While the desire for such cooperation has been expressed on various occasions, there is little evidence of any steady development. The 'Cinemas of the Maghreb' is a regional category which makes more sense in relation to Europe and more particularly France, be it in terms of production, circulation of personnel, or distribution, because most films are produced as co-productions with at least one European country, most often France, and the films are more likely to be distributed in France than in the other countries of the Maghreb or Europe.

'Cinemas of the Maghreb' is thus rooted in a cultural rather than filmic understanding of the Maghreb. It is associated with representations of the Maghreb in films rather than with shared conceptions of film aesthetics, industrial practices and film culture. In other words, the notion of 'Cinemas of the Maghreb' is largely based on the implicitly shared understanding that the three cinemas provide their audiences with access to a common culture based on a series of common themes: the condition of women, migration, the rise of Islamic fundamentalism. These recurrent motifs that characterize the 'Cinemas of the Maghreb' are read in promotional and critical discourses as being fraught with the tension between tradition and modernity which Lila Abu Lughod has argued is an 'organizing trope' that emphasizes the failures of the East modernize through imitation of the West. This reductive view has been detrimental to an understanding of the complex ways in which aspirations to modernity operate in diverse national projects, projects that are the outcome of specific local religious, cultural and economic contexts and specific relations with the West.[7]

---

7   L. Abu-Lughod (ed.), *Remaking Women: Feminism and Modernity in the Middle East* (Princeton: Princeton University Press, 1998), 18.

Films have notably been treated as a privileged medium for representations of the 'condition of women' in Arab Muslim countries or in the Maghreb, and numerous books about these regional cinemas devote a chapter or even a whole section to the issue.[8] But it is important to dissociate films on the 'condition of women' that are a constitutive feature of 'Cinemas of the Maghreb', from films directed by women. While the two categories often overlap because numerous long-feature fiction films by women represent women and have female protagonists, films 'on the condition of women' have been directed by men as well. Some of the films directed by women have enjoyed international critical and box-office success, particularly Moufida Tlatli's *Silences du Palais* (1993) and *Saison des hommes* (2000), Yamina Bachir-Chouikh's *Rachida* (2002) and Raja Amari's *Satin Rouge* (2002), but a close look at box-office receipts and releases on VHS or DVD shows that such successes have overall been matched by Ferid Boughedir's *Halfaouine, l'Enfant des terrasses* (1990), and to a lesser extent by Nadir Moknèche's *Viva Laldgérie* (2004) or *Delice Paloma* (2007). The interest in films on the 'condition of women' as it can be traced in promotional, critical and academic discourses confirms the extent to which the 'Cinemas of the Maghreb' operate as representations of a Maghrebi culture rather than as a questioning of the ways in which Maghrebi filmmakers make films, think about form or film aesthetics.[9]

In this chapter, I will examine the ways in which the confusion and slippage between films on the condition of women and films by women filmmakers (with the issue of the representation of the condition of women becoming more important than women filmmakers) contributes to the construction of the regional Maghrebi dimension and works against interrogating the specificity of women filmmaking in the 'Cinemas of the Maghreb'. In order to do so, I will look into the relationships between the

---

8   Brahimi, *50 ans de cinéma maghrébin*; O. Leaman, *Companion Encyclopedia to North African and Middle Eastern Cinemas* (London: Routledge, 2001).

9   It is quite striking that in *L'image de la femme au Maghreb, une étude Méditerranéenne*, edited by Khadija Moshen-Finan (Paris: Actes Sud/Bakhzach, 2008), the author of the third essay is a well-known Tunisian critic and writer, Hedi Khélil, who devotes his article to the representations of women in Tunisian cinema.

funding and international circulation of films in different markets as these can help us understand how films on the condition of women participate in the construction of the 'Cinemas of the Maghreb'. More importantly, I will show how processes of valorization inherent in the economic and cultural power relations between the Maghreb and Europe (mainly France) promote films by women filmmakers, erase their specificity and, paradoxically, fail to contribute to their international reputation. In order to do so, I will first provide a brief survey of the films directed by Maghrebi women or women of Maghrebi origin in each national industry in the Maghreb or in Europe, and then I will proceed to explore the patterns of production and circulation of these films, before reflecting on the difficulties and limits inherent in trying to assess the visibility and/or reputation of a set of films. Women filmmaking in the countries of the Maghreb is a fairly recent phenomenon. Two films were made in the 1970s, three in the 1980s, eight in the 1990s and twenty-six in the 2000s. We should thus be wary of premature generalizations about patterns of development. Nevertheless, a number of emerging trends can be identified. In a fourth part, I will interrogate the potential impact on the Maghreb region of the gradual phasing out of national filmmaking policies within Europe, increasingly superseded by the development of a common European policy in support of international film production and its subsequent reconfiguration of regional programmes. In the context of this major transformation, I will speculate on the visibility and recognition of women filmmakers of the Maghreb and of Maghrebi descent.

Examining patterns of funding in a transnational context is not easy: information about film production, distribution and exhibition often remains sketchy, partial and contradictory. A certain amount of information about film production is kept secret. Another difficulty lies in the very different systems and tools that each country has devised to record data on the production, distribution and exhibition of films on the domestic or international markets which makes any comparison quite hazardous. Third, data about the circulation of films is intimately tied to the promotion of the films in question which means that some sources for the collection of data outside strict institutional control are liable to biases. Despite these limitations, an examination of the patterns of funding and the various contributions to the financing of film through cross-checking data available from various databases and film credits constitutes an interesting angle for approaching the contribution of women filmmakers to the construction of a regional category as well as the significance of films by women in this category.

An Overview of the Production of Films by Women Filmmakers

The corpus used here consists of twenty-eight filmmakers and thirty-nine long-feature fiction films made by women working in the national film industries in the Maghreb, women from the Maghreb or of Maghrebi descent working in Europe and representing the Maghreb in their films, and women acknowledging a relationship with at least one country in the Maghreb through their work.[10]

Moroccan Women Filmmakers in Morocco or in the European Diaspora up to 2008[11]

In 1982, Farida Bourquia directed *La Braise*, the first film by a woman in Morocco. In 2009, there were ten women out of eighty-nine filmmakers in the official database of the Centre Cinématographique Marocain

---

10 All the data concerning film releases in France and in Europe is based on a comparison between the figures available on the Chiffres du Box-Office database and the Lumière database available on the website of the Observatoire Européen de l'Audiovisuel.

11 In the use of the term 'diaspora' here, I wish to follow Avtar Brah's idea that 'diasporic journeys are essentially about settling down, about putting roots "elsewhere". These journeys must be historicized if the concept of diaspora is to be used as a useful historic device', 'Diaspora, Border and Transnational Identities', in *Cartographies of Diaspora* (London: Taylor & Francis, 1996), 186.

(CCM) who had directed fifteen long-feature films between them (out of the 176 films included in the official film catalogue). Farida Benlyazid made four films, Narjiss Nejjar three, and Farida Bourquia two. Eight of these films feature a woman as the main protagonist, but films by Moroccan women filmmakers include a wider range of themes, genres, narratives and aesthetic experiments. Proportionally, these films do not get distributed as widely on the European national markets as films by their Tunisian or Algerian counterparts. Moroccan cinema is by far the most dynamic film industry in the Maghreb, but also the most 'insular'. In comparison with Algeria and Tunisia, films by Moroccan filmmakers have had the most limited international circulation in Europe and France. Out of these fifteen films, five are national productions and ten are international co-productions. All of the international co-productions include funding from at least one European country, and seven of them feature French funding. Six films by women filmmakers were released in France with an average of 56,000 tickets sold per film in France. Cinema audiences for films by Moroccan women are on average lower than those of their female colleagues from Algeria or Tunisia, but much higher than those for their male Moroccan counterparts. Out of the thirty-six films with Moroccan funding distributed in Europe over the last fifteen years, seven of them are by women filmmakers (all of them with women as protagonists struggling in diverse ways for freedom in Morocco) and five among them have been distributed in France. In other words, one in five Moroccan films distributed in Europe has been directed by a woman whereas films by women filmmakers in Morocco do not add up to one tenth of the national catalogue. This means that films by Moroccan women filmmakers are comparatively more present on European screens than films by their male counterparts, with films by women representing 25 per cent of the box-office receipts for the films released in Europe. This does not mean that all these films were successful at the box office – quite the contrary. Furthermore, only four of all these films have been available at some point on VHS or DVD.

Algerian Women Filmmakers in Algeria or in the European Diaspora up to 2008

Since *La Nouba des Femmes du Mont Chenoua* by Assia Djebar (1978), there have been ten women filmmakers in Algeria who have directed a total of twelve feature films. Three of these films are national productions, four are French-Algerian productions and five are French productions. There is a clear distinction between the films directed by women most often in France that tend to focus on identity, be it sexual, religious or cultural, in a postcolonial metropolitan France – *Inch'Allah Dimanche* by Yamina Benguigui (2001); *Un fils* by Amal Bedjaoui (2003); *La Petite Jérusalem* by Karin Albou (2005) – and films about Algeria's painful national history, its blindspots, the unspoken aspects of official histories and their effect on women's lives – *Rachida* by Yamina Bachir Chouikh (2002) or *Barakat!* by Djamila Sahraoui (2006). The economic dependence on France is striking. First, even though the filmmakers acknowledge ties to Algeria, the films are French and listed as French national film productions. The Lumière database lists twenty-one films with Algerian funding released in Europe since 1996, twenty of them have been released in France. Among these films, there are only three films by women, the other films included in our corpus being listed as French productions. Films by Algerian women filmmakers or women of Algerian descent are fewer but they do well in France where seven of them have been distributed. They drew on average 58,500 spectators, a higher figure than the average film with Algerian funding. Eight films are available on VHS or DVD. In spite of the faltering state of the Algerian film industry, films by Algerian women filmmakers are proportionally much more present than those by Moroccan women filmmakers. A woman is the main protagonist in eight of these films.

Tunisian Women Filmmakers in Tunisia or in the European Diaspora up to 2008

The first long-feature film by a Tunisian woman filmmaker was *Fatma 75* (1978). Tunisia counts eight women filmmakers and twelve long-feature films, all of them being international co-productions; one is a Maghrebi film production and all of them are co-produced with at least one European country, nine among them are co-produced with France. Twenty-four films by Tunisian filmmakers have been released in Europe since 1994, and amongst these films by women filmmakers have been fairly successful on the European market; six films have been released on European screens, five of these in France. Six have been available at some point on VHS or DVD. The average box-office receipts for films directed by Tunisian women are much higher than for those for their Moroccan or Algerian counterparts, with on average 92,600 tickets sold per film. It is noteworthy that films with women protagonists fare better than others whether they are made by male or female filmmakers. Nor have all the films made by women been successful abroad; Selma Baccar's period dramas and Kalthoum Bornaz's films have not found a European distributor. Nadia El Fani's *Bedwin Hacker* (2003), with its narrative centred on women involved in a chase for the freedom of speech via the Internet on both sides of the Mediterranean, received very little attention in France. The films that have done well constitute a restrictive range of film productions by women: they are dramas about the intimacy of two generations of Muslim women in Tunisia after World War Two who are confronted with sexual and cultural oppression. Successful films by male filmmakers fit within this pattern as well.[12] In other words, films by women filmmakers portraying struggling women in Tunisia (often read as the Maghreb) have greatly contributed to the international reputation of Tunisian cinema.

12 P. Caillé, 'Figures du féminin et cinéma national tunisien', *Actes du Colloque de l'AFECCAV – La Fiction Eclatée: Etudes socioculturelles*, ed. J.P. Bertin-Maghit and G. Sellier (Paris: L'Harmattan, 2007). There is one exception: Karin Albou's *Chant des Mariées* (2008) represents the relationship between two female teenagers, one is Jewish the other is Muslim. As the film is seen from the point of view of the Jewish teenager, the film departs from the standard narrative of oppression to reinscribe it within a more historical, albeit problematic, framework.

Trends in Patterns of Production and Circulation of Films by Women Filmmakers

Analyses of film representations are certainly very useful for tracking shifts in the construction of national, regional, ethnic and gender identities, but a closer look at the film selection processes, be they economic, political and/or cultural, reveals the extent to which it is necessary to take into account the grounds upon which corpuses of films are constituted. An enquiry into the circuit of production, circulation and reception of films is therefore important for interrogating the ways in which films reach viewers, whilst it is nevertheless necessary to recognize the limitations which the researcher faces when seeking to grasp these largely invisible encounters between films and their audiences and consequently the meanings that viewers give to films.

The integration of women into the film industry varies from country to country. In a booming Moroccan film industry, there has been a relatively large influx of women filmmakers – the number of women listed in the database of the CCM rose from six out of sixty-one filmmakers in 2007 to eleven out of eighty-nine at the end of 2009. However, the place of women filmmakers raises certain questions. Some of them, such as Narjiss Nejjar or Leila Kilani (who is turning from documentary to fiction), have already obtained funding from the CCM for new films. But the vast Ali'N Prod programme run by Nabil Ayouch and the CCM has not contributed to the training of any new female directors. The objective of this new Moroccan film production company – which has already produced thirty films – is to train new personnel in the film industry, to develop a popular Moroccan cinema based on the production of genre films, and to strengthen the film industry through the development of very close ties between film production and television. The Algerian film industry came to a halt during the 1990s because of the war and the collapse of the national industry that had organized film production. Since then, 'Alger, Capitale de la culture Arabe 2007' constituted an unexpected boon for the film industry. It enabled the production of *L'Envers du Miroir* (2006) by

Nadia Chérabi, but there is no sign of more stable and regular production taking root. Films about Algeria or its diasporic communities are produced primarily from France and/or with French funding by Djamila Sahraoui, Yamina Bachir-Chouikh or Yamina Benguigui. The Tunisian film industry no longer produces as many films as it used to. In recent years, only a few Tunisian women filmmakers have been making films, such as Kelthoum Bornaz (*L'Autre moitié du ciel*, 2006) or Raja Amari (*Les Secrets*, 2008), and numerous young women filmmakers have not been able to move on to the production of long-feature films. Such discontinuous developments highlight the extent to which what may be read as women's growing access to filmmaking should be treated with caution.

A close examination of film production shows that the circulation of these films follows the same paths as their funding. In my corpus, there are four national Moroccan productions, three Algerian productions and four national French productions. Even though some of these national productions have been shown in international festivals, overall they have had more specifically national trajectories. Maghrebi co-productions are very few and irregular; there have been two Maghrebi co-productions between Morocco and Tunisia, *Porte sur le ciel* by Farida Benlyazid (1988) and *Bedwin Hacker* by Nadia El Fani (2002). Films are much more likely to circulate between one country in the Maghreb and Europe or France than across the three countries in the Maghreb. Unsurprisingly, films are more likely to be seen in the countries that contributed to their production: co-productions with Germany, Belgium or the Netherlands have led to releases of the films in the countries where they got funding, although this is not automatic. Twenty films are international co-productions with French funding, and twelve of them have been released in France. French funding was minor for six out of the eight films that did not get a release. One of them, Fatma Zohra Zamoum's *Z'Har* (2008) is an experimental film shot in HD with very little chance of ever achieving commercial distribution; another one, Djamila Sahraoui's *Ouardia avait deux enfants*, was still in post-production at the time of writing.

The regional Maghrebi dimension from a European / French vantage point thus confirms the privileged relationship between each country in the Maghreb and one European country, most often France, in the production and international circulation of films. It also highlights the regularity of the total number of films with Moroccan, Algerian or Tunisian funding released in France, the number of films by women filmmakers, and the number of tickets sold for these films, independent of the actual number of films being produced and the diversity of themes, narratives, formal and aesthetic experiments in each of these national industries. We may infer from this that demand constitutes a much stronger determinant than supply with regard to the number of Moroccan, Algerian and Tunisian films released in Europe and France, European markets making room for only so many films regardless of the variety of films that are available. From my initial corpus of thirty-nine films, only twenty-one films were distributed in Europe (eighteen in France), which means that almost half of these films were not. Considering the importance of the international (i.e. European and primarily French) market for these films, the 'commercial' selection operates as a clear narrowing down of the diversity of the filmic representations available. To a certain extent, we could regard international commercial distribution as a two-tier system; it zooms in on a limited set of films that gain greater visibility, and within this group, filmgoers adhere by their choice of films to an even smaller group of films which have become emblematic.[13]

In this way, films directed by women often participate unwittingly in the construction of a Maghrebi dimension through the selective process at work in commercial distribution and through the prevalence of the cultural over the filmic dimension in the promotional discourse and in the critical reception of their work. The commercial successes that ensue contribute to the prominence of a very specific type of film. In this respect, films by Tunisian filmmakers that have fared much better internationally have been emblematic. They are predominantly auteur films, dramas with women protagonists that are centred on the quest for freedom or on the

---

13   It is not within the scope of this chapter to focus on an in-depth analysis of the questions raised by the relationship between economic and marketing dimensions and the sociological dimension of the cultural practices associated with specific film cultures.

domination or oppression of women in a patriarchal context, with the domination constructed in national and/or regional (i.e. cultural) terms. This type of film, which is part of a larger tradition of Arab social films, contributes to the promotion and questioning of regional (i.e. cultural) Maghrebi representations of the domination of women, at the expense of specifically national and historical contexts, no matter how present such contexts may be in the films themselves. This is particularly striking in the reception of Moufida Tlatli's *Silences du Palais* (1993) in France, despite the fact that the film's narrative is founded on the parallel between the ongoing oppression of women and the struggle for national Tunisian independence. Films with women protagonists have generally done better on European and French screens than other films: period films (Farida Benlyazid's or Selma Baccar's), popular genre films and films made within the industry (Farida Bourquia's and Nadia Cherabi's films have not been released) and Zakia Tahiri's latest comedy *Number One* (2006) drew limited audiences in France.

From Production to Recognition: The Problems with Assessing Visibility and Recognition

Is the growing number of women filmmakers and films by women of the Maghreb or of Maghrebi descent a sign of their integration into national industries or even into an international film culture? There are good reasons to doubt this. First, some of the women filmmakers included in our study do not believe that women filmmakers constitute a relevant or useful category for analysing film production in the Maghreb. A colloquium, 'Maghreb, des réalisatrices et leurs films' was organized as part of a research project 'Maghreb et cinémas' on 9–10 April 2010 in Saint-Denis in order to discuss the position of women in the film industry from a regional perspective.[14] Nine women filmmakers working in Morocco, Algeria, Tunisia or France had been invited. The colloquium revealed that no one questioned the legitimacy of women filmmakers in the industries but a few were ambivalent with regard to raising the question of filmmaking in the Maghreb in terms of gender. They may even have been implicitly calling into question the often challenged legitimacy of research produced in the West on women as a universal category of analysis as Chandra Talpade Mohanty did in 1988. The discussions exposed the discrepancy between, on the one hand, the large number of films by women that had narratives centred on women protagonists and, on the other, the very wide range of views these filmmakers presented in describing their position in the film industry. The heated debates brought to light the rejection by some, at least publicly, of a struggle against what may be regarded as discrimination. This resistance was related to their emphasis on the point that making films is difficult for both male and female filmmakers. Such a position is also related to their desire to be recognized as filmmakers without any consideration of their sex or of the geographical/cultural area where they make films. To a certain extent, the fear that some women filmmakers have of being trapped in a gender and/or regional ghetto reinforces the idea that women filmmakers do not constitute a distinct category.

Such a state of affairs is puzzling because, as I showed earlier, the increase in the number of women filmmakers and films directed by women filmmakers may be interpreted in different ways. The filmographies of women filmmakers are still much shorter than those of their male counterparts in the three countries, even for filmmakers whose careers have spanned over two decades. Only Farida Benlyazid has been able to make four long-feature fiction films, while Selma Baccar and Farida Bourquia, who also belong to the first generation of filmmakers, have made three and two films, respectively. Among the women who directed their first long-

---

14 'Maghreb et Cinémas: Circulation des Films, Production des Savoirs et Constitution d'un Patrimoine' is a research project funded by the Agence Universitaire de la Francophonie.

feature films in the late 1990s or early 2000s, Narjiss Nejjar, Zakia Tahiri, Karin Albou, Moufida Tlatli, Kalthoum Bornaz, Raja Amari and Djamila Sahraoui have so far been able to make two films each; a few of them have a third one in production. By comparison, there are at least seven filmmakers in Morocco credited with at least six long-feature films, some of them with as many as ten. Whereas women working in Algeria have not yet been able to make a second long-feature film, Mohamed Lakhdar-Hamina, Merzak Allouache, Mohamed Chouikh, Amar Laskri or Rachid Benhadj and others have long filmographies, and the same is true in Tunisia. With two fairly successful first films, it took Moufida Tlatli seven years and Raja Amari eight to complete a second one.

Accessing European Support Systems for African, Middle Eastern and South-East Asian Films

Sources of Funding for Internationally Dominated Cinemas in Europe

Global patterns of film aid cannot be considered unilaterally, simply in terms of French or European aid being distributed to various regions in the world, but need to be integrated into an analysis of the relationships of forces at work within the media industries. Thus, film and audiovisual industries worldwide have been trying to maintain their competitiveness, to find outlets for their own productions, and all this is fought out on economic, cultural and ideological grounds. The blurring of the distinction between film and audiovisual media, related in part to the sprawling development of digital media as well as the shift from a national to a regional conception of film and audiovisual production, has obviously had a significant impact on conceptions of film. While a thoroughly developed analysis of the successive European MEDIA programmes is beyond the scope of this chapter, it is important to highlight the different forms that European cooperation has taken in relation to the media industries.

Since 1991, there have been four different European MEDIA Programmes run by the European Commission and aimed at boosting cooperation among the members of the European Union (and a few others) to foster the training of personnel, stimulate film and audiovisual production and the circulation of these productions within the European Union and beyond. The latest programme runs from 2007 to 2013 with a budget of 755 million euros.

In parallel to these successive MEDIA Programmes, the European Commission has also run two distinct programmes of European aid that are much more modest in scale. The outcome of the Barcelona Process that officially initiated an economic, political and cultural partnership in 1995 around the Mediterranean, Euromed Audiovisual, is meant to foster cooperation between the EU and the MEDA countries (twelve countries in the eastern and southern Mediterranean), and automatically reintegrates countries of the Maghreb within a larger Mediterranean area. There have been two successive programmes since 2000 with 18 and 15 million euros each, and a much less visible third phase was launched in 2009. These Euromed Audiovisual programmes are articulated around a series of projects whose description highlights the extent to which gender constituted – initially at least – a significant preoccupation. Here as well, the extension of support to television programmes confirms the blurring of the distinction between film and television, and the prominence of audiovisual media over film. It implies an industrial rather than an artistic conception of film that may contribute to foregrounding the question of the representation of women in audiovisual media, rather than the position of women in the film industry. In descriptions of the projects, the question of images of women in the audiovisual industry seems to override and sideline questions about film, the position of women in the film industry and about film aesthetics – even though, in communicating about its projects, Euromed often mentions films by women filmmakers.

This issue of the representation of women in such programmes is associated with an Arab cultural dimension while the Maghrebi dimension, subsumed within a larger Arab or Mediterranean one, is non-existent. Part of Euromed Audiovisual I, Elles ... Pionnières was devoted to the production of a series of twelve documentaries on Arab women whose political, cultural or artistic contribution to the world deserved attention.[15] In Euromed Audiovisual II, the Caravane du Documentaire Arabe and Medscreen promoted the Euro-Arab dimension and a greater visibility for Arab cinema; it also devoted much attention to documentaries by Arab women filmmakers without promoting them as such. Promoting the representation of women through the support provided by this programme appears to be a double-edged sword. It constitutes a widely shared and consensual concern but it is difficult to evaluate the extent to which it contributes to the production and the visibility of the films by women filmmakers. Another European programme also serves as a substitute for French national programmes, among others the *Plan Afrique Images* and the *Fonds Image Afrique*, suspended in 2008. The EU-ACP project (2007–9), which excluded the countries of the Maghreb, was aimed at contributing to the development of audiovisual industries in the seventy-nine ACP (African, Caribbean and Pacific) countries.

Both these European programmes risk being phased out and replaced by a new cooperation programme underway, MEDIA Mundus (2011–13), a larger project-based programme that will be awarded 13.5 million euros over three years. Emphasizing the need for cultural diversity and presented as an alternative to an extension of Euromed Audiovisual and EU-ACP, the programme will support consortia based in Europe and Media Mundus third countries. Its objective is to boost international cooperation with countries outside the EU regarding training, and the production and distribution of films and audiovisual productions.[16]

The growing importance of Europe, which is concomitant with a reduction in national aid, is the outcome of different sets of preoccupations; the first being to rationalize the contribution of the EU and to strive to articulate a more coherent international media policy.[17] There

---

15   The documentaries produced were shown in film festivals and broadcast on Arte in 2005, but it is difficult to find any evidence of the legacy of these documentaries produced within the scope of this programme.
16   http://ec.europa.eu/information_society/media/docs/mundus/mundus_proposal_en.pdf (accessed 1 October 2011).
17   T. Hoefert de Turégano, 'Sub-Saharan African Cinemas: The French Connection', *Modern and Contemporary France*, 13.1 (2005), 71–83.

was a growing sense of the culturally inadequate ways in which film aid had been granted. Strict regulations governing the use of aid awarded to regionally specific film and audiovisual projects meant that much of the subsidies had to be spent in part with the goal of (indirectly) supporting the French film industry. The new programmes have been less restrictive and allow for more of the money to be spent in the countries of production. The shift to a wider interest in audiovisual production is also concomitant with a reduced interest in the artistic dimension associated with celluloid. Other transformations characterizing the evolution of these programmes involve the development of aid attached to the distribution and circulation of films, and the move away from a nationally centralized French organization to a more decentralized network of programmes. Thus French regions and, to a lesser extent, cities have become significant agents in film production.

International film funding is not limited to these large European programmes and I cannot here hope to cover the history of the intricate network of foundations and festivals in France and Europe that have contributed to film production and sometimes distribution as well, but I will instead focus on some significant international or national state-run programmes such as the *Fonds Sud* in an analysis of women filmmakers' access to funding.

Women Filmmakers' Access to Funding within Patterns of Aid for Maghrebi Films

While funding from Europe continues to rise in importance, it has not completely ended the intervention of other international institutions, state-run programmes or initiatives. Exploring the effects of the meshing together of these different sources of funding and its effect on the production and circulation of Maghrebi films will enable us to speculate on the different meanings associated with the geographical, cultural, institutional and/or political category of the 'Maghreb'.

European Aid

It is clear that the regional Maghrebi dimension is neither significant nor visible in relation to international or national aid programmes aimed at supporting the production and/or distribution of Maghrebi films. The Maghreb is thus re-inscribed within a larger Mediterranean region covered, on one hand, by the Euromed Audiovisual Programmes and, on the other, by Euromed Cinemas within it, which is run by the CCM in Morocco and the EU. The Mediterranean region is thus re-inscribed within a larger Euro-Mediterranean area through a European policy aimed at fostering the circulation of cultural products. Euromed Cinemas seeks to facilitate the distribution and exhibition of films from the MEDA region, and to a lesser extent from the EU, across the Mediterranean. Between 2006 and 2009, it supported 107 films, among them twenty films with funding from one of the countries in the Maghreb or directed by a Maghrebi filmmaker, but only two films by Moroccan women filmmakers.[18] The overall outcome of this programme is difficult to assess: it is the result of local initiatives rather than national policy and some of it is used to help national releases.[19] But a closer look at these figures highlights the ways in which the three countries of the Maghreb do not clearly privilege films produced in the other two countries when claiming aid for distribution or exhibition of films.

Within this tension between the erratic existence of a regional Maghrebi category and the contradictory signals of its construction, Maghrebi cinema is subsumed within a larger political and economic European-Arab and/or Mediterranean category. Out of twenty-seven international productions in my corpus, twenty are international co-productions that include French funding which means that twenty-four out of the thirty-nine films directed by Maghrebi women filmmakers or by women filmmakers of Maghrebi descent working in the Maghreb or in Europe rely partially or

---

18 Yasmine Kassari's *L'Enfant Endormi* (2004) and Zakia Tahiri's *Number One* (2007).
19 For the list of films supported by Euromed Audiovisual, see http://www.euromed-cinemas.org/supported-films.php?menu=FR#FR (accessed 1 October 2011).

completely on French funding. Looking at sources of funding available outside European programmes described earlier in the light of the circulation of the films, I will examine the ways in which films by women have access to such sources. Such a question raises financial as well as symbolic issues; obtaining funding is a means to make a film, to obtain a certain legitimacy and thereby to aspire to greater visibility.

Francophone Aid

The Fonds Francophone de Production Audiovisuelle du Sud (FFPAVS), created by the Agence Intergouvernementale de la Francophonie (AIF) and the Conseil International des Radios et Télévisions Francophones (CIRTEF), is run by the Organisation Internationale de la Francophonie (OIF) and funded to the order of about 2.5 million euros a year. It covers thirty-seven countries – out of the fifty-six members of the OIF – and supports the writing, production, postproduction and distribution of film and audiovisual programmes – both fiction and documentary – with aid ranging from 8,000 to 110,000 euros. Between 1990 and 2007, the FFPAVS has been fairly generous to Maghrebi filmmakers,[20] even though it seems to privilege certain countries much more evidently than it privileges the funding of films by women filmmakers. Figures show that Tunisia benefited more from the FFPAVS, with support for fifty-two long feature films and among them eight films by women, while it contributed to twenty-seven Moroccan films, among them only three films by women filmmakers.

The 'Cinemas of the Maghreb' have been the object of festivals and sporadic or recurrent thematic screenings at larger festivals in France, Morocco, Switzerland, Tunisia, and those working in the film industries in the Maghreb have expressed the desire for developing such cinemas, but

20   Since its creation, the FFPAVS has supported twenty-seven Moroccan and forty-two long-feature Tunisian films. Algeria does not belong to the OIF. Among them, there are three Moroccan and eight Tunisian films by women. It has also contributed to the making of sixteen Moroccan audiovisual productions, forty-three Tunisian audiovisual productions, five Moroccan productions and among them four were directed by women.

there is still little evidence of their coming into being within programmes for the development of film and audiovisual production, distribution, exhibition and broadcast. The African dimension is more present and stronger with relatively well-known websites devoted to film criticism that have produced much more systematic knowledge about films.[21] The most recent initiatives for the creation of an aid programme emerged from the Panafrican Festival which took place in Algiers in the summer of 2009. Funded by the Algerian Government, and aimed at supporting the production of four long-feature and four short African films, the programme has already selected two Maghrebi (i.e. Tunisian) films, but it foregrounds the African dimension of these films.[22] In the same way, during the 2010 Cannes Film Festival the Organisation Internationale de la Francophonie (OIF) with the Fédération Panafricaine des Cinéastes (FEPACI) announced a new Panafrican Fund for film production.[23] African cinema as a regional category has worked as the promotion of a collective effort to support dominated and barely visible films and/or national cinemas. The objectives of such initiatives, whether devoted to film criticism or to film production, is to counteract what is perceived as excessive reliance on a faltering Europe. The development of resources devoted to African cinema has produced much more systematic knowledge about film.

French Aid

Created in 1984, the prestigious *Fonds Sud* devoted exclusively to film and run jointly by the French Ministry of Culture and Communication

---

21 See, for instance, *Africiné, le site de la Fédération Africaine de la Critique Cinématographique* (www.africine.org), *Africultures, le site de référence des cultures africaines* (www.africultures.com), or *Clap Noir, cinémas et audiovisuels africains* (www.clapnoir.org). These websites have a continental understanding of Africa and include the countries of the Maghreb.
22 Selected films were announced during the seminar held during the Panafrican film festival that focused on potential patterns of development for African cinemas, among them a short by Ala Eddine Slim and a long-feature Tunisian film by Nawfel Saheb Ettaba.
23 The official OIF press release can be read at http://www.francophonie.org/La-Francophonie-s-associe-a-la.html.

and the Ministry of Foreign Affairs, has supported over 400 films from almost seventy countries. In 2004, it extended the geographical scope of its aid to include South-East Asia and South America without any increase in its budget of 2.5 million euros, much to the dismay of many African and Maghrebi filmmakers. Over the course of its existence, the *Fonds Sud* has contributed to the production of seventy-four films from the three countries of the Maghreb; twenty-two Moroccan, eighteen Algerian, and thirty-four Tunisian, sixteen of which were released for commercial distribution in France. Among these films, thirteen were directed by women: four from Morocco (*Nos Lieux Interdits* (2008) by Leila Kilani, *Les Yeux Secs* (2004) by Narjiss Nejjar, *L'enfant endormi* (2004) by Yasmine Kassari, *Une porte sur le ciel* (1988) by Farida Benlyazid); three from Algeria (*Ouardia avait deux enfants* (in postproduction) and *Barakat!* (2006) by Djamila Sahraoui, *Rachida* (2002) by Yamina Bachir-Chouikh); and ten from Tunisia (including three films by Raja Amari, *Les Secrets* (2009), *Lambeaux* (listed as in production), *Satin rouge* (2002), two films by Moufida Tlatli, *Les Silences du Palais* (1993), *La Saison des hommes* (2000), and *Keswa, le fil perdu* (1997) by Khaltoum Bornaz). Eight of these Maghrebi films were released on French screens. In 2009, only one film – Sahraoui's latest film – received funding to the tune of 150,000 euros. The proportion of films by women filmmakers supported by *Fonds Sud* is larger than the proportion of films directed by women in the national industries. A closer look also shows that *Fonds Sud*, FFPAVS and other international sources of complementary funding contribute to the development of film industries that have a regular production and an international development, which explains how the Tunisian film industry, which was stronger in the 1990s, benefited the most.

Much more recent, the *Fonds de la Diversité* (CNC/Acsé) created in 2007 has also provided aid for production of film or audiovisual programmes, in the form of complementary funding for films that have already been selected either by the CNC or the Acsé (*Agence Nationale pour la Cohésion Sociale et l'Égalité des Chances*). It is aimed at promoting audiovisual works that challenge cultural stereotypes and promote diversity, and consequently privileges representations of cultural diversity in metropolitan France. It has supported a number of filmmakers from the Maghreb or of Maghrebi descent, contributing to the release on DVD of Yamina Benguigui's latest film productions, *Aïcha* (2008), *9/3*, *Mémoire d'un territoire* (2006), and *Plafond de verre, les défricheurs* (2006), and to the production of documentaries such as *Lamine La France* (2009) by Samia Chala, *Les imams vont à l'école* (2010) by Kaouther Ben N'hia and *La Cuisine en heritage* de Mounia Anna Meddour, as well as one long-feature fiction film, *Française* (2008), by Souad El Bouhati.

Aid coming from Conseils Régionaux, Conseils Généraux and French cities is more difficult to track, but these institutions have also contributed to the making of films by Maghrebi women filmmakers and the diasporas.

To conclude, a systematic look at these various sources of funding highlights some regularity in the patterns of development. While there is no clear sign that the development of a European policy with respect to film and audiovisual production, distribution and exhibition has contributed to the support and visibility of women filmmakers, it is quite clear that the promotion of women's issues and images of women is more important.[24] Francophone and French aid over recent years has contributed to a shift in the relationship between institutions and representations through moving images. There is a clear move away from the promotion of film as art, which still characterizes such funds as the *Fonds Sud*, toward the promotion of cultural representations of oneself as vehicles for cultural change. This move away from film as art is cultural, technological and economic. It highlights the extent to which digital production has become significant and concomitantly contributed to the blurring of the distinction between film and audiovisual production and/or to the decline of film. Considering that out of our corpus of thirty-nine fiction films, only two were shot in HD, the strong attachment of women filmmakers to film and their desire to be recognized as filmmakers in film as well as the reluctance to take on the struggle for recognition in terms of gender may delay or work against their

---

24 It is interesting to note here that Euromed has initiated a Euromed Gender Equality programme aimed at promoting gender equality and curbing violence against women. Within it, a project is also devoted to research on the images of women in the media. See http://www.euromedgenderequality.org/template.php?menu_secondaire=17&code_menu=17 (accessed 1 October 2011).

integration. It is clear that European policy in relation to the Mediterranean does not pay much attention to film.

National Patterns of Funding

From looking at our corpus of films, it appears that women's relationships to national film industries follow similar patterns across the Maghreb. The early films by women filmmakers are generally made within national film industries whereas later films are international co-productions, with Morocco remaining more 'national'. Women anchored in any of the national industries, whether in the Maghreb or to a lesser extent in Europe, seem to get funding more regularly than women filmmakers whose careers span both sides of the Mediterranean. The sustainability of the national industry and the integration of women filmmakers within it have a more significant impact on their access to production than the contribution of symbolically important international funds. Strangely enough, the international reputation of a film does not seem to guarantee greater access to future national (or international) funding for its director. In this respect, the comparison between Farida Benlyazid's and Moufida Tlatli's trajectories is suggestive. The former has been able to produce very diverse films in spite of the quasi absence of international distribution while Moufida Tlatli's highly acclaimed films have not enabled her to produce a third film.[25] International funding, while promoting a certain type of film by women filmmakers, has failed to account for the diversity of films directed by women filmmakers.

International Francophone and French funding for the production and distribution of film looks favourably upon films about women and/or films directed by women from the Maghreb or from the diasporas. While foregrounding a concern for women's issues, the effects of the shift to European funding upon the production and circulation of films by women seems more ambivalent. These additional sources of funding and support

---

25 Moufida Tlatli's third film, *Nadia et Sarra* (2003), was produced for television.

to national funding are all the more significant because the figures (when they are available) indicate that, for all the countries concerned, women's film productions are much cheaper than average.[26] It is clear, however, that the number of films that get support follows the patterns of development of national industries, which implies that such funds cannot guarantee the production of a larger number of films by women and that they cannot replace the funding offered within national film industries. Patterns of development for women working in the three countries of the Maghreb in relation to their respective national industries appear to be identical to those for women working in France with respect to the French film industry.

Conclusion:
The Mediterranean, Maghrebi Women Filmmakers and their Films in the Construction of a Regional Cinema

At this point, tracing the presence of films by women filmmakers from the Maghreb and the diasporas or of Maghrebi descent in national and/or transnational regional contexts means understanding the meshing together and overlap of various regional dimensions, each of them constituting a specific economic, cultural and political construction of a particular regional cinema. Defining the visibility and the significance of this presence therefore involves an examination of the articulation between national and international funding of films by women filmmakers, the legitimatizing role of funds that constitute significant sources of complementary finance, the circulation of films directed by women filmmakers from Morocco, Algeria, Tunisia and the diasporas, the availability of films on different media, and, last but not least, the various promotional, critical and scholarly discourses they generate.

---

26 CBO (Les Chiffres du Box-Office) records film budgets, albeit unsystematically.

In this respect, the Mediterranean dimension meshes together the political and economic agenda of the EU with regard to the Mediterranean, i.e. the MEDA countries, and the mythic dimension of a heterogeneous regional entity characterized by a history of migrations. The Mediterranean dimension is constructed geographically as a privileged area for the circulation of films, thereby becoming a means of bringing peace through a better understanding of its cultural diversity, whereas references to an Arab region have more straightforward cultural overtones. In this second case the promoted aim is better knowledge of a common Arab-Muslim culture in the southern Mediterranean. Lately, much more of the aid aimed at the production and circulation of film and audiovisual production in the Mediterranean has been awarded by the European Union based on the funding of specific projects. Within this Mediterranean or Euro-Arab understanding of film and audiovisual media, women filmmakers from the Maghreb and their films dissolve within a larger concern about images of women in the media and the promotion of positive images of women which is the outcome of a policy conceived in relation to audiovisual production rather than in relation to film, combined with a policy designed to enhance gender equality. Neither of these policies seems to have significantly contributed to film production by Maghrebi women filmmakers or to the circulation of their films, nor does it seem to be their priority.

The 'Cinemas of the Maghreb' do not constitute a meaningful region with regard to film or audiovisual media from an economic standpoint. They refer primarily to a set of representations of an elusive common Maghrebi culture rather than to a region sharing similar conceptions of film production, film aesthetics or constituting a privileged area for the circulation of the films. The implicit foregrounding of a cultural Maghrebi dimension constructs the 'Cinemas of the Maghreb' as enabling (international) audiences to obtain access to a better understanding of the Maghreb rather than as a region whose cinematic specificity and whose conception(s) and production(s) of films need to be explored. It presupposes the Maghreb as a regional and meaningful entity thus obscuring the cultural location of this construction and the circulation of capital, people, films and other cultural productions that generated it. It highlights the extent to which this regional dimension is primarily postcolonial as it is largely related to a history of immigration embedded in a history of French colonization at the expense of an interest in a history of film or filmmaking.

Within this context, women filmmakers have not been the object of the production of as much knowledge as their films and the portraits of women in them.[27] However, their films have overall produced more reliable returns. Looking at patterns of funding in relation to the circulation of films by women filmmakers from the Maghreb and/or of Maghrebi descent brings to light the economic and cultural postcolonial tensions at work. Until very recently, filmmakers who have worked consistently on representations of women have fared better than their male counterparts within existing Francophone and French programmes, both in terms of international funding and the circulation of films, even though films on the condition of women obviously constitute a restrictive category that is distinct from films directed by women. This does not mean that all these films have done well: looking at the figures clearly highlights the extent to which historical dramas and genre films have not been successful. Films by women that have been successful are films representing women in the Maghreb since independence struggling against oppression. What is the result of such an interest which is not so much in film as in women in the Maghreb? We can only speculate on such a question, but the tendency to consider the Maghreb as the object to be recovered from films has been an issue hampering the recognition of women filmmakers from the countries of the Maghreb and from the diasporas. The category that has enabled their ascent may also serve as a trap that limits their recognition.

In spite of the significance of the term 'Cinémas du Maghreb' in France and the desire to promote such cinemas in the Maghreb, there are no grounds for affirming that the Maghreb constitutes a significant region nor does it ever appear to be with respect to film production, industrial cooperation, film distribution or even film heritage. There is no fund for the production and/or distribution of films that is specifically attached

27 Florence Martin's *Veils and Screens: Maghrebi Women's Transvergent Cinema* (Bloomington: Indiana University Press, 2011) is a forthcoming contribution.

to the region, only limited initiatives coming essentially from Morocco to promote a greater visibility for films across the three countries.

The same absence of the Maghreb can be observed on the most significant websites dedicated to the promotion of these films. Websites actively involved in the development of a film culture that takes into account Moroccan, Algerian and Tunisian films are usually national, such as the CCM website or the website promoting Tunisian cinema that devotes part of its coverage to Maghrebi cinema.[28] Others focus on the more ambiguous African dimension that tends to obscure the Maghreb because the term 'African cinemas' may refer either to the whole continent and include the Maghreb or may be limited to Sub-Saharan Africa. Unlike websites promoting African cinemas and cultures, a few Maghrebi websites devoted to culture contribute to the circulation of significant information about events and may offer interesting analyses but their purpose is primarily to inform rather than build thorough and systematic resources.[29] Somewhere in between, the project of Babelmed is devoted to the construction of a Mediterranean identity and covers a much wider range of preoccupations that include social and economic issues.[30]

At this point in time, to contend that the growing number of women filmmakers in the three countries of the Maghreb and the diasporas combined with the relative success of some of their films are the unequivocal signs of a slow but irreversible process of integration may be too presumptuous. Women filmmakers, primarily in the French diasporas, refuse to be defined in relation to such a region, and some of them in the Maghreb also refuse to be included in any category foregrounding gender. Furthermore, it is impossible to deny that the women filmmakers who have fared well are also those who are integrated in national industries (Morocco, Algeria, Tunisia, France and Belgium). So far there has not been any completely transnational production for women, nor has an international career

---

28 See www.cinematunisien.com (accessed 1 October 2011).
29 See, for instance, Maghrebarts, http://www.maghrebarts.ma/ (accessed 1 October 2011).
30 See, Babelmed, le site de cultures méditerranéennes, http://www.babelmed.net/ (accessed 1 October 2011).

guaranteed easier and more regular access to funding for women, quite the opposite. The concern is that shifts in the articulation of the fragile balance between the national, regional and international dimensions may not serve women in the long term. One would need to look much more closely into the terms of the relationship between the interest in the Maghreb and Maghrebi women and the interest in Maghrebi films by women filmmakers in order to understand what critics, experts and audiences in French/European cultures appropriate and appreciate about these films. From a European standpoint, the growing emphasis on the Mediterranean and/or Arab dimensions at the expense of the Maghreb may well work against the success that films by Maghrebi women have enjoyed in the last twenty years.

Bibliography

Abu-Lughod, L. (ed.), *Remaking Women: Feminism and Modernity in the Middle East* (Princeton: Princeton University Press, 1998).
Armes, R., *Cinémas du Maghreb: Images postcoloniales* (Paris: L'Harmattan, 2006).
Bensmaïa, R., *Experimental Nations: The Invention of the Maghreb* (Princeton: Princeton University Press, 2003).
Berrah, M., 'Cinémas du Maghreb', *Cinémaction*, 14 (1981).
Brah, A., 'Diaspora, Border and Transnational Identities', in *Cartographies of Diaspora* (London: Taylor & Francis, 1996).
Brahimi, D., *50 ans de cinéma maghrébin* (Paris: Minerve, 2009).
Caillé, P., 'Figures du féminin et cinéma national tunisien', in J.P. Bertin-Maghit and G. Sellier (eds), *Actes du Colloque de l'AFECCAV – La Fiction Eclatée: Etudes socioculturelles* (Paris: L'Harmattan, 2007).
Caillé, P., 'Le Maroc, l'Algérie, la Tunisie des réalisatrices ou la construction du Maghreb dans un contexte postcolonial', in P.-N. Denieuil (ed.), 'Socio-Anthropologie de l'image au Maghreb', *Maghreb et Sciences Sociales* (2009–10).
Caillé, P., '"Cinemas of the Maghreb": Reflections on the transnational and polycentric dimensions of regional cinemas', *Studies in French Cinema* (2012).

Chamkhi, S., *Cinéma tunisien nouveau: Parcours autres* (Tunis: Sud Editions, 2002).

Chamkhi, S., *Le Cinéma tunisien à la lumière de la Modernité: 1996–2006* (Tunis: Centre de Publication Universitaire, 2009).

Dwyer, K., *Beyond Casablanca: M.A. Tazi and the Adventure of Moroccan Cinema* (Bloomington: Indiana University Press, 2004).

Gabous, A., *Les Femmes et le cinéma en Tunisie* (Tunis: Cérès Editions/Crédif, 1998).

Hoefert de Turégano, T., 'Sub-Saharan African Cinemas: The French Connection', *Modern and Contemporary France*, 13/1 (2005), 71–83.

Ismaël, *Cinéma en Tunisie, Kaléidoscope d'une saison 2007/2008* (Tunis: Arts Distribution, 2008).

Jaïdi, M.D., 'Le Cinéma maghrébin: la quête vers un idéal', *Wachma*, 3–4 (2008), 19–28.

Jaïdi, M.D., *Le cinéma colonial: Histoire du cinéma au Maroc* (Rabat: Aljamal, 2001).

Khalil, A. (ed.), 'Through the Lens of Diaspora: North African Cinema in a Global Context', *The Journal of North African Studies*, 12.3 (2007), 273–5.

Khélil, H., *Abécédaire du cinéma tunisien* (Tunis: Hédi Khélil, 2007).

Khélil, H., *Le Parcours et la Trace, Témoignages et documents sur le cinéma tunisien* (Tunis: Éditions Médiacom, 2002).

Khlifi, O., *Histoire du cinéma en Tunisie* (Tunis: Société Tunisienne de Diffusion, 1970).

Leaman, O., *Companion Encyclopedia to North African and Middle Eastern Cinemas* (London: Routledge, 2001).

Martin, F., *Veils and Screens: Maghrebi Women's Transvergent Cinema* (Bloomington: Indiana University Press, 2011).

Talpade Mohanty, C., 'Under Western Eyes: Feminist Scholarship and Colonial Discourses', *Feminist Review*, 3 (1988), 65–88.

Nacify, H., 'For a Theory of Regional Cinemas: Middle Eastern, North-African and Central Asian Cinemas', *Early Popular Visual Culture*, 6/2 (2008), 97–102.

Serceau, M. (ed.), 'Les Cinémas du Maghreb', *Cinémaction*, 111 (2004).

MARY BEN BONHAM

Hybrid Cultural Significance for the Bibliothèque Municipale à Vocation Régionale de Marseille

This chapter explores the connection between a building's form and its cultural significance.[1] Based on my analysis, the Bibliothèque Municipale à Vocation Régionale de Marseille is representative of a hybrid Mediterranean culture. Through functional aesthetics, the library's architecture bridges times, cultures and places.[2]

---

1   The author extends sincere thanks to: the Miami University School of Fine Arts for a scholarship and teaching grant, which supported this research; the Association for the Study of Modern and Contemporary France for providing the context for dialogue at their 2009 conference 'France and Mediterranean' at the University of Portsmouth – the conference discussions gave added dimension to my analysis of architecture and culture; the many individuals in France who contributed their time and provided information about the library, including Corinne Ezavin of Adrien Fainsilber Associés, Frédéric Bertrand of Marseille Aménagement, and Carole Grousset-Jeanjean and Patrick Michel of BMVR, Marseille.
2   In 'The Open Book: Laminated Translucent Stone Façade at Marseille's Alcazar Library', published in *Material Matters: Making Architecture*, proceedings of the 2008 Association of Collegiate Schools of Architecture West Fall Conference (Los Angeles: University of Southern California, 2008), the author has described the material and qualitative effects of the translucent stone façade as it relates to the building's site and physical context. This text chapter extends that work to explore the cultural context of the building with an emphasis on French and Mediterranean identities.

BMVR, Marseille: view of glass-marble façade including Alcazar name on historic entrance.

L'Alcázar

Commonly called l'Alcazar by Marseille residents, the library is located on the former site of a famed entertainment venue of the same name. The former theatre and music hall's name remains imprinted in gold leaf over the restored public entrance, facing the tree-lined Cours Belsunce near the Vieux-Port, between the mountains and the sea. The name maintains the former building's reference to the majestic *mudéjar*-style Alcázar palace in

Seville, Spain, where Islamic influence is found in the light-filtering, intricately carved stone.³ *Alcázar* is a word derived from the Arabic meaning 'palace'. Now a palace of books opened for the people, Marseille's library features a screen of glowing marble, and the library's patrons are its main attraction.

In one respect, the library embraces the dispersed centrality of French cultural identity. It was completed in 2004 by the Parisian architectural firm Adrien Fainsilber Associates. As a Bibliothèque Municipale à Vocation Régionale (BMVR), it is among the first group of twelve regional libraries in France funded conjointly with state, regional and local resources, closely associated to the Bibliothèque nationale de France in Paris. Marseille's BMVR library is a thoroughly modern construct – first in its mission to redefine what a contemporary library should offer its patrons, and second in its execution featuring an inventive use of structures and materials. The building technology used in the design follows a tradition of technological innovation that can be traced back to Paris and the ambitions of the Mitterrand-era Grands Projets. These associations firmly implant the project in contemporary French culture. At the same time, the library's design skilfully absorbs qualities and proportions of the local building style and integrates influences of Spanish and North African architectural heritage. The library responds to its climate by integrating classic regional methods to ventilate the building and protect it from the sun. The association with traditional design principles from other Mediterranean countries resonates here in Marseille, a city with many immigrants from the lands that share Marseille's hot and dry climate. Further on, we will

---

3   *Mudéjar* is a specific type of Islamic-style architecture originating in Spain: 'The less well-known term *Mudéjar* refers to architecture carried out for Christian patrons by Muslim craftsmen. *Mudéjar* architecture uses many of the most characteristic features of Islamic architecture including Arabic calligraphy and the horseshoe arch.' 'The best example of Mudéjar architecture in Seville is the Alcázar which was rebuilt as the palace of Pedro the Cruel in the fourteenth century [...] The palace contains a series of courtyards or patios which are decorated with intricate carved stonework arcades and polychrome tile dadoes.' A. Petersen, *Dictionary of Islamic Architecture* (London: Routledge, 1996), 265, 257.

explore how these associations have been merged in terms of the aesthetics and functionality of the architecture. We will first, however, consider the broader context of the library – the significance of its placement, origins and institutional mission.

Mise en place

On the north shore of the Mediterranean Sea, Marseille has been continuously occupied since ancient times. In terms of size today, France's 'third city' could be called the Euro-Mediterranean's 'first port'. Peripheral in location to her nation's centre, Marseille is a mainstay of the international network of *pôles* now defining the rim of the Mediterranean Sea. To build in this place is to confront many worlds – past and present, central and peripheral, local and global, French and Mediterranean.

'Place' can be construed in both geopolitical and socio-cultural terms. An important public edifice built in contemporary Marseille might put up a façade addressing any one of these geopolitical identities: [*French*] [*European*] [*Euro-Mediterranean*] [*Mediterranean*]. Of course, these monikers are not mutually exclusive. For example, a city can be easily considered simultaneously French and European. A city can also be both central and peripheral. A city in the centre, Marseille anchors the Provence – Alpes – Côte d'Azur region. Marseille is also a city on two cusps. It is on the southern shore of the European continent, strategically located midpoint on the coastal arc of European Union nations fronting the Mediterranean Sea – the grouping loosely defined as *Euro-Mediterranean*. At the same time, Marseille is on the northern shore of the Mediterranean international agglomeration. Here, the place of Marseille is further defined by special trajectories connecting it to other nations' ports across the sea. Some connections, as in Marseille's inextricable alignment with the port of Algiers, remain significant in contemporary as well as colonial contexts. Place-labels become ambiguous when they extend to socio-cultural identities

> *Marſeilles* is ſituate at the Foot of a Hill, which riſes in the Form of an Amphitheatre in Proportion to its Diſtance from the Sea.
>
> Excerpt from 'A Description of Marseilles' from *Marseille. Chambre du Conseil de l'Hôtel de Ville* (London, 1722).

attributed to these regions. People within the boundaries of a geopolitical area commonly hold multiple and sometimes merged cultural identities that defy simple boundary lines. A resident of Marseille may identify herself as [*marseillaise*] [*française*] [*algérienne*] [*musulmane*] [*juive*]... as one, several or none of these. These affiliations elude clear definitions (are they legal classifications, perceived identities, familial lines?) and naturally blur as applied to individuals, neighbourhoods and whole societies. We accept that architecture in form and materials can hold meaning as a manifestation of culture, as *patrimoine*. How then can architecture speak to multiple or merged cultural identities without becoming a pastiche of styles and symbols?[4]

4   *Patrimoine culturel* refers to cultural heritage of national significance. Alone, the word *patrimoine* is frequently used to refer to significant public buildings. The meaning of *patrimoine* as inheritance or property reflects how closely the arts and technology of a culture become intertwined with issues of national and individual identity. Exploration of the question of how architecture can speak to multiple or merged cultural identities, as pertains to the discussion of French and Mediterranean cultural hybridity, is by nature embedded in repercussive colonial contexts. Author Sheila Crane examines themes of transnationalism and interdependence between significant works of French and Algerian architecture in 'Architecture at the Ends of Empire: Urban Reflections between Algiers and Marseille', in G. Prakesh and K.M. Kruse (eds), *The Spaces of the Modern City: Imaginaries, Politics, and Everyday Life* (Princeton: Princeton University Press, 2008), 99–143. In this chapter, Crane presents the historical example of a late colonial-era mass housing project in Algiers, the Diar el-Mahsul designed by French architect Fernand Pouillon, among other 'complex interactions and hybrid constructions'. The Diar el-Mahsul is notable in a discussion of cultural identities, in part due to its separate districts for Europeans and Algerians with differently appointed apartments, and its repossession after independence by

A possible answer is found by returning to the root *geo-* found in geopolitical. Now add 'geophysical' as another lens through which to view 'place'. Any particular site has a concomitant terrain and climate, not to be overlooked in the architect's response lest the building function poorly. The Mediterranean region's signature sunlight and hot, arid weather patterns are nonplussed by geopolitical boundaries. A building design with a focused response to climate is called 'bioclimatic'.[5] Bioclimatic design places emphasis on the human occupants' physical comfort, with emphasis on the person as biological entity more so than a socio-political one. This approach allows a release from an architectural response focused solely on cultural significance and culture's inherent multiplicity and contradictions. The bioclimatic design approach can be the basis of an apolitical architecture that is fitted to a broad geographic region such as the Mediterranean.

Certainly, architecture and sites with a strong sense of place are formed in response to culture as well as climate. Attention to both is required. In the process of design, the two concerns (one as complex as humanity, one as dependably constant as the sun) may be artificially separated, allowing deeper consideration of each. But in the synthetic act of building they are reconstituted in architecture that resonates with people and place. Ironically, this type of adroit architectural response may be achieved by temporarily suspending consideration of the socio-cultural and political identities of the very people the building is intended to speak to. In other words, the building ends up speaking to people by speaking to the common language of their place.

Algerians. Crane writes about the post-colonial decades: 'In the face of migrations, buildings from the colonial era in Algiers remained as significant symbolic presences within the urban landscape. At the same time, the pressing need to accommodate waves of *repatriés* and Algerian immigrants had important effects on the architecture of Marseille.'

5 B. Meersseman and A. De Herde, *Bioclimatic Architecture* (Louvain-la-Neuve: Centre de Recherches en Architecture, Cellule Architecture et Climat, Université Catholique de Louvain, 1992).

Origins and Operations

Just as surely as the context of place is many-layered, the context of client and constituency eschews simple definition. The origins and operational mission of the Bibliothèque Municipale à Vocation Régionale de Marseille can be understood by retracing the origins of the project and the objectives of its multi-tiered client.

The library's client was in fact a multi-tiered collection of state, regional and local politicians along with cultural ministers and library institutional administrators at all of these levels. Add to this the customary interdisciplinary project team of project manager, contractor, architect, engineers and a host of other building consultants.[6] The diversity of the clients and actors, each holding discrete goals and bringing unique expertise to the project table, led to the fitting and innovative architectural response.

The BMVR programme began with legislation passed by France in 1992 to encourage and fund the construction of regional libraries. A main objective of the measure was to stimulate the *lecture publique*, part of a generalized effort by the French government to increase citizen participation in cultural activities within an array of institutional networks. Another objective was to protect the nation's rare books and decentralize some of the functions of the *dépôt légal*, wherein copies of published works are collected in the national coffers – an institutionalized method of building up the national patrimony begun when the Bibliothèque nationale de France (BNF) was still the Bibliothèque du Roi.[7] The law that created the BMVR designation shares some of the same spirit as the decentralization legisla-

---

6 *Appropriate Sustainabilities* lists the library's project team members in addition to Adrien Fainsilber Associates: Clients – Municipality of Marseille, Marseille Development Corporation; Associated architect – Didier Rogeon; Lighting design – Speeg & Michel; Engineering – Beterem. The project area is noted as 21,000 square metres. M. Emery, *Appropriate Sustainabilities: New Ways in French Architecture* (Basel: Birkhäuser, 2002), 143.
7 B. Blasselle, *Bibliothèque nationale de France: L'esprit du lieu* (Paris: Editions Scala, 2001), 6–7.

tion of 1985, which had far-reaching effects bolstering the autonomy and identity of regional governments. The City of Marseille, its General and Regional councils, and the French State shared the financing of the Alcazar project. Both the high profile and tiered support created the context that made the library possible.[8]

BMVR libraries by definition must have large collections and serve a broad regional population of at least 100,000 inhabitants. The Marseille BMVR easily met this requirement – the population of the commune of Marseille alone numbers 800,000.[9] The library's patrons are simply residents of an area, not people defined by other socio-economic or political means. This is now an expected fact for a public project in contemporary France, but it underscores the evolution of the French library from centralized exclusivity (in its earliest origins as a library within the Louvre palace for the King and court) to global-reaching inclusivity (in the current framework of the BNF and its dispersion in the form of the BMVR

---

8   Since the 1980s regional governments in France have had increased authority to commission and build civic buildings, but lack of financial wherewithal prevented broad developments in cultural institutions. In 1992, with a focus on improving the quality of life in the regions, France approved the construction of twelve regional libraries with close ties to the BNF. This act enabled the construction of large regional libraries meeting size, collection and constituency requirements. Planning for the initial twelve began between 1993 and 1997, and BMVR libraries have opened in many cities including Marseille, Lyon, Rennes and Montpellier. These libraries not only share access to research databases and other shared resources, as do other large libraries in France. Significantly, each BMVR project would also receive State financing amounting to forty per cent of construction costs including furnishings and equipment. See A. Fritzinger, 'Bibliothèque municipale a vocation régionale (BMVR): histoire et contexte d'un projet – L'exemple de la future bibliothèque publique de Marseille', *Colloquium Public Libraries and New Technologies: How to Fight Info-poor exclusion?* (Lisbon: Department Cultural da Camara Municipal de Lisboa, 2002), http://www.citidep.pt/papers/acf/acfbnt.html (accessed 1 October 2011). In the case of the Marseille BMVR, the state's contribution was 20 per cent of the total project investment, while the city of Marseille's portion was 52 per cent. C. Grousset-Jeanjean, *Fiches Techniques Alcazar* (Marseille: Bibliothèque Municipale à Vocation Régionale de Marseille, 2008), 11.
9   Grousset-Jeanjean, *Fiches Techniques*, 3.

and other libraries with access to its collections). This modern form of library is actually planned for continued evolution. Another requirement of BMVR designation is that the building structure must be adaptable to ongoing technological and institutional change.

Beyond its defined role providing modern multimedia library services and cultural resources for the region, the BMVR project was of key relevance to many groups at the municipal level. The library's site near the Vieux-Port was strategically selected and promoted by Marseille's mayor, Jean-Claude Gaudin. The choice of the abandoned Alcazar site allowed for a nostalgic revival of memories of the old music hall, nationally cherished for its role as a bastion of French *chanson*. The Belsunce neighbourhood of the city had been in economic decline for some time, and has seen much urban renewal efforts by the city of Marseille in partnership with building owners and developers. The library's project manager and contracting firm had already been engaged in many historic preservation and renewal projects in the Marseille region. They brought expertise in renovating the area's historic structures to the team. During the course of the library's construction, archaeological remains from a Roman settlement were uncovered. While this delayed the project for a year and increased its cost, these discoveries only added to the project's rich cultural story. The library's arrival in this old but important neighbourhood, fronting a busy boulevard and tramway line, created a vitalizing pole for the urban renewal efforts. The site's unique historical features and the library's programmatic role as a cultural institution serving all residents may have allowed this particular project to achieve broad acceptance. This is in contrast to other redevelopment in the area that has been more contested and seen by some as a process of gentrification.

The city of Marseille placed a number of provisions in the project brief that figured significantly in the architectural response. These included a height restriction in keeping with the typical buildings in the district, preservation of the original stone façade on the outer corner of the building site, building around a mid-block alley, and use of clay-tile roofing on most of the roof plane.[10] Within these constraints, architects Adrien Fainsilber Associates conceived of a thoroughly contemporary response that respects the site's historic context while still placing emphasis on the library's civic prominence and functional needs. This balance is particularly evident in the dramatic use of the translucent marble-glass façade and central atrium. These bioclimatic design strategies were employed by the architect to meet functional as well as aesthetic programmatic objectives.

The translucent marble-glass wall is integrated into the library's west-facing entrance façade. The outermost surface of this wall is an expansive curtain of thin translucent marble and glass. A clearly modern intervention to the streetscape, its visual impact is quieted by its alignment in height and width with the adjacent traditional stone façades that characterize the neighbourhood. This dimensional conformance to physical context suppresses an ostentatious reading of the façade, even while the marble material denotes a sense of civic prominence. The marble-glass façade is suspended like a curtain at a height one storey above the street, carefully hung on the main building structure by a lightweight metal truss about one metre in depth. A conventional clear glass wall framed in aluminum forms the innermost layer that extends down to meet the street. In this inner plane, the library's public entrance doors lead to an entrance foyer that is visually completely open to passers-by. Together with the air space in between the two layers, this composition forms what is called a 'double façade'. Natural sunlight is diffused as it penetrates the façade layers during the day – perfect for the reading rooms that are located here at each level. Artificial lights installed within the space of the doubled layers create a screen-like glow directed to the street at night. Stone is a traditionally dignified choice of material for a library façade. The translucent stone application here delivers a symbolic message of dignity combined with contemporary openness that traditional opaque stone could not. In this way the façade meets the library's internal operational needs (the provision of well-lit, acoustically isolated reading rooms above, a secure yet inviting entrance below) while

---

10   Grousset-Jeanjean, *Fiches Techniques*, 3.

BMVR, Marseille: view of atrium 'inner street'.

also supporting its mission to encourage public awareness and participation in the cultural institution.

Like the façade, the library's central atrium performs on multiple levels beyond pure functionality. It is a space for circulation and exchange, like another street knit into the urban grid. Oriented along an east–west axis, it turns and extends the Cours Belsunce pedestrian zone several blocks deep into the neighbourhood. Open to four public levels and anchored by two pairs of glass elevators, the atrium brings freely flowing organization to the library's contents. The open exposed structure visibly expresses the BMVR's capability to adapt to changing technology.

The building structure is a reinforced concrete frame and slab system. The proportions of its column grid are distinctly native to the historic neighbourhood. The historic Cours Belsunce area is largely made up of a mixture of nineteenth-century three-bay and four-bay buildings. Terraced houses adjoin into blocks that form the proportions of the three-dimensional streetscape. Called *maison marseillaise* or *trois-fenêtres*, the buildings are generally six to seven metres wide overall, featuring three window bays of about two metres wide between thick stone walls. The modules are based on the spanning ability of timber sizes available to builders in this era of the port city. The architects adopted the timber module as the proportional system for the library, effectively weaving the larger scale of the library to be harmonious with the neighbourhood's smaller scale. Open public areas such as the atrium, book stacks and reading rooms are generally sized to be four modules across (about eight metres), while adjacent open hallway areas are one module wide (two metres). These areas are defined by a *tram escosse* or 'scotch plaid' column grid. In this system, pairs of concrete columns clearly define two-metre wide walkways at the floor level. Overhead, running lines of building utilities are organized, further delineating the main paths used by library patrons. Ceiling areas are generally left open for ease of access to communication and other networks, which the library expects to periodically upgrade. This open network of spaces and routes gives a framework to a very large interior while maintaining proportions and human scale native to the neighbourhood.

The library site actually spans two rectangular blocks in the urban fabric behind the Cours Belsunce. A small but vibrant neighbourhood street continues to pass between the blocks. Here, the library's structural grid and the long atrium space are interrupted, resuming on either side of the street. The building stays functionally unified by a continuous basement below street level and a pedestrian bridge at the third level. The ensemble is figuratively unified by the atrium's sun-shade louvres spanning high overhead the alley.

The theme of openness and adaptability extends in the institution's mission to support use by persons of all ages and physical abilities. Although accessibility in design has now become commonplace in public buildings in the United States, due to the Americans with Disabilities Act of 1990, this was not the case in France at the time the Marseille library was built. When the building opened in Marseille in 2004, it ambitiously demonstrated accessibility to its patrons by use of Braille signage, floor guidance markers and comprehensively accessible routes. These features complement the library's programmatic mission to increase access of the collections to the visually impaired, as seen in the creation of the *Lire Autrement* service department. Other library constituent groups are overtly served with named departments or collections: *Le Public Jeunesse, Le Patrimoine, La Civilisation* among other categories. Specific service to multiple cultural constituencies, for example the ethnically diverse residents of the Cours Belsunce neighbourhood, is not explicit in the library's dossier of departments, services and lists *par les chiffres*. Instead the library proposes to offer service to '*tous publics, tous usages*'.[11] As they are primarily defined as denizens of a broad BMVR region, the library patrons of the most modern of French libraries remain ostensibly uncategorized. Yet the diversity of their patronage is reflected in the library architecture that does not homogenize in its response to place.

The Architectural Agenda

The architects clearly embraced the project's precondition of sensitivity to urban context and use. But they integrated the project even further into its setting by allowing the Mediterranean climate of Marseille to inform the architectural solution. While welcomed by the client, neither the bioclimatic response nor the resultant energy conservation was a requirement

---

11   Grousset-Jeanjean, *Fiches Techniques*, 1–13.

of the project brief. Adrien Fainsilber Associates proposed the west-facing translucent double façade and sun-shaded atrium as strategies that manage solar heat gain while admitting useful daylight. The architectural devices work together in the design to provide solar protection, increase thermal inertia, assist with convective envelope cooling and reduce internal lighting gains – all primary principles of bioclimatic architecture.[12]

The library's most striking architectural features represent both climatic and cultural concerns. The west-facing entrance wall described earlier is clad beautifully with book-matched sheets of *Arabescato Carrara* marble layered between clear glass panes. The two layers of the double façade form a wide, naturally ventilated buffer zone of air that shields the library reading rooms from direct western sun and heat while admitting diffuse daylight for the readers. The double façade performs functionally similar to two common features found in vernacular Mediterranean architecture – *mulqaf* wind-catching towers and *mashrabiyya* lattice-like screens.[13] A central atrium space in the library forms an inner 'street' into which all the floors open, extending the urban pedestrian experience. This large, communal space is also shielded from direct daylight by a screen of large louvres delicately balanced on the rooftop. There is more to the role of the atrium than managing air movement and bringing controlled daylight into the core of the building. Similar to the open courtyards and domed halls of Islamic palaces and mosques, the library's atrium functions as a place of central focus, for passage and for gathering, for bringing people together in search of commonly shared goals.[14]

12 Meersseman and De Herde, *Bioclimatic Architecture*.
13 Petersen, *Dictionary of Islamic Architecture*, 277–8, 298.
14 The plan of the Great Mosque of Timnal, Algeria, is an example of a large open gathering space that is similar to Marseille's library in these aspects: a large rectangular columned space used for gathering is oriented with the long side following the east–west axis; auxiliary spaces to one side of the open space share the same column grid; the ensemble is surrounded by thick, insulating mass – masonry walls in the case of the ancient mosque, concrete walls and double-façade air space in the case of the modern library. Petersen, *Dictionary of Islamic Architecture*, 18.

The design of the building also integrates climatic and cultural concerns in less obvious ways. Space for the library in this densely built area of the city was found by carefully demolishing portions of blocks and buildings rather than by the wholesale clearing that often accompanies urban renewal. The new site retains a prominent original stone façade corner at the main street frontage, and literally bridges over a pre-existing small but active alleyway deeper into the block. The neighbourhood's narrow streets and consistent building heights allow adjacent structures to shade one another and the pedestrians for much of the day. The thick stone masonry of the older surrounding buildings acts like thermal insulation, slowing the transfer of outdoor heat to the interior. To similar effect, the modern library conforms to the existing urban fabric and uses high-mass concrete for its structure. As mentioned earlier, even the proportions of the building frame are derived from the buildings native to the historic neighbourhood. Like the surrounding buildings, large sections of the new library roof are roofed with clay tile. These high-mass building materials, which have stored the heat during the day, begin to release the heat to the cooler sky at night. This diurnal cycle readies the spaces for comfortable occupation the next day with reduced use of mechanical heating and cooling systems. In this manner, the context of the city is not merely preserved but in fact woven into the fibre of the new building.

How is it that the bioclimatic design response described thus far, naturally Mediterranean (because it is fitted to its climate), naturally Marseillaise (because it is fitted to its city), is also representative of French culture in a national sense? The answer can be found through a study of the architect's previous work. Adrien Fainsilber Associates' history with bioclimatic design goes back over twenty years to this firm's pioneering work on the Musée des Sciences et de l'Industrie at Parc de la Villette, one of the first in the series of France's Grands Projets. For this project, completed in 1986, the architects conceived of the Grandes Serres – three wide south-facing multi-storey glass greenhouses that were intended to visually and symbolically connect the science and nature themes of the museum to the controlled natural landscape of the park. In the quest to construct a glass enclosure with the utmost transparency, the architects teamed with the British firm Rice Francis Ritchie Engineers. The team made further developments to the existing technology of frameless 'point-supported glass', a façade system in which laminated glass is supported on structural points at the corners of glass panes rather than by continuous frames. For the greenhouse structure at La Villette they innovatively developed a special bolt for the façade system that allowed specially strengthened glass to bear the weight of the wall while wind forces were transferred though the bolt to the backing structure. As a result of this modification, the tensile cable backing structure behind the glass could be lightweight and highly transparent.[15]

For the temperate climate of Paris, the greenhouses serve as a thermal buffer with dual functions depending on the season. In summer, warm air drawn into the glass greenhouses rises by convection and is vented to the outside. In winter, cold air entering the greenhouses is warmed by the sun and held, insulating the adjacent building from the cold outside as well as delivering warm air to the building interior. In this way the greenhouses protect the main building area from the extremes of heat and cold. They attach to the main public spaces of the museum – large sky-lighted areas surrounding a multi-storey atrium. The greenhouse and atrium volumes are both naturally lit; the atrium is protected with shading devices on the roof as would become the model for the Marseille library. The greenhouses and the atrium enable the museum to 'breathe naturally' in a sense. In this way, the bioclimatic design supports the symbolic objective of the museum to interconnect science and nature.[16]

The frameless structural system used at La Villette was applied to another Grand Projet soon thereafter by a different architect. For the Pyramide du Louvre, architect I.M. Pei employed the highly transparent system to stunning material and symbolic effect. At the Louvre, the system was not formulated to conserve energy but rather for its transparence and crystalline properties. The system, so publicly featured in two of the promi-

---

[15] A. Fierro, *The Glass State: The Technology of the Spectacle, Paris, 1981–1988* (Cambridge: MIT Press, 2003), 182–7.

[16] A. Fainsilber, 'The Metamorphosis of a Monumental Building', in S. Fachard (ed.), *Paris 1979–1989* (New York: Rizzoli, 1988), 151.

Museum of Science and Industry: view of greenhouses.

BMVR, Marseille: view of interior of double façade.

nent Grands Projets, came to represent the high levels of technical innovation achieved by France. Architectural materials gave a tangible nature to two national ambitions – dominant leadership in art and technology on the one hand, desire for transparency between people and their cultural and political institutions on the other.[17]

This type of glass wall has since become used extensively throughout Europe and now the world in applications both grand and commonplace. Its significance has been reduced to imply a ubiquitous 'European' or 'high-tech' design aesthetic. But for Adrien Fainsilber Associates, the concept behind their use of the façade has remained one of aesthetics combined with bioclimatic functionality. Subsequent to designing the museum at La Villette, the architects have made a practice of evolving the bioclimatic concept, applying thermal buffer zones in the form of double façades and atriums to numerous public and commercial buildings in different parts of France. For the Alcazar project in 2003, they adapted the concept again in a way that was suited to the library use, the hot-dry climate and the dense urban context of Marseille. Here, the façade's translucent outer skin diverts unwanted heat from the western exposure but elegantly admits useful diffuse light. Unlike a greenhouse, the cavity of a double façade is narrower, the width of a service catwalk only. The stone and glass layers along with the air mass create the buffer that mitigates the flow of heat to the interior. In Marseille's climate, the need is almost always for cooling; therefore the cavity is permanently open at the top and bottom. Natural convection, or airflows induced by the effect of hot air rising within the cavity, aids cooling day and night. The atrium concept as adapted to the

---

[17] In the preface to a book about the architecture of the Grands Projets, former French president François Mitterrand concludes his introduction to the projects with this statement of intent: 'Beauty excites curiosity, responds both to the needs of the heart and mind and has an educative and training value. The Grands Projets will, I hope, enable us to understand our roots, our history. They will allow us to foresee and conquer the future. This is the meaning of the *grands projets*: to illustrate the continuing ambition of the country as a whole.' 'Preface', in Fachard (ed.), *Paris 1979–1989*, 9. For a developed discussion of the concept of *transparence* relative to architecture in France, see the previously cited work of Fierro, *The Glass State*.

Marseille library is more visually open than its predecessor at La Villette. It is flowing with people, lined with conveniences of a new information era. Shiny metallic frames guide glass elevators to each level of the library; floor slabs are like neatly indexed trays holding books. Thanks to the use of contemporary technology and new material combinations, the lineage of which is being traced here as having a distinctively French pedigree, the library's double façade and atrium function effectively together in ways distinctly Mediterranean: simultaneously ventilating, protecting and lighting the reading rooms within. The elements work passively with the local climate as do their counterparts across the water – a shaft to catch wind, a courtyard to shade and pull breezes through, a screen to filter light.

Material Precedents

Adaptation of the greenhouse concept to Marseille was accomplished in part by innovations in the materials used. Here, composite stone and glass panels were introduced into the frameless point-supported structure, a refreshing update of the system developed nearly two decades earlier for La Villette. The panel manufacturing process merged laminated glass technology with stone veneering craft for a new product. The assembly appears to be proprietary to a company in Villeurbanne, near Lyon, France. Their fabricators painstakingly laminated thinly sliced book-matched *Arabescato Carrara*, a delicately veined marble, between two glass layers creating large (up to 2,600mm by 1,400mm) façade panels. The composite sheets were laid out on the façade in book-matched sets of four panels by mirroring the material across both horizontal and vertical axes.[18] This arrangement forms a quadrant, not unlike a diamond or reverse diamond veneer pattern in wood. The smooth, modern exterior appearance of the library façade

18 Fiberstone, 'Fitting Technics – AEG Façades (Attached External Glass)' http://www.fiberstone.com/anglais/meo/meo_facVEA.html (accessed 1 October 2011).

owes much to this minimalist system, where only small, flush bolt caps are visible near each panel's four corners. The backing structural frame of steel and tensile cables secures the outer façade to the main building structure and resists wind loads. Like the veins protruding from the underside of a leaf, the structural lines can only be seen in glimpses from oblique viewing angles, by looking upward from the sidewalk underneath the double façade, or from within the library. Perhaps it is more than a linguistic coincidence that laminated glass in French is called *verre feuilleté*, the descriptor beautifully derived from the word *feuille*, or leaf.

In another play on words, the natural veining pattern of the marble, when mirrored or 'book-matched' about a panel edge, visually recalls the pages of an open book. The reference of the wood furniture veneering term to the turning over of book pages is poetically appropriate for the public face of a building dedicated to reading. The assembly of the façade panels reminds us of the care and craft used in assembling a fine piece of furniture or interior panelling. The attention to detail and near luxury of the workmanship is certainly fitting for a public institution of this stature. Yet the modernity of the composite materials and technologies used handily avoid any associations with stuffiness or exclusiveness such workmanship might otherwise imply.

Inside at daytime, diffuse natural light creates visual quietness for library reading rooms. This quality is paired with acoustical buffering provided by the outdoor curtain of air. I have seen the reading rooms in daytime use by groups of students studying for exams. The sunlight passing through the façade silhouettes them dramatically against the naturally formed Arabesque patterns of the marble in the glass wall. The power of natural light to define gathering spaces becomes nearly tangible.

Again in daytime, when viewed from the street the appearance of the façade in direct sunlight is that of honed stone rather than glass. This condition of apparent opacity resonates with the introversion of the surrounding stone-clad façades of the quarter. At night, the building is lit from the interior, glowing like the traditional alabaster light fixtures that would have been found in the old theatre. I like to think of the luminous façade as a second marquée for the building, subliminally advertising that this venue is open for all to enter.

The use of translucent stone as a contemporary façade material with qualitative effects is not without precedent: In the Beineke Rare Book and Manuscript Library at Yale University (c. 1963), thick translucent stone panels set into a concrete frame protectively wrap the whole perimeter of the building, admitting sombre diffuse light to the interior. In a Parisian building whose mission was to promote understanding of the Arab world in France, the walls of the mysterious inner courtyard of the Institut du Monde Arabe (c. 1987) are lined with tiny translucent stone tiles. These modern-day examples are furthermore indebted to ancient use of stone to cool and beautify interiors in hot climates around the world. A beautiful example is found at the *Mughal*-era Red Fort in Delhi, India, where thick marble was carved to a thin transparent membrane above carved open marble lacework. Returning to the Mediterranean region, in Seville, Spanish stone is also carved delicately in ornate, repetitive patterning above the stone arches and light-diffusing window screens of the Alcázar palace. My admiration for the use of translucent stone material in the Marseille library is not to suggest this is a new idea, but rather a brilliant adaptation of an ancient practice.

Hybrid Cultural Significance

In talking with the staff and building engineer of the Marseille library, the building functions as a library with a high level of success. In the press, upon opening to the public and in blogs near that time, the library appears to be earning its rank as public *patrimoine*. For example, on 22 November 2004, *Le Monde* ran an article entitled 'Popular success for Marseille's Alcazar' noting that the library sees a broad spectrum of patrons, between 3,000 and 8,000 people daily.

The design approach applied in Marseille's library undoubtedly allows a vernacular response to climate to come into the modern technological age, both aesthetically and functionally. But this title holds a loftier claim:

hybrid cultural significance for the Bibliothèque Municipale à Vocation Régionale de Marseille. Is the combination of discrete architectural strains into a harmonious whole in the BMVR a form of universalism? A universalist view might imply that architectures and cultures arriving from the south (Spain and North Africa) have been assimilated by the French architect into the design. Or is it that views brought by the architect from the north (Paris, representing France and Europe) were assimilated into the local architectural tradition? This line of thinking becomes circular, placing emphasis on defining Marseille's position relative to geopolitical boundaries. Can the architecture of the BMVR, so closely engineered to place, instead be called 'hybrid'? Mixed sources form a hybrid product that is unique from the original sources and fitted to thrive in a particular place. In positing that the architecture of Marseille's library has hybrid cultural significance, the analysis ultimately shifts the focus from origins of place and identity and allows us to focus on the local performance and human qualities of the result.

Bibliography

Blasselle, B., *Bibliothèque nationale de France: L'esprit du lieu* (Paris: Editions Scala, 2001).

Bonham, M.B., 'The Open Book: Laminated Translucent Stone Façade at Marseille's Alcazar Library', *Material Matters: Making Architecture*, proceedings of the 2008 Association of Collegiate Schools of Architecture West Fall Conference (Los Angeles: University of Southern California, 2008).

Crane, S., 'Architecture at the Ends of Empire: Urban Reflections between Algiers and Marseille', in G. Prakesh and K.M. Kruse (eds), *The Spaces of the Modern City: Imaginaries, Politics, and Everyday Life* (Princeton: Princeton University Press, 2008).

Emery, M., *Appropriate Sustainabilities: New Ways in French Architecture* (Basel: Birkhauser, 2002).

Fainsilber, A., 'The Metamorphosis of a Monumental Building', in S. Fachard (ed.), *Paris 1979–1989* (New York: Rizzoli, 1988).

Fierro, A., *The Glass State: The Technology of the Spectacle, Paris, 1981–1988* (Cambridge, MA: MIT Press, 2003).

Grousset-Jeanjean, C., *Fiches Techniques Alcazar* (Marseille: Bibliothèque Municipale à Vocation Régionale de Marseille, 2008).

Meersseman, B., and De Herde, A., *Bioclimatic Architecture* (Louvain-la-Neuve: Centre de Recherches en Architecture, Cellule Architecture et Climat, Université Catholique de Louvain, 1992).

Mitterrand, F., 'Preface', in S. Fachard (ed.), *Paris 1979–1989* (New York: Rizzoli, 1988).

Petersen, A., *Dictionary of Islamic Architecture* (London: Routledge, 1996).

PART 3

Politics of Ethnicity and Postcolonial Memory

BRUNO LEVASSEUR

French *Artistes Engagés* Transcending the Politics of 'Race' and 'Ethnicity' in the Republic (1980s–2000s)[1]

In recent decades, following the process of decolonization, new stigmatizing and threatening images of 'race' and 'ethnicity' have appeared in popular culture, particularly in art.[2] The culture of protest and its political dimension and specific aesthetic has offered an original counter vision of the new 'racist' discourse that, in Western nations, associates immigrants with job snatchers, welfare scroungers or threats to security. In France, the mobilization of French *artistes engagés* has been highly significant over the last thirty years and can be located within the development of this aggravating rhetoric of the 'new racism' defined by Martin Barker.[3] Here, my purpose is to analyse representations of the 'ethnic other' in the artistic culture of *engagement* and the alternative perception of the French Republic which can be found in this artistic culture from the mid-1980s.

Since the end of the 1970s, depictions of France have increasingly intersected with the traditional pattern of the postcolonial narrative based

---

1   I wish to thank Dr Karima Laachir, Dr Béatrice Dammame-Gilbert and Dr Angela Kershaw for their support, comments and advice while preparing this chapter. I am also indebted to the challenging and inspiring contributions made by the participants of the ASMCF conference held in Portsmouth in September 2009.
2   P. Gilroy, *Between Camps* (London: Allen Lane, 2000), 214–15.
3   M. Barker, *The New Racism* (London: Junction Books, 1981), 20–4. In a nutshell, the 'new racism' emphasizes the fantasies of the bond of blood and belonging, and concentrates more on the homogeneity of the nation and its authentic cultural life endangered by 'newcomers'.

on 'race', 'ethnicity' and 'frontiers'.[4] The issue of 'racism' in the French Republic, and the correlative discrimination against (North) African immigrants and their children, has become one of the most significant topics in the nation.[5] Their deprivation and marginalization have called into question the fundamental principles of 'liberty, equality and fraternity' at the basis of the Republic. They have also brought into question the issues of 'assimilation' into the national community and the acquisition of citizenship. As a specific category, white *artistes engagés*[6] and their emphasis on tolerance and justice have played a vital role in stimulating debates on issues of 'difference' and 'equality'. In this sense, they have proven highly significant in short-circuiting, since the end of the Algerian war (1954–62), the demonizing images of the 'ethnic other'. They have also succeeded in showing the persistence of a paradoxical 'evolutionary racism' in contemporary France.[7]

The main object of this chapter is to demonstrate how recent, cultural representations of protest made by white French left-wing artists have, since the 1980s, provided alternative ways to oppose and resist the *grands discours* promoting both 'difference' and 'intolerance' in the Republic. For leading British scholars such as John Solomos and Les Back,[8] 'popular culture [should not be considered] as uniform and fixed' since it 'populariz[es] racial imagery' in various subtle ways. Oscillating between the author's art and interests,[9] the work of French *artistes engagés* is concomitant with

---

4 E. Balibar, *La Crainte des masses* (Paris: Galilée, 1997) and E. Balibar 'La forme nation: histoire et idéologie', in E. Balibar and I. Wallerstein (eds), *Race, Nation, Class* (Paris: La Découverte, 1997), 117–43.
5 A. Hargreaves, *Multi-Ethnic France* (London: Routledge, 2007).
6 H. Dufournet, L. Lafargue de Grangeneuve, A. Schvartz and A. Voison, 'Art et Politique sous le regard des sciences sociales', *Terrains et Travaux*, 13 (2008), 3–12.
7 On this concept, see P.-A.Taguieff, 'Universalisme et racisme évolutionniste: le dilemme républicain hérité de la France coloniale', *Hommes et migrations*, 1207 (Mai–Juin 1997), 90–7.
8 J. Solomos and L. Back, *Racism and Society* (Basingstoke: Macmillan, 1996), 157.
9 J. Street, 'The Politics of Popular Culture', in *The Blackwell Companion to Political Sociology*, ed. K. Nash and A. Scott (Oxford: Blackwell, 2004). Theorizing the politics of protest, Street observes that within the politics of popular culture, culture

certain aesthetic and political features distinct from the dominant 'racialist' discourse on 'otherness'. Here, I shall propose that the decentered portrayals of the 'ethnic other' in artistic *culture engagée* offer a complex outlook on the construction of the 'immigrant problem' in France.[10] Through a comparison between hegemonic and protest discourses, my goal will be to show how these portrayals, which are commonly divided into the '*zoufri*' and the '*beur*',[11] do not only demythologize the egalitarian discourse of the French Republic, but also challenge ethnic absolutism and national belonging.

My study is grounded on selected literary, cinematic and musical artefacts produced over the last thirty years by some leading French *artistes engagés*. It concentrates on the following works: Didier Daeninckx's detective novel, *Le Bourreau et son double*; Bertrand Tavernier's documentary, *De l'autre côté du périph'*, and Les Ogres de Barback's rock song, *Pardon Madjid*.[12] I consider the militant contribution of these artists to be vital within the French artistic sphere. Drawing on Jacques Derrida's work on hospitality, I maintain that these neglected productions have testified to the existence of distinct modes of depicting the French Republic away from the prevalent 'racial' and 'ethnic' determinants of 'intolerance' and 'closure'. Following Derrida's assertion that 'Western discourses' have served to forge both centrisms and determinisms, I argue that the subversive creations by French *artistes engagés* are suited for the purpose of deconstructing fixed ideas and beliefs. Through their de-essentialization of the 'ethnic other', they generate a 'supplement' which not only supplies us with a 'new political

certainly expresses political views. Subsequently, artists take part in conventional politics (304–11).

10 M. Silverman, *Deconstructing the Nation* (London and New York: Routledge, 1992), 94.

11 A. Boubeker, *Les Mondes de l'ethnicité* (Paris: Balland, 2003). Boubeker lists two figures of ethnicity in French society: the '*zoufri*', 'tour à tour guerrier, travailleur, immigré', and that of the '*beur*', their descendants incarnating 'l'évasion des cités et invasion de l'espace public' (ibid., back cover).

12 D. Daeninckx, *Le Bourreau et son double* (Paris: Gallimard, 1986). B. Tavernier and N. Tavernier, *De l'autre côté du périph'* (1997). Les Ogres de Barback, *Pardon Madjid* (2007).

horizon' but also works towards a reconfiguration of 'friendship' and 'hospitality'. In this chapter, my ultimate goal will be to show how French *artistes engagés* have contributed, since the mid-1980s, to challenging the politics of 'race' and 'ethnicity' while urging for a redefinition of the Republic.

Contemporary French Politics and Artistic Representations of 'Race' and 'Ethnicity'

Contemporary France:
Republicanism, the 'Immigrant Problem', Moral Panic

Republicanism in France is grounded on egalitarian principles designed to ensure social justice and equality, underline the supremacy of the people and foster their participation in the public sphere.[13] The core of French republicanism, which suppresses all privileges linked to rank, blood or birth, is its emphasis on citizenship rather than 'race' or 'ethnicity' as the defining condition of membership of the national community. These elements combine to represent a 'model' or a 'fiction'. Traditionally, France is considered to be different from most Western nations in what can be seen as its long history of immigration and capacity to integrate foreigners.[14] However, the 'manufacturing' of the presence of immigrants as a 'problem', in the 1980s and early 1990s, provided the basis for a drastic reframing of national ideals and restored a long political culture of discrimination and segregation.[15] To put it briefly, over several years, in the context of France's 'great mutation', the reconfigured portrayal of immigrants and their descendants as

---

13   J. Hayward, *Governing France* (New York: Norton, 1983), 88.
14   G. Noiriel, *Le Creuset français* (Paris: Seuil, 1988).
15   M. Silverman, *Deconstructing the Nation* (London and New York: Routledge, 1992), 94. J. House, 'Anti-Racism and Anti-Racist Discourse in France from 1900 to the Present Day', unpublished PhD thesis (University of Leeds, 1997), 42.

culturally, ethnically and religiously different came to emblematize the fears and fantasies of the break-up of the national community.[16] At the heart of public discourse, this prompted the implementation of 'national republicanism',[17] followed by political violence and repressive legislation.[18] In republican France, signs of this 'moral panic' concerning immigration also manifested themselves in popular culture, most obviously in artistic productions.[19]

'Alternative Depictions of the 'Ethnic Other':
French Artistes Engagés as New Bearers of 'Francité'

Questions of 'race' and 'ethnicity' in France have been a much developed topic within 'art for the masses'.[20] More than white *artistes engagés*, creators of foreign origin were seen as the main contributors to a dissident representation of the new 'immigrant problem'. As amply demonstrated, the cultural contribution of minority artists, born and raised in the '*banlieues*', is perceived as having been greatly significant in changing depictions of

---

16 M. Wieviorka, *La France raciste* (Paris: Seuil, 1992), 25. See also: C. Wihtol de Wenden, 'North African Immigration and the French Political Imaginary', in M. Silverman (ed.), *Race, Discourse and Power in France* (Aldershot: Avebury, 1990), 99, 101.

17 E. Balibar, 'Le droit de cité ou l'apartheid?', in E. Balibar (ed.), *Sans papiers: l'archaïsme fatal* (Paris: La découverte, 1999), 89–116. To quote Balibar, the 'national républicanisme' refers to the 'gestion autoritaire de l'immigration' based on 'des discours répressifs et des pratiques qui tendent à les légitimer' (92). A notorious example remains the highly mediatized 'affaire des sans-papiers' in the summer of 1996.

18 For a historical overview of 'l'identité nationale', see G. Noiriel, *A quoi sert l'identité nationale?* (Paris: Seuil, 2007). For details about the new 'Ministère de l'immigration et de l'identité nationale' (2007), read J.-F. Bayart, 'Il n'y a pas d'identité française', *Le Monde* (6 June 2009).

19 S. Cohen, *Folk Devils and Moral Panics* (London: MacGibbon and Kee, 1972).

20 R. Pouivet, 'Qu'est-ce que l'art de masse?', http://philosophie.scola.ac-paris.fr/C603-04Pouivethtm (accessed 1 May 2010).

'ethnic' and 'racial' matters in France since the early 1980s.[21] Tyler Stovall, among others, has provided a fine analysis of works by filmmakers, writers and singers who were descendants of African immigrants and whose work represented the intensification of the suburban crisis.[22] In a more specific way, Michel Laronde, Carrie Tarr and Hadj Miliani[23] have particularly insisted on the crucial part played by *'beur'* culture while detailing the importance of their 'infiltration' in literature, cinema and music.[24] Overall, I entirely subscribe to the idea that minority and *beur* artists have questioned the significance of racial and ethnic divisions within French society'. However, I shall maintain here that they have not been the only ones to have undertaken a criticism of French republicanism[25] and de-essentialize the image of the 'ethnic other'.

In contemporary France, the changing depiction of the 'immigrant' – even if the point seems to have been overlooked[26] – is also connected to the significant contribution of a culture of *engagement* reminiscent of a strong republican tradition. Cutting across the artistic sphere, this contribution has supplied a considerable quantity of alternative depictions of

21   A. Hargreaves, 'The Contribution of North and Sub-Saharan African Immigrant Minorities to the Redefinition of Contemporary French Culture', in C. Forsdick and D. Murphy (eds), *Francophone Postcolonial Studies* (London: Arnold, 2003).
22   T. Stovall, 'From Red Belt to Black Belt: Race, Class, and Urban Marginality in Twentieth-Century Paris', *L'Esprit créateur*, XLI/3, 9–23.
23   M. Laronde, *Autour du roman beur* (Paris: L'Harmattan, 1993). C. Tarr, *Reframing Difference: Beur and Banlieue Filmmaking in France* (Manchester: Manchester University Press, 2005). H. Miliani, 'De la culture d'exil au syncrétisme culturel', in H. Gafaïti (ed.), *Cultures transnationales de France* (Paris: L'Harmattan, 2001).
24   J. Forbes and M. Kelly, *French Cultural Studies* (London and New York: Routledge, 1995), 269–73. 'Beur' culture refers to the explosion of cultural practices and art of protest by Franco-Maghrebin painters, writers, filmmakers, musicians and comedian. See also A. Begag and A. Chaouïtte, *Ecarts d'identité* (Paris: Seuil, 1990).
25   N. Guénif-Souilamas (ed.), *La République mise à nu par son immigration* (Paris: La Fabrique, 2006).
26   The scarcity of analysis attempting to articulate the promotion of tolerance and hospitality by French *artistes engagés* across culture is striking. To my knowledge, no scholarly study of the trans-artistic contribution of white French *artistes engagés* between the 1980s and the 2000s exists.

the 'ethnic other', alongside those circulated by minority artists. Affiliated with the development in France of a *'culture rebelle'*,[27] it has proven particularly dynamic in supplying alternative fictional, cinematic and musical representations of stigmatized immigrants in the Republic.

Historically, French intellectuals and artists in the Republic have offered alternative and dissident views, in the shade of the dominant discourse. Referring to the Algerian war, historian Michel Winock[28] has demonstrated how intellectuals and artists contributed, through their decentred vision, to raising public attention on both sides of the Mediterranean to France's violence and discrimination against colonial subjects. In addition, Winock has shown how intellectuals and artists fought actively along with the 'African Philosopher' Jean-Paul Sartre for Algeria's independence.[29] As the origin of the changing depiction of the 'other', I argue that the commitment of French *artistes engagés* and their support for the 'wretched of the Earth' is still alive. Fighting the 'French colour line,'[30] Daeninckx's detective novel *Le Bourreau et son double*, Tavernier's documentary film *De l'autre côté du périph'* and Les Ogres de Barback's song *Pardon Madjid*[31] appear as exemplars of this established republican trend. Here, I shall suggest that their creators acting as protectors of *égalité* and *fraternité* can be regarded as new *porteurs de 'francité'* (bearers of *'francité'*).

As indicated, I am concentrating in this chapter on the provision of a different insight into the evolution of 'racism' and 'ethnicity' in contemporary France; one that shows how, between the 1980s and 2000s, white French militant artists proposed, through *art engagé*, a significant change of perspective on the Republic. I shall suggest that within the 'art [of]

27 X. Crettiez and I. Sommier (eds), *La France rebelle* (Paris: Michalon, 2002).
28 M. Winock, *Les Voix de la liberté* (Paris: Seuil, 2001).
29 R. Young, 'Preface', in Jean-Paul Sartre, *Colonialism, Neo Colonialism* (London: Routledge, 2006), vii.
30 For a brief historical outlook on 'racisme républicain', see P. Tévanian, *La République du mépris* (Paris: Seuil, 2007), 6–7.
31 D. Daeninckx, *Le Bourreau et son double* (Paris: Gallimard, 1986) ; B. Tavernier, *De l'autre côté du périph'* (1997); Les Ogres de Barback, *Pardon Madjid* (2007).

protest',[32] the focus by white left-wing committed artists on interstitial strategies primarily based on an alternative imaginary, together with the reversal of standpoints, deconstructed stereotypical portrayals of the immigrants while opening up 'supplementary' perspectives within 'the order of discourse'. Focusing on the memory of exploitation, violence and exclusion (*Le Bourreau et son double*); dealing with the denaturalization of suburban space (*De l'autre côté du périph'*); and concentrating on the promotion of repentance and openness (*Pardon Madjid*), I argue that these artistic creations have not only participated in fighting against the recent abusive politics of 'race' and 'ethnicity' in France, but have also fostered new 'games of truth'[33] concerning the ideals and reality of the French Republic.[34]

French *Artistes Engagés* Challenging 'Race' and 'Ethnicity' in the Republic (1980s–2000s)

Although the question of speaking for others remains a complex and problematic issue,[35] it is important to note that French *artistes engagés* bearers of '*francité*' have greatly contributed through literature, cinema and music to the opening of new 'horizons' about 'difference' and discrimination in

32 J. Balansinski and L. Mathieu (eds), *Art et contestation* (Rennes: Presses Universitaires de Rennes, 2006).
33 M. Foucault, *L'Ordre du discours* (Paris: Gallimard, 1971); M. Foucault, *Dits et écrits, 1954–1988* (Paris: Gallimard, 1994), 1524. Foucault's notion of 'jeux de vérité' refers to 'd'autres atouts dans le jeu de vérité' and sheds light on 'd'autres possibilités rationnelles, enseignant aux gens ce qu'ils ne savent pas' (1543).
34 G. Raymond, 'The Republican Ideal and the Reality of the Republic', in A. Cole and G. Raymond (eds), *Redefining the French Republic* (Manchester: Manchester University Press, 2006), 5–24.
35 For a stimulating approach to this issue, which raises ethical and political questions, see L. Alcoff, 'Speaking for the Others', *Cultural Critique*, 92 (Winter 1991), 5–32. Alcoff states 'as a type of discursive practice, speaking for others has come under increasing criticism, and in some communities rejected' (6).

France.³⁶ Fighting their way through the 'misery of anti-racism'³⁷ their humanist voice has often echoed and supported republican principles defended by the 'new intellectuals'.³⁸

Beyond the fact that Daeninckx, Tavernier and Les Ogres de Barback all share a deep interest in postcolonial racism, the cultural artefacts under scrutiny are characterized by the similarities that they share with the post-1968 productions of protest. Representative of a mistrust of institutional discourses, these cultural productions respectively affiliated with 'neo polar', the '*nouvelle école documentaire*' and alternative rock trends, show a leftist bias which reflects, first and foremost, a desire to subvert the dominant rhetoric on 'race' and 'ethnicity' in France.³⁹ In this political and artistic framework, a change of perspective fostered by these leftist *créations engagées*

---

36 B. Vercier and D. Viart, *La Littérature francaise au présent* (Paris: Bordas, 2005), 245–62; G. Hayes and M. O'Shaughnessy, *Cinéma et engagement* (Paris: L'Harmattan, 2001); B. Lebrun, *Protest Music in France* (Aldershot: Ashgate, 2009).

37 R. Gallissot, *Misère de l'antiracisme* (Paris: Marcantère, 1985).

38 On the 'new intellectuals', see F. Hourmant, *Le Désenchantement des clercs* (Rennes: Presses Universitaires de Rennes, 1997). To cite a few examples on the activism of French artists and intellectuals: in 1983, singer Renaud (Séchan) and P. Bourdieu both joined the final steps of the 'Marche contre le racisme et pour l'égalité' – see respectively D. Gaster, *L'intégrale de Renaud* (Paris: City Editions, 2006), 102; and R. Schor, *Histoire de l'immigration en France* (Paris: Armand Colin, 1996), 306. In 1996, following the Debré Laws, filmmaker C. Lanzmann, rock band Rita Mitsouko and novelist J.-M. Le Clézio protested in support of the 'sans papiers' along with Jean Baudrillard and Derrida. More recently, the Ministère de l'immigration et de l'identité nationale (2007) saw the resistance and opposition of filmmaker R. Guédiguian, singer Dominique A and historians G. Noiriel and P. Weil.

39 M. Witt, 'On Documentary Filmmaking in France in the 1990s', conference paper delivered at the Association for the Study of Modern and Contemporary France annual conference ('Reinventing France: Towards the New Millennium'), University of Cardiff, 1999. D. Looseley, *Popular Music in Contemporary France* (Oxford: Berg, 2003), 44–50. This desire to subvert the dominant rhetoric on 'race' and 'ethnicity', is discernable not just in the artists' background (at various levels, Daeninckx, Tavernier, Les Ogres de Barback have all been involved in militancy), but also in the way they are critical of those who are in power.

is certainly offered by Didier Daeninckx's treatment of republicanism and racism in his detective novel, *Le Bourreau et son double*.

Secret and Memory of Colonial and Postcolonial Racism in *Le Bourreau et son double* (1986)

The French Republic with its egalitarian and humanitarian principles has contributed to making a name for France as a 'land hospitable to migrants'.[40] However, the return in the 1980s of the 'tyranny of the national'[41] proved highly controversial and anti-republican. For French scholars, the reason concerns France's pact of oblivion regarding its colonial past.[42] As a means of reflecting on society, creations from the *littérature engagée* certainly propose a different outlook on dominant representations of 'race' and 'ethnicity' in the Republic. Symbolic of what Sartre famously called a 'sword', Didier Daeninckx's pen in *Le Bourreau et son double* not only distorts the construction of immigrants and their descendants as a 'problem', but also provides a base for rethinking and transcending 'racism' and 'ethnicity' since the Algerian War.

In French history, many studies on immigration have emphasized the special status of the '*zoufri*' and the '*beur*'.[43] Initially, the brief openness of the Republic in the early 1980s symbolized a strong hope for young Franco-Maghrebis, but certainly never erased the *distance raciale* that made diasporic representatives 'outsiders'.[44] Part of a long series of militant anti-

---

40 C. Wihtol de Wenden cited by M. Rosello, *Postcolonial Hospitality* (Stanford: Stanford University Press, 2001), 3.
41 G. Noiriel, *La Tyrannie du national* (Paris: Calman Levy, 1991).
42 N. Bancel and P. Blanchard, *Cultures Postcoloniales de France* (Paris: Autrement, 2006).
43 Boubeker's analysis of the fate of the *zoufri* and *beur* is eloquent and demonstrates full well the fear of immigrants and their children in France. See A. Boubeker, *Les Mondes de l'ethnicité* (Paris: Balland, 2003).
44 K. Laachir, 'The Ethics and Politics of Hospitality in Contemporary French Society', unpublished PhD thesis (University of Leeds, 2003), 14.

racist 'romans noirs',[45] *Le Bourreau et son double* (which represents one of Daeninckx's first testimonies of support for immigrants)[46] fosters a stimulating reflection on the evolution of institutional racism in the Republic between the 1960s and 1980s.[47] One of the first notable features of *Le Bourreau et son double*, written after the *Marche pour l'égalité et contre le racisme* in 1983, is perhaps how much Daeninckx strives to unveil the 'great secret'[48] regarding the historical enforcement of a closed republicanism in France.[49] Many elements in this novel, which deals with the murder of a trade unionist by the chief of a plant security, work to identify and explain the secret links existing between colonial and postcolonial racism in contemporary France. Thus, scenes of abusive and discriminatory treatment of the immigrant workers, put in lorries or boats[50] to come and work in France, regularly appear in Daeninckx's novel and contribute to countering the fear of the '*zoufri*'.[51] Similarly, Daeninckx also dwells on 'the Arab communities who were harnessed to chains' as a symbol of the postcolonial workers.[52] As a motif, the novelist writes about the local police who spread terror in Courvilliers and the 'Hotch factory' in the name of safety.[53] Reintroducing, in the context of the industrialization, the 'recolonization

---

45 N. Compard, 'L'image des immigrés dans les romans noirs des années 50 à nos jours', unpublished PhD thesis (Université de Franche-Comté, Besançon, 2008).

46 Described as a postcolonial writer, Daeninckx's testimony of support for the '*zoufri*' but also to the '*beur*' is manifold. For instance, *Lumière Noire* (1987) is dedicated to Abdel (Benyahia) assassinated by a drunken policeman in Pantin in December 1986; 'Cité perdue' alludes to another '*arabicide*', Taoufik Ouannès, killed in La Courneuve in 1983.

47 P.-A. Taguieff, 'Universalisme et racisme évolutionniste: le dilemme républicain hérité de la France coloniale', *Hommes et migrations*, 1207 (Mai–Juin 1997), 90–7.

48 P. Verdaguer, *La Séduction policière* (Birmingham: Summa, 1999), 251.

49 M. Winock, *Nationalisme, antisémitisme et fascisme en France* (Paris: Seuil, 1990), 11.

50 T. Ben Jelloun, *L'Hospitalité française* (Paris: Seuil, 1984), 47.

51 M. Lamont, 'Immigration and Racial Boundaries among French Workers', in H. Chapman and L. Frader (eds), *Race in France: Interdisciplinary Perspectives on the Politics of Differences* (Oxford and New York: Berghahn, 2004), 144.

52 L. Pitti, 'Algériens à Boulogne Billancourt', *Rouge* (5 mai 2010).

53 D. Daeninckx, *Le Bourreau et son double* (Paris: Gallimard, 1986), 10, 16–17, 73.

of the colonial subject in France', Daeninckx's *Le Bourreau et son double* not only sheds a new light on postcolonial immigration, but brings to the fore violent forms of neo-colonialism.[54] In this respect, Daeninckx's brief but forceful comments on 'the slave market organised by Hotch', 'sending the cattle' and 'the manipulation of the herd' to satisfy both 'the planning of production' and 'the volume of demand'[55] are certainly typical of the neo-polar,[56] but they also illustrate the racism perpetrated by the 'powerful' in the postcolonial era.

In Daeninckx's *Le Bourreau et son double*, examples of the secret exploitation of a cheap foreign workforce appear regularly, and in very different ways. They therefore contribute to disrupting the general discourse of violence and danger on the part of immigrants living in France during the 1970s.[57] Yet this is not to say that the secret practices against immigrant workers is solely limited to the economic sphere. Despite the fact that the republican model of integration prohibits 'difference' because it leads to divisions in society, Daeninckx demonstrates how residential segregation is actually strictly enforced in this period, and further isolates the '*zoufris*' and their children. By insisting on the specificities of a relegated type of housing of the Courvilliers *cité* – ironically named 'Cité République' – where 'concrete was incessantly thrown up to [...] provide a workforce, hostels, studios, to house armies of single people',[58] Daeninckx highlights the 'unsaid of history' and simultaneously unveils how insidiously immigrants have been kept at arm's length within the Republic.[59]

---

54 K. Ross, *Fast Cars, Clean Bodies* (Cambridge, MA: MIT Press, 1995). See also R. Gallissot, 'Le Mouvement ouvrier s'inscrit dans une situation coloniale', *El Watan* (26 February 2007).
55 D. Daeninckx, *Le Bourreau et son double* (Paris: Gallimard, 1986), 57, 92, 127.
56 J.-N. Blanc, *Polarville* (Lyon: Presses Universitaires de Lyon, 1991).
57 M. Lamont, 'Immigration and Racial Boundaries Among French Workers', in Chapman and Frader (eds), *Race in France, Interdisciplinary Perspectives on the Politics of Differences* (Oxford and New York: Berghahn, 2004), 144.
58 D. Daeninckx, *Le Bourreau et son double* (Paris: Gallimard, 1986), 10.
59 P. Lorcin, *Imperial Identities* (London: I.B. Tauris, 1995).

In this period marked by the rise of physical violence against the 'Beurs' and their invasion of public space,[60] another peculiarity of *Le Bourreau et son double* concerns Daeninckx's emphasis on the memory of the Algerian War. Linked to the 'great secret', this emphasis is at the core of colonial and postcolonial racism, and accounts for the fast diffusion in France of xenophobia and racism during the 1970s and 1980s. In his classic work on the legacy of the Algerian War, *La Gangrène et l'oubli*, Benjamin Stora argues that the so-called 'war without a name' is still with us today. 'The repression of its memory', writes Stora, 'continues to haunt the very foundations of French society'.[61] Situating the murder of trade unionist Claude Werbel in the context of the Algerian War, Daeninckx's *Le Bourreau et son double* challenges what Charles Forsdick has called 'the dungeons of history'[62] while providing the reader with a platform of reflection on long-term memory. In this respect, Daeninckx's insistence on the institutionalization of violence and the horrors of the Algerian War is cardinal. In the French (prisoners) camp of 'Souk Lémal', whose name sounds like the commerce of evil, Daeninckx writes that 'the recruits were like in a zoo' and the smells of vomit, dejection and blood 'hit the stomach'. Mainly made up of 'fellahs', the prisoners were forced to 'cut down the bush' and 'lay down a network of barbed wire'. In the meantime, others were tortured in the 'bathtubs', 'where the water had run red'. Daeninckx's incorporation of these '*trous de mémoire*' helps to unearth racist and ethnic memories of discrimination. Enhanced by the complex structure of a novel based on a subtle *dédoublement* and *enchevêtrement* of histories,[63] reinforced by the suggestive metaphor of cancer,[64] they force

60 S. Bouamama, *Dix ans de marche des beurs* (Paris: Desclée de Brouwer, 1994).
61 B. Stora, *La Gangrène et l'oubli* (Paris: Seuil, 1998), 13.
62 C. Forsdick, '"Direction les oubliettes de l'histoire": Witnessing the Past in the Contemporary French Polar', *French Cultural Studies*, 12 (2001), 333–50.
63 D. Daeninckx, *Le Bourreau et son double* (Paris: Gallimard, 1986), 7, 22, 109, 131, 132, 159, 191.
64 The image of the (black) cat named Chatka (after crabmeat), and its strategic location at the opening (ibid., 9) and epilogue of the novel (ibid., 201) are certainly telling.

the reader to ponder on the ongoing republican racism afflicting (North) African immigrants and their descendants.[65]

As we have just seen, French *littérature engagée* in the 1980s provides an original account of the colonial and postcolonial racism afflicting diasporic communities from North Africa residing in France. By focusing on the secret and the memory 'of this Republic which forgets all too often its founding ideas',[66] Daeninckx's *Le Bourreau et son double* participates to disrupt the traditional portrayal of the immigrant as a 'problem', but also allows the reader to question the racist legacies affecting the '*Beurs*' in the 1980s. In stark opposition to the growth of xenophobia and 'difference', the author of *Meurtres pour mémoire*[67] defies the 'evolutionary racism'[68] and ultimately tries to promote a way of life together 'in the Cité'.[69] In the French Republic, the politics of 'race' and 'ethnicity' underwent a dramatic aggravation in the 1990s. The return of a republican consensus towards immigration,[70] followed by the implementation of the 'national republicanism',[71] contributed to reinforcing in France the stigmatization of the immigrant perceived as the new internal enemy.[72] The contribution of white, left-wing filmmakers during this period represented a key phase

---

[65] For colonial and postcolonial violence, see R. Branche, *La Torture et l'armée pendant la guerre d'Algérie (1954–1962)* (Paris: Gallimard, 2001).On the violence against the '*zoufris*' near Marseille in 1973, see F. Gaspard and C. Servan-Schreiber, *La Fin des immigrés* (Paris: Seuil, 1985), as well as Ed Naylor's chapter in this volume.

[66] N. Bancel, P. Blanchard and F. Vergès, *La République coloniale* (Paris: Albin Michel, 2003), 25.

[67] D. Daeninckx, *Meurtres pour mémoire* (Paris: Gallimard, 1984).

[68] P.-A. Taguieff, 'Universalisme et racisme évolutionniste: le dilemme républicain hérité de la France coloniale', *Hommes et migrations*, 1207 (mai–juin 1997), 90–7.

[69] H. Van Effenterre, 'La Cité grecque, modèle de la République des Républicains', in S. Bernstein and O. Rudelle, *Le Modèle républicain* (Paris: Presses Universitaires de France, 1992), 13–50.

[70] D. Blatt, 'Immigrant Politics in a Republican Nation', in A. Hargreaves and M. McKinney (eds), *Postcolonial Cultures in France* (New York and London: Routledge, 1996), 46–50.

[71] E. Balibar, 'Le droit de cité ou l'apartheid?', in E. Balibar (ed.), *Sans papiers: l'archaïsme fatal* (Paris: La Découverte, 1999), 92.

[72] M. Rigouste, *L'ennemi intérieur* (Paris: La Découverte, 2009).

in the evolution of the *culture artistique de l'engagement*. As we shall see, filmmaker Bertrand Tavernier and his acclaimed *Au-delà du périph* is undeniably a prominent representative of commitment in this new decade of resistance and opposition.

Immigration and 'Intégration' in Bertrand Tavernier's *Au-delà du périph* (1997)

Republican France has been confronted by a continuous and expanding crisis, which has been attributed to the presence of (North) African immigrants and their alleged non-integration in the national community. Concomitant with the supposed loss of Frenchness, cardinal aspects of the crisis have been linked to the control of the nation's borders and the issue of the *banlieues immigrées*.[73] Thus, the 1990s coincided with a dramatic aggravation in the perceptions of the 'ethnic other' together with a reorientation of the nation's discourse on the city margins. As an alternative art form,[74] documentary filmmaking provides a different image of the nation and participates to reframe, in the context of the new inhospitality laws,[75] the 'moral panic' concerning immigration and the suburbs. Away from the 'cinema of the *banlieue*',[76] Bertrand Tavernier offers in *Au-delà du périph* a stimulating critique which challenges through the real the intensive racialization of France in the mid-1990s.

Tavernier's documentary creation of 1997, which evokes the hardships and the difficulties as well as the courage and the joys of the inhabitants of a Parisian *cité* (the Grands-Pêchers in Montreuil), directly engages with

---

73 F. Dubet and D. Lapeyronnie, *Les quartiers d'exil* (Paris: Seuil, 1992), 58.
74 S. Bruzzi, *New Documentary: A Critical Introduction* (London and New York: Routledge, 2000).
75 I am referring more specifically to the Pasqua and Debré laws passed in the mid-1990s. On this, see D. Fassin, A. Morice and C. Quiminal, *Les Lois de l'inhospitalité* (Paris: La Découverte, 2006).
76 T. Jousse, 'Le banlieue film existe-t-il?', *Cahiers du Cinéma*, 492 (juin), 37–9.

the danger posed by the so-called non-integration of immigrants and the '*banlieue* problem'.[77] In a similar way to the focus on secrets and memories, the focus by Tavernier on integration and belonging contradicts and calls into question the 'repressive practices' connected with the French politics of 'race' and 'ethnicity' in the mid-1990s. Inspired by the support of French filmmakers for the *sans papiers*[78] and a provocative response by the former Ministre de la Ville, Eric Raoult, who stated that 'integration is not cinema', a significant aspect of Tavernier's *Au-delà du périph'* concerns the stress placed on ordinariness which proposes a sharp counterpoint to the fear of the immigrant and the *banlieues*. The manner in which Tavernier works to render the atmosphere of the peripheral territory by stressing the banality and insignificance of the multi-ethnic Grands-Pêchers effectively helps to break down spectacular and threatening images of immigrant violence and suburban danger.[79] Examples in the film are numerous. Taking us around the estate, Tavernier's camera alludes to the mediocrity of daily life in this space by lingering on buildings and public places, and following residents in their peregrinations. As a recurrent device suggestive of the banality of the site, Tavernier films children playing, cycling, running or kicking a ball outside the tower blocks. Similarly, he shows us residents of different origins shopping, watching television, reading the news, etc. The most revealing example of the 'normality' of existence in the ethnically diverse Cité des Grands-Pêchers concerns the shot of an old woman relaxing on a bench in front of a billboard advertising the Debré laws. Capturing the plain reality of the 'immigrant banlieues',[80] this shot contributes to providing a vivid contrast with the new irrational fears and prejudices against immigrants in the Republic.

77 M. Dikeç, *Badlands of the Republic* (Oxford: Blackwell, 2007).
78 For information on the *Manifeste des 66* written in 1997 by filmmakers A. Desplechin and P. Ferran, see C.-M. Trémois, *Les Enfants de la liberté* (Paris: Seuil, 1997), 9.
79 J.-M. Stebé, *La Crise des banlieues* (Paris: Presses Universitaires de France, 1999), 5.
80 'Moi je m'en vais, malheureusement!'; 'Pour moi, la cité, c'est un lieu normal, ce qui ne va dire qu'il n'y a pas de problème'; 'C'est pas plus mal qu'ailleurs', B. Tavernier and N. Tavernier, *De l'autre côté du périph'* (1997).

If the emphasis of the documentary on the real participates to reframe the image of the French '*banlieues*' as '*hors-lieu de la République*', one should not deduce that Tavernier deliberately omits difficult and spectacular episodes taking place in the stigmatized town of Montreuil. As in any other multi-ethnic *cité*, problems in the Grands-Pêchers exist and do appear on screen. To cite only one example, the moment when a young resident fixes his moped and gets shot at by a neighbour is relevant and suggests the existence of episodic scenes of violence on the estate. However, because such a moment seized by Tavernier's continued filming is replaced in a context of normality, it passes relatively unnoticed. Therefore, this 'immigrant banlieue'[81] stigmatized by Raoult himself appears in the movie like many other territories of the Republic – banal, ordinary, integrated.

Another characteristic of *Au-delà du périph'*, and the way it denaturalizes received ideas about the threat posed by immigration and suburbs in the mid-1990s, concerns the emphasis given by the filmmaker to the level of 'institutional racism'[82] which runs against egalitarian and humanitarian values of the French republican. Whereas *communautarisme* (communitarianism) imposed itself during this period as a new buzzword justifying the implementation of new racial politics, Tavernier contradicts the institutional discourse and shows how immigrants and foreigners have been confined to the margins of French society. Resorting to the specificities of the documentary in terms of time and research,[83] Tavernier's focus on the 'differentialist management of immigration'[84] is, in this regard, eloquent. First, the reduction in schooling subsidies (confirmed, in the documentary, by a representative of the Montreuil School Authorities) provides the spectator with a perfect insight into the ways children of African descent are neglected by an education system which displays the sort of institutional

---

81 F. Dubet and D. Lapeyronnie, *Les quartiers d'exil* (Paris: Seuil, 1992), 58.
82 D. Sibony, 'Institution et racisme', in M. Wieviorka (ed.), *Racisme et modernité* (Paris: La Découverte, 1993), 141.
83 Y. Jeanneau, 'Sur la définition du documentaire de création', in R. Copans and Y. Jeanneau, *Filmer le réel* (Paris: La bande à Lumière, 1987), 13–22.
84 R. Castel, *La discrimination négative* (Paris: Seuil, 2007), 41.

racism that the Republic claims to reject.[85] In the same way, Tavernier's treatment of employment discrimination on racial grounds shows the economic hardships experienced by the young underprivileged, and suggests that this form of 'workplace racism'[86] reinforces their social marginalization. The comment made in front of the camera by a Montreuil councillor is revealing and testifies to the existence in the Republic of a growing internal racism: 'People are bitter [...] I admire their capacity to stick together!'

The most significant change of image offered by Tavernier's documentary regarding the questions of immigration and integration in the context of the new *banlieuephobia* of the mid-1990s is probably expressed through the political commitment of the nationals and non-nationals living in the Grands-Pêchers. In this respect, the filmmaker's negotiation of the plain reality[87] proposes a significant rupture with the enduring demonization of immigration and criminalization of the '*Beurs*'. If Tavernier's focus on everyday life of the *cité* gives many adults the opportunity to vent out their discontent, frustration and anger regarding the 'imaginary racism' pervading French society during this period, the filmmaker's camera also shows how demonized teenagers of foreign origin can express and manifest themselves on issues of 'race' and 'ethnicity', 'inclusion' and 'exclusion' as part of a 'democratic' process. Evoking certain aspects of Jacques Rancière's work,[88] the following comments made by one young adult of sub-Saharan descent are in this respect eloquent, and serve to demonstrate their belonging to France while condemning the racist use of integration: 'For me, integration means to be somewhere and to live there! Integration doesn't exist! Integration is a load of rubbish invented to make people think that immigrants are bad boys!'

85 F. Dubet, 'Le Racisme à l'école', in S. Paugam (ed.), *L'Exclusion: l'état des savoirs* (Paris: La Découverte, 1996), 298.
86 P. Bataille (ed.), *Le Racisme au travail* (Paris: La Découverte, 1997).
87 S. Bruzzi, *New Documentary. A Critical Introduction* (London and New York: Routlege, 2000).
88 J. Rancière, *La Mésentente* (Paris: Galilée, 1995) or J. Rancière, 'CNDP 2001: Lycée/Iatable ronde pédagogique: l'exclusion existe-t-elle? Les réponses de Jacques Rancière', http://www.cndp.fr/tr_exclusion/rep_ranc.html (accessed 1 April 2010).

The contribution of French *cinéma engagé* in fighting racism during the mid-1990s is certainly essential,[89] and supplies the spectator with an original gaze on the exponential fear of the immigrant and the *banlieue* in republican France. In *Au-delà du périph'*, Tavernier's treatment of the problematic issues of immigration and integration in France effectively illustrates during this period the profound feeling of belonging among *cité* residents in general and adolescents in particular. Poles apart from the degrading discourse on the alterity of immigrants and alien nature of the suburbs, Tavernier's commitment stresses, in a very different manner, the evolution of 'race' and 'ethnicity' politics in the 1990s and in so doing questions both the validity and legitimacy of French national republicanism. In France, the start of the new millennium certainly represented another major turning point regarding the mutations of 'evolutionary racism'. What Sarah Gibson has called, in relation to Western nations, the new 'politics of deterrence'[90] has contributed to the diffusion in France of racism, resentment and hatred. During the 2000s, the impact of white French left-wing artists can be affiliated with the rich and diverse contribution of the *musique engagée*. As heirs to a long tradition of protest, the rock band Les Ogres de Barback and their song *Pardon Madjid* can serve to illustrate another interesting facet of the French *artistes engagés*' battle against institutional racism in the Republic.

Opening up the Republic to the 'Ethnic Other': Les Ogres de Barback's 'Pardon Madjid' (2007)

Since the mobilization of committed filmmakers in the mid-1990s, the official discourse about immigration in the French Republic has become even more demonizing. The development of republican nationalism, which

---

[89] G. Hayes and M. O'Shaughnessy, *Cinéma et engagement* (Paris: L'Harmattan, 2001).

[90] S. Gibson, "'Abusing Our Hospitality": Inhospitableness and the Politics of Deterrence', in J.G. Molz and S. Gibson (eds), *Mobilizing Hospitality: The Ethics of Social Relations in a Mobile World* (Aldershot Ashgate, 2007), 166.

saw the creation in 2007 of the Ministry for Immigration, Integration, National Identity and Co-development, has been the object of numerous debates, mainly singling out the alarming regression of republican principles. In a period characterized by unprecedented levels of state racism, French *artistes engagés* have severely criticized and strongly opposed the 'ethnicization of the Republic'.[91] The main interest of French *musique engagée* is, in this case, to resist the implementation by political authorities of a '*république du mépris*' to quote Pierre Tévanian.[92] In *Pardon Madjid*, Les Ogres de Barback's demonstrate how, by way of protest, committed musicians have contributed to opening up the Republic while transcending the image of the 'other'.

Let us note for a start that French rock music is traditionally known for its capacity for resistance and opposition.[93] More commonly called 'Les Ogres', the band certainly appear somewhat different (apart from their art) within the spectrum of artists already considered. Part of French 'protest music',[94] this band, composed of four members of the same family (the Burguières), belongs to a new generation of artists who, despite being born after decolonization, have actively promoted the moral ideals of solidarity and fraternity at the heart of the Republic. Released in 2007 at the time of the movement against the Ministry of immigration and National Identity,[95] *Pardon Madjid* and its focus on immigration has certainly gained greater resonance in the Republic since the recent debate on national identity.

In the context of the new *république du mépris* promoting prejudice and 'difference', a primary feature of *Pardon Madjid* 'is its memorial dimension linking together racism, the 'confession' and the 'forgiveness'. Theorizing apartheid, Sophie Pons has argued that the 'confession' and

---

91   C. Bertossi, 'L'Ethnicisation et la République', *Accueillir*, 251 (septembre 2009), 7–8.
92   P. Tévanian, *La République du mépris* (Paris: Seuil, 2007).
93   D. Looseley, *Popular Music in Contemporary France*, 44–50.
94   B. Lebrun, *Protest Music in France* (Aldershot: Ashgate, 2009).
95   See the petition circulated in June 2007 and signed by Les Ogres: 'Non au ministère de l'immigration et de l'identité nationale', http://www.gisti.org/spip.php?article939 (accessed 1 October 2011).

'forgiveness' could be the first steps towards a policy of reconciliation, thus allowing us to 'break the debt and the forgetting'.[96] Apart from its title, many lines of this song featuring on the album, *Du simple au néant*,[97] intersect with Pons's works and refer explicitly to political and historical landmarks related to immigration so as to encourage forgiveness: from the Algerian war where the '*zoufris*' fought for France, to the threshold of intolerance with the thematics of the '*bruit et l'odeur*'.[98] Relocating the question of the '*zoufri*' and their descendants in a new temporality, these lines contribute to placing republican racism in a historical continuity that makes it part of a collective past. In so doing, it provides a basis for '*une politique du pardon*'[99] which can help with examining the past that remains associated with the Republic.

Beyond this historical dimension, the interest of Les Ogres de Barback in opening the *république* and breaking the *mépris* is also manifest in their focus on postcolonialism and French culture. As noted earlier, the 'engagement' of intellectuals fighting against the strengthening of French inhospitality has been highly apparent since the 1980s. To cite only two examples, French sociologist Alain Touraine[100] has asked whether we could live together being equal and different at the same time. In a more recent contribution, Saïd Bouamama has argued that the Republic needed to accept diversity and to evolve.[101] In 2007, Les Ogres's *Pardon Madjid* addresses the same question raised by many intellectuals. Countering the new closed republicanism, they try to go beyond the fear of *métissage* and

---

96 S. Pons, *Apartheid: L'aveu et le pardon* (Paris: Bayard, 2000). See also A. Olivier (ed.), *Le Pardon: Briser la dette et l'oubli* (Paris: Seuil, 1998).
97 Les Ogres de Barback, *Du Simple au néant* (2007).
98 A direct quote from a speech made by Jacques Chirac as mayor of Paris in 1991, 'le bruit et l'odeur', has become a catch-phrase symbolizing a form of ordinary racism. Zebda had a hit in 1995 with *Le bruit et l'odeur* on their eponymous album. The expression is also used in *Pardon Madjid*.
99 S. Lefranc, *Politiques du pardon* (Paris: Presses Universitaires de France, 2002).
100 A. Touraine, *Pouvons-nous vivre ensemble égaux et différents?* (Paris: Fayard, 1997), 11.
101 S. Bouamama, *La France* (Paris: Larousse, 2009), 109.

loss of identity still so prevalent in post-colonial France.[102] Including rap singer Magyd Cherfi as a guest, *Pardon Madjid* seeks to promote conviviality in France and overcome the fear of regression of French culture. In that respect, besides the composition of the band (the Burguières and Cherfi), hybridity is very much prevalent in the lyrics, melody and musical composition. Alternating with Burguière, *Pardon Madjid* incorporates some Arabic lyrics sung by Cherfi. This specificity is also to be associated with hybrid sounds combining rock and classical music but also North African folkloric tunes. The blending of the lyrics combined with the mixing of the music provides a new space for reflection that not only acknowledges the evolution of France, but also testifies to the realities of its cultural evolution and hybridity. Consequently, while mirroring the fear of losing one's identity, Les Ogres help us to reconsider the '*proximité coloniale*' and '*distance nationale*'.[103]

Overall, this song promotes openness and tolerance and seeks to promote an utopian hospitality. If we know that inversion or reversal is at the basis of the concept of 'utopia',[104] the idea of hospitality is a more complex issue. Derrida claims that hospitality bears an aporetic dimension, but the Derridean concept of a 'to-come' proves helpful for deconstructing and transcending the notions of 'camp entrenchment' typical of the postcolonial world.[105] In *Pardon Madjid*, Les Ogres de Barback clearly aim to build on, through utopia, a possible hospitality. Thus the song invites us to 'welcome our brothers by letting our imagination define who they may be' (*imaginer nos frères serait plus accueillant*) a powerful counter-point to readymade images and prejudices. Like the apologies contained in *Pardon Madjid*, these lyrics show by the process of inversion the support of the band for a more tolerant approach to immigration. More significantly, it is

---

102   N. Bancel, P. Blanchard and F. Vergès, *La République coloniale* (Paris: Albin Michel, 2003).
103   S. Lamri, 'Entre Maghreb et France: Proximité coloniale et distance nationale', *Brèves du Casnav* (2003), 4.
104   A. Pessin, *L'Imaginaire utopique aujourd'hui* (Paris: Presses Universitaires de France, 2001).
105   P. Gilroy, *Between Camps* (London: Allen Lane, 2000).

clearly the inversion of the autochthonous / migrant – 'all the white immigrants pressed at the gates of the desert' which is the most significant part of the song. Symbolic of a utopian dream, it forces the listener to revise its preconceptions about migratory flows and ethnicity. Breaking down discourses on 'racism' and 'ethnicity', it also contributes to thinking about ways of a hospitable world-to-come.

Conclusion

In France, *French artistes engagés* have reflected in different ways on the politics of 'race' and 'ethnicity' and have offered, between the 1980s and the 2000s, a significant renewal of the traditional depictions of the Republic. Generally speaking, *artistes engagés* departed from fixed and determined discourses, and projected a political view on the Republic. Heirs to the protectors of republican principles, the *artistes engagés*, with their interest in the 'other', showed a radically different image of immigrants and their descendants. As *porteurs de 'francité'* in the context of the gradual closure of the French nation, they succeeded in circulating, along with minority artists, 'supplementary' representations which challenged representations of immigrants in contemporary France.

By evoking racial and ethnic markers in a different way, French *artistes engagés* have successfully transmitted a different image of immigrants which are a far cry from the simplistic depictions communicated in the more commercial productions. Dictated by a trans-artistic discourse combining revolt and idealism, *artistes engagés*, with their emphasis on breaches and alternative imaginaries together with the reversal of viewpoints, have participated in the re-evaluation of the question of immigration and short-circuited most of the negative stereotypes usually associated with their representations. By reflecting on the usual 'racist' and 'ethnic' portrayals, they have supplied the public with other games of truth (*jeux de vérité*),[106] thereby challenging through literature, cinema and music the 'manufacturing' of immigration as a 'cultural problem'. Similarly, they have uncovered the long-term inconsistencies of the Republic towards immigration from *outre-Méditerranée*.

Although frequently branded as 'anarchists', French *artistes engagés* have certainly contributed to modifying opinions about contemporary France. Their stress on the history of 'republican racism', their deconstruction of integration along with their use of forgiveness to promote a more open society and a more welcoming nation, have surely demonstrated the mythology surrounding French discourse on the 'other'. Yet, this dissident outlook on the politics of 'race' and 'ethnicity' possesses certain limitations.[107] At first, the way they use such concepts to discuss immigration is certainly illuminating and of great interest, but it only appeals to a limited proportion of the population. Consequently, the possibility of influencing and changing discourse in the long term remains limited. Moreover, the supportive attitude expressed towards immigrants bears signs of a longstanding solidarity among French leftist militants. However, their discourse can be judged as somewhat reductive: although the message it broadcasts favours an opening of France to multiculturalism, it does not always stress enough what *is* different.

In *Contre-feux*, Pierre Bourdieu developed the notion of an engaged knowledge (*savoir engagé*) to counter a depoliticized culture and strongly encouraged scholars and artists to invest their talent in critical struggles.[108] Overall, it can be said that some committed French *artistes* have made a significant contribution to the debate on immigration, race and ethnicity in the Republic. Even though their critical discourse has not reached a wide audience, they have certainly help feeding the debate about the nature of the Republic today by articulating complex, creative and challenging

---

106   M. Foucault, *Dits et écrits, 1954–1988* (Paris: Gallimard, 1984), 1524.
107   This idea and the second remark formulated below are inspired from B. Lebrun 'Banging on the Wall of Fortress France: Music for *Sans-Papiers* in the Republic', *Third Text*, 20/6 (2006), 711–21 (719).
108   P. Bourdieu, *Contre-feux* (Paris: Liber, 2001).

ideas, and images on immigration. In so doing, and in parallel to dissident creations of minority artists, such artists have revealed themselves to be a significant component of popular culture that deserves attention. On the threshold of a new decade where the 'ethnic other' continues to be seen as the 'enemy', their committed creations constitute impressive pacifying weapons of mass construction '*armes de constructions massives*' in the context of France's contemporary politics of 'race' and 'ethnicity'.[109]

Bibliography

Abel, O. (ed.), *Le Pardon: Briser la dette et l'oubli* (Paris: Seuil, 1998).
Alcoff, L., 'Speaking for the Others', *Cultural Critique*, 92 (1991), 5–32.
Balansinski, J., and Mathieu, L. (eds), *Art et contestation* (Rennes: Presses Universitaires de Rennes, 2006).
Balibar, E., 'Le droit de cité ou l'apartheid?', in E. Balibar (ed.), *Sans papiers: l'archaïsme fatal* (Paris: La Découverte, 1999), 89–116.
Balibar, E., *La Crainte des masses* (Paris: Galilée, 1997).
Balibar, E., 'La forme nation: histoire et idéologie', in E. Balibar and I. Wallerstein, *Race, nation, classe* (Paris: La Découverte, 1997).
Bancel, N., and Blanchard, P., *Cultures Postcoloniales de France* (Paris: Autrement, 2006).
Bancel, N., Blanchard, P., and Vergès, F., *La République coloniale* (Paris: Albin Michel, 2003).
Barker, M., *The New Racism* (London: Junction Books, 1981).
Bataille, P. (ed.), *Le Racisme au travail* (Paris: La Découverte, 1997).
Bayart, J.-F., 'Il n'y a pas d'identité française', *Le Monde* (6 June 2009).
Begag, A., and Chaouïtte, A., *Ecarts d'identité* (Paris: Seuil, 1990).
Ben Jelloun, T., *L'Hospitalité française* (Paris: Seuil, 1984).
Bertossi, C., 'L'Ethnicisation et la République', *Accueillir*, 251 (septembre 2009), 7–8.

---

109 B. Tavernier, 'Le cinéaste Bertrand Tavernier ni brumeux ni électrique ...', *Nice Matin* (20 May 2009).

Blanc, J.-N., *Polarville* (Lyon: Presses Universitaires de Lyon, 1991).
Blatt, D., 'Immigrant Politics in a Republican Nation', in A. Hargreaves and M. McKinney (eds), *Postcolonial Cultures in France* (New York and London: Routledge, 1996), 46–50.
Bouamama, S., *Dix ans de marche des beurs* (Paris: Desclée de Brouwer, 1994).
Bouamama, S., *La France* (Paris: Larousse, 2009).
Boubeker, A., *Les Mondes de l'ethnicité* (Paris: Balland, 2003).
Bourdieu, P., *Contre-feux* (Paris: Liber, 2001).
Branche, R., *La Torture et l'armée pendant la guerre d'Algérie (1954–1962)* (Paris: Gallimard, 2001).
Bruzzi, S., *New Documentary: A Critical Introduction* (London and New York: Routledge, 2000).
Castel, R., *La discrimination négative* (Paris: Seuil, 2007).
Cohen, S., *Folk Devils and Moral Panics* (London: MacGibbon and Kee, 1972).
Compard, N., 'L'image des immigrés dans les romans noirs des années 50 à nos jours', unpublished PhD thesis (Université de Franche-Comté, Besançon, 2008).
Crettiez, X., and Sommier, I. (eds), *La France Rebelle* (Paris: Michalon, 2002).
Daeninckx, D., *Meurtres pour mémoire* (Paris: Gallimard, 1984).
Daeninckx, D., *Le Bourreau et son double* (Paris: Gallimard, 1986).
Dikeç, M., *Badlands of the Republic* (Oxford: Blackwell, 2007).
Dubet, F., and Lapeyronnie, D., *Les quartiers d'exil* (Paris: Seuil, 1992).
Dubet, F., 'Le Racisme à l'école', in S. Paugam (ed.), *L'Exclusion: l'état des savoirs* (Paris: La Découverte, 1996).
Dufournet, H., Lafargue de Grangeneuve, L., Schvartz, A., and Voison, A., 'Art et Politique sous le regard des sciences sociales', *Terrains et Travaux*, 13 (2008).
Fassin, D., Morice, A., and Quiminal, C., *Les Lois de l'inhospitalité* (Paris: La Découverte, 2006).
Forbes, J., and Kelly, M., *French Cultural Studies* (London and New York: Routledge, 1995).
Forsdick, C., '"Direction les oubliettes de l'histoire": Witnessing the Past in the Contemporary French Polar', *French Cultural Studies*, 12 (2001), 333–50.
Foucault, M., *L'Ordre du discours* (Paris: Gallimard, 1971).
Foucault, M., *Dits et écrits, 1954–1988* (Paris: Gallimard, 1994).
Gallissot, R., *Misère de l'antiracisme* (Paris: Marcantère, 1985).
Gallissot, R., 'Le Mouvement ouvrier s'inscrit dans une situation coloniale', *El Watan* (26 February 2007).
Gaspard, F., and Servan-Schreiber, C., *La Fin des immigrés* (Paris: Seuil, 1985).
Gaster, D., *L'intégrale de Renaud* (Paris: City Editions, 2006).

Gibson, S., "'Abusing Our Hospitality": Inhospitableness and the Politics of Deterrence', in J.G. Molz and S. Gibson (eds), *Mobilizing Hospitality: The Ethics of Social Relations in a Mobile World* (Aldershot: Ashgate, 2007).
Gilroy, P., *Between Camps* (London: Allen Lane, 2000), 214–15.
Guénif-Souilamas, N. (ed.), *La République mise à nu par son immigration* (Paris: La Fabrique, 2006).
Hargreaves, A., 'The Contribution of North and Sub-Saharan African Immigrant Minorities to the Redefinition of Contemporary French Culture', in C. Forsdick and D. Murphy (eds), *Francophone Postcolonial Studies* (London: Arnold, 2003).
Hargreaves, A., *Multi-Ethnic France* (London: Routledge, 2007).
Hayes, G., and O'Shaughnessy, M., *Cinéma et engagement* (Paris: L'Harmattan, 2001).
Hayward, J., *Governing France* (New York: Norton, 1983).
Hourmant, F., *Le Désenchantement des clercs* (Rennes: Presses Universitaires de Rennes, 1997).
House, J., 'Anti-Racism and Anti-Racist Discourse in France from 1900 to the Present Day', unpublished PhD thesis (University of Leeds, 1997).
Jeanneau, Y., 'Sur la définition du documentaire de création', in R. Copans and Y. Jeanneau, *Filmer le réel* (Paris: La bande à Lumière, 1987), 13–22.
Jousse, T., 'Le banlieue film existe-t-il?', *Cahiers du Cinéma*, 492 (juin 1995), 37–9.
Laachir, K., 'The Ethics and Politics of Hospitality in Contemporary French Society', unpublished PhD thesis (University of Leeds, 2003).
Lamont, M., 'Immigration and Racial Boundaries among French Workers', in H. Chapman and L. Frader (eds), *Race in France: Interdisciplinary Perspectives on the Politics of Differences* (Oxford and New York: Berghahn 2004).
Lamri, S., 'Entre Maghreb et France: Proximité coloniale et distance nationale', *Brèves du Casnav*, 4 (2003).
Laronde, M., *Autour du roman beur* (Paris: L'Harmattan, 1993).
Lebrun, B., 'Banging on the Wall of Fortress France: Music for *Sans-Papiers* in the Republic', *Third Text*, 20/6 (2006), 711–21.
Lebrun, B., *Protest Music in France* (Aldershot: Ashgate, 2009).
Lefranc, S., *Politiques du pardon* (Paris: Presses Universitaires de France, 2002).
Looseley, D., *Popular Music in Contemporary France* (Oxford: Berg, 2003).
Lorcin, P., *Imperial Identities* (London: I.B. Tauris, 1995).
Miliani, H., 'De la culture d'exil au syncrétisme culturel', in H. Gafaïti (ed.), *Cultures transnationales de France* (Paris: L'Harmattan, 2001).
Noiriel, G., *Le Creuset français* (Paris: Seuil, 1988).
Noirel, G., *La Tyrannie du national* (Paris: Calman Levy, 1991).

Noiriel, G., *A quoi sert l'identité nationale?* (Paris: Seuil, 2007).
Ogres de Barback, Les, *Pardon Madjid* (2007).
Ogres de Barback, Les, *Du Simple au néant* (2007).
Pessin, A., *L'Imaginaire utopique aujourd'hui* (Paris: Presses Universitaires de France, 2001).
Pitti, L., 'Algériens à Boulogne Billancourt', *Rouge* (5 mai 2010).
Pons, S., *Apartheid: L'aveu et le pardon* (Paris: Bayard, 2000).
Pouivet, R., 'Qu'est-ce que l'art de masse?', http://philosophie.scola.ac-paris.fr/C603-04Pouivethtm (accessed 1 May 2010).
Rancière, J., 'CNDP 2001: Lycée/latable ronde pédagogique: l'exclusion existe-t-elle? Les réponses de Jacques Rancière', http://www.cndp.fr/tr_exclusion/rep_ranc.html (accessed 1 April 2010).
Rancière, J., *La Mésentente* (Paris: Galilée, 1995).
Raymond, G., 'The Republican Ideal and the Reality of the Republic', in A. Cole and G. Raymond (eds), *Redefining the French Republic* (Manchester: Manchester University Press, 2006).
Rigouste, M., *L'ennemi intérieur* (Paris: La Découverte, 2009).
Rosello, M., *Postcolonial Hospitality* (Stanford: Stanford University Press, 2001).
Ross, K., *Fast Cars, Clean Bodies* (Cambridge, MA: MIT Press, 1995).
Schor, R., *Histoire de l'immigration en France* (Paris: Armand Colin, 1996).
Sibony, D., 'Institution et racisme', in M. Wieviorka (ed.), *Racisme et modernité* (Paris: La Découverte, 1993).
Silverman, M., *Deconstructing the Nation* (London and New York: Routledge, 1992).
Solomos, J., and Back, L., *Racism and Society* (Basingstoke: Macmillan, 1996).
Stebé, J.-M., *La Crise des banlieues* (Paris: Presses Universitaires de France, 1999).
Stora, B., *La Gangrène et l'oubli* (Paris: Seuil, 1998).
Stovall, T., 'From Red Belt to Black Belt: Race, Class, and Urban Marginality in Twentieth-Century Paris', *L'Esprit créateur*, 41/3 (2001), 9–23.
Street, J., 'The Politics of Popular Culture', in K. Nash and A. Scott (eds), *The Blackwell Companion to Political Sociology* (Oxford: Blackwell, 2004).
Taguieff, P.-A., 'Universalisme et racisme évolutionniste: le dilemme républicain hérité de la France coloniale', *Hommes et migrations*, 1207 (mai–juin 1997).
Tarr, C., *Reframing Difference: Beur and Banlieue Filmmaking in France* (Manchester: Manchester University Press, 2005).
Tavernier, B., and Tavernier, N. (dirs), *De l'autre côté du périph'* (1997).
Tavernier, B., 'Le cinéaste Bertrand Tavernier ni brumeux ni électrique ...', *Nice Matin* (20 May 2009).
Tévanian, P., *La République du mépris* (Paris: Seuil, 2007).

Touraine, A., *Pouvons-nous vivre ensemble égaux et différents?* (Paris: Fayard, 1997).
Trémois, C.-M., *Les Enfants de la liberté* (Paris: Seuil, 1997).
Van Effenterre, H., 'La Cité grecque: modèle de la République des Républicains', in S. Bernstein and O. Rudelle (eds), *Le Modèle républicain* (Paris: Presses Universitaires de France, 1992), 13–50.
Vercier, B., and Viart, D., *La Littérature francaise au présent* (Paris: Bordas, 2005).
Verdaguer, P., *La Séduction policière* (Birmingham: Summa, 1999).
Wieviorka, M., *La France raciste* (Paris: Seuil, 1992).
Wihtol de Wenden, C., 'North African Immigration and the French Political Imaginary', in M. Silverman (ed.), *Race, Discourse and Power in France* (Aldershot: Avebury, 1990).
Winock, M., *Nationalisme, antisémitisme et fascisme en France* (Paris: Seuil, 1990).
Winock, M., *Les Voix de la liberté* (Paris: Seuil, 2001).
Witt, M., 'On Documentary Filmmaking in France in the 1990s', conference paper delivered at the Association for the Study of Modern and Contemporary France annual conference ('Reinventing France: Towards the New Millennium'), University of Cardiff, 1999.
Young, R., 'Preface', in *Jean-Paul Sartre: Colonialism, Neo Colonialism* (London: Routledge, 2006).

ED NAYLOR

'A system that resembles both colonialism and the invasion of France': Gaston Defferre and the Politics of Immigration in 1973

The early 1970s was a key period in the forging of contemporary immigration politics in France. July 1974 has traditionally been seen as a watershed, with the decision of the new Chirac Government to officially halt non-EEC immigration. Bringing down the curtain on an era of *laissez-faire* policy associated with the post-war *Trente Glorieuses*, the move marked the transition to a new paradigm which has more or less endured to the present day. Neither large-scale labour importation nor notional full employment has ever returned.

However, as Laurens has convincingly shown, the original decision was a provisional suspension and the product of conflicting considerations.[1] Only retrospectively did it come to be seen as a logical (and definitive) response to prolonged economic recession. There were also antecedents well before the oil crisis of late 1973.[2] Ministerial circulars in 1968 and 1972 (Marcellin-Fontanet) significantly restricted the conditions under which migrants could legally reside and work in France. These changes have been interpreted as part of a shift towards a 'new immigration policy' which sought to limit extra-European immigration on assimilationist rather than economic grounds.[3] The growth of migration from the Maghreb during

---

1   S. Laurens, '1974 et la fermeture des frontières: analyse critique d'une décision érigée en *turning-point*', *Politix*, 82 (2008), 67–92.
2   Itself a rather arbitrary proxy for recession since unemployment in France did not rise significantly until the third quarter of 1974.
3   M. Silverman, *Deconstructing the Nation: Immigration, Racism and Citizenship in Modern France* (London: Routledge, 1992), 72–6.

the 1960s was a particular concern among policymakers. An attempt to privilege certain migration flows over others can be seen in the contrast between an annual quota of 65,000 workers agreed with Portugal in 1971 and the reduced quota of 25,000 that the Algerian Government was forced to concede the following year. From this perspective the 1974 'halt' looks less like a turning point than the culmination of a control agenda established in a period of strong economic growth.

At the same time, the fact that these restrictive measures came in the form of administrative circulars rather than legislation was indicative of a reticence on the part of the government. In marked contrast to the 1980s, senior centre-right politicians were reluctant to engage in political debate on the terrain of immigration, and reforms tended to be couched in the technocratic language of rationalization. On the one hand this reflected the widespread consensus on the utility of foreign labour to the French economy, a view still publicly defended by major private sector employers in late 1973 and expressed in the UDR election manifesto of April 1974. On the other hand, public discussions of immigration at this time were also characterized by a new solicitude for the plight of immigrant workers. In the wake of May '68, far-left groups, immigrant protests and a range of civil society organizations helped to popularize themes of exploitation and racism.[4] The latter came in response to a growing number of violent attacks on Maghrebis and the rising profile of far-right groups such as Ordre Nouveau, with its slogan 'France for the French'. Widely relayed by a sympathetic mainstream press and a partially liberalized television network, desperate housing and working conditions together with the issue of racism became firmly established in the public arena. These concerns saw the passage of France's first anti-racist legislation in 1972, along with a number of modest government initiatives in the domain of housing and employment rights for foreign workers.

It was in this context that in the late summer of 1973 Marseille was engulfed by a wave of racist polemic and violence directed at the city's

---

4   See G. Noiriel, *Immigration, antisémitisme et racisme en France (XIXème–XXème siècle): discours publics, humiliations privées* (Paris: Fayard, 2007), 562–5.

Algerian population. The episode met with widespread condemnation locally and nationally, but also drew some more mixed reactions. The French authorities maintained a dogged scepticism towards claims of racism, preferring to emphasize the presence of foreigners as a source of social tension. More surprising still was the response of the city's socialist mayor, Gaston Defferre.

Mayor of Marseille since 1953 and president of the socialist group in the National Assembly, Defferre was a leading figure in the Parti Socialiste and a national political personality. Through his media group he also owned *Le Provençal*, the pre-eminent newspaper in the region, whose editorial line and political coverage were tightly integrated into the political machine which governed Marseille. Largely unremarked by biographers and historians, in 1973 Defferre took his cue from events in the city to publicly develop a series of unorthodox proposals on immigration which this chapter aims to explore.[5] We will begin by surveying the events of 1973 in Marseille, before examining Gaston Defferre's public statements and how they might be interpreted. Though redolent of political opportunism in a local context, it will be argued that his stance also offers some wider insights into the politics of postcolonial immigration in 1970s France.

The *Rentrée* of 1973 in Marseille

In June 1973, heavily publicized clashes between militants of Ordre Nouveau and far-left opponents in Paris led the Interior Ministry to ban both the extreme right group and the Ligue Communiste. Over the course of that summer, a series of local incidents across the Midi began to take on the appearance of an anti-immigrant *psychodrame*. In July, police violently

---

[5] Gastaut briefly cites Defferre's February article in *Le Provençal* as an example of the 'tolerance threshold' idea in the 1970s. Y. Gastaut, *L'immigration et l'opinion en France sous la Vème République* (Paris: Seuil, 2000), 473.

dispersed *sans-papiers* demonstrators in Grasse with the public approbation of the mayor, and the same month saw a series of racially motivated attacks reported on the Côte d'Azur. On 10 August, the Municipal Council of Toulon unanimously adopted a motion which called for 'constant surveillance of estates and neighbourhoods where North Africans are heavily concentrated', and ended with the slogan 'Toulon wants to remain Toulon'.[6]

The wave of xenophobia reached its apogee in Marseille during the last week of August. A dramatic local crime, in which a bus driver was stabbed to death by a mentally disturbed Algerian man, was rapidly exploited by a racist campaign directed at the city's Algerian population. Local activists of the Front National, founded the previous autumn, began distributing tracts proclaiming 'Marseille is afraid!', and announced the formation of a 'Comité de Défense des Marseillais' (CDM).[7] Within twenty-four hours the CDM launched a petition which condemned 'l'immigration sauvage'[8] and called for tighter controls. The region's right-wing daily newspaper, *le Méridional*, published an infamous editorial on 26 August entitled 'Enough, enough, enough!' An extraordinary diatribe in a newspaper with a circulation of some 85,000, the piece began 'We've had enough of Algerian thieves, enough of Algerian vandals, enough of Algerian syphilitics, enough of Algerian rapists ...'[9] In common with a number of interventions, the polemic had a distinctly *revanchard* flavour, intoning that 'contrary to what they were led to believe, independence has only brought them misery'. Both the editorial's author, Gabriel Domenech, and the newspaper itself were later prosecuted under the 1972 anti-racist law. *Le Méridional* also turned

---

6 Resolution of the Toulon Municipal Council, 10 August 1973. This and all subsequent translations are my own unless otherwise specified.
7 Note from the Commissariat Central de Marseille to the Prefect of Police, 28 August 1973. Archives Départementales des Bouches-du-Rhône (henceforth ADBdR), 135 W 52.
8 Literally translating as 'wild immigration', the term was more generally understood to mean unregulated or unfettered immigration.
9 *Le Méridional* (26 August 1973).

over its pages to a number of the region's centre-right politicians who were quick to jump on the bandwagon.

The local section of the Union des Jeunes pour le Progrès condemned 'the North African, anti-French mob', and was promptly disowned by its national committee in Paris.[10] Local UDR deputy Marcel Pujol denounced the killing in an open letter to the Interior Minister linking immigrants to prostitution and organized crime, while a UDR mayor from northern Marseille penned an article entitled 'Workers yes, assassins ... no!' in which he complained of a 'reign of terror'.[11] Rhetorical denunciations soon gave way to violence and vigilante attacks. In less than a month, six Algerian men and one Tunisian man were killed in Marseille alone.[12]

Local parties of the left and trade unions were quick to denounce the racist campaign, and along with a range of local dignitaries issued repeated calls for calm. Defferre's *Le Provençal* newspaper and the communist *Marseillaise* both echoed these sentiments, with the *Marseillaise* accusing the Government of complacency towards the far right.[13]

In the face of violent attacks, a series of strikes and protests were organized by immigrant workers in conjunction with the far-left Mouvement des Travailleurs Arabes (MTA).[14] Anti-racist associations, religious leaders and much of the national press soon began to echo this alarm. Reporters dispatched from Paris gave prominent coverage to the 'wave of racism' said to be engulfing Marseille.[15] A burgeoning anti-racist consensus culminated

10   *Le Méridional* (26 August 1973).
11   *Le Méridional* (28 August 1973).
12   Gastaut, *L'immigration et l'opinion en France*, 291.
13   *La Marseillaise* (29 August 1973).
14   The group came to prominence in Marseille during the protests against the Marcellin-Fontanet circulars in late 1972. On the MTA, see A. Hajjat, 'Des comités Palestine au mouvement des travailleurs arabes (1970–1976)', in A. Boubeker and A. Hajjat (eds), *Histoire politique des immigrations (post)coloniales: France 1920–2008* (Paris: Editions Amsterdam, 2008), and R. Aissaoui, *Immigration and National Identity: North African Political Movements in Colonial and Postcolonial France* (London: Taurus, 2009).
15   Gastaut, *L'immigration et l'opinion en France*, 292–3.

in the announcement of a 'National Day Against Racism' by the CGT, CFDT and parties of the left, scheduled for the last week of September.

A confidential report on these events, sent to the Interior Ministry by the prefect of the Bouches-du-Rhône, offers some insights into the perspective of the French authorities.[16] Markedly different from the interpretation prevailing in the mainstream press, it was an official view which mirrored that of local police reports and, as will be seen, was also reflected in the statements of the most senior government figures. In analysing the violence in Marseille, a binary distinction was made between two 'communities', one 'North African', the other 'European'.[17] Crimes involving these two groups were then listed in the ten days after the bus killing. Those imputed to 'North Africans' (and where 'Europeans' were considered to be the victims) ranged from petty theft and stone-throwing by children, to bar brawls and two stabbings. Set against these were crimes of which 'North Africans' had been victims, which included four murders, a Molotov cocktail attack and a series of shootings. In an odd equivalence this was taken as evidence of tensions between 'two communities'. Summarizing the 'underlying causes' in his briefing, the prefect literally underlined '*the geographical concentration of Maghrebi immigrants, and notably Algerians, in certain cities and in certain neighbourhoods*'. By way of conclusion he suggested that 'in this context, with each community simultaneously manifesting signs of aggression and self-defence, the murder of the Marseille bus conductor and its immediate repercussions [the violent deaths of six Algerians and a Tunisian] do not really come as a surprise.' This was the 'seuil de tolérance' (tolerance threshold) thesis in all but name. The theory held that when the proportion of foreign residents in an area exceeded a certain level,

---

16 'Rapport sur les problèmes de l'Immigration Nord-Africaine Dans les Régions de Marseille, Nice et Toulon Suite aux Evénements Récents', Jean Laporte, Prefect of the Bouches-du-Rhône and Provence-Côte d'Azur, to the Cabinet du Ministère de l'Intérieur, 25 September 1973. ADBdR 135 W 52.

17 Here the prefect was drawing on an earlier report from the Renseignements Généraux which used the same terms. Report from the Service Régional des Renseignements Généraux (SRRG) to the Prefect of the Bouches-du-Rhône, 6 September 1973. ADBdR 135 W 52.

dangerous social tensions would inevitably result. Crudely mechanistic and devoid of an empirical basis, the notion of a 'tolerance threshold' had the advantage of precluding any substantial reflections on racism. Though rapidly discredited in academic circles, its influence on political and media discourse grew during the 1970s and into the 1980s.[18]

'Another problem about which very little has been said or written until now': Defferre's Public Interventions

In line with most local politicians of the left, Gaston Defferre publicly condemned racist exploitation of the bus killing. His loyal lieutenant, secretary of the Bouches-du-Rhône section of the PS Charles-Emile Loo, rapidly penned an article for *Le Provençal* entitled 'Justice yes, racism no'.[19]

However, as reports of attacks on Algerians emerged Marseille's mayor was unsurprisingly more reluctant than national newspapers or anti-racist associations to accuse his local electorate of a collective xenophobic reaction. In a radio interview with France-Inter on 30 August he instead hailed the calm and sang-froid of the *Marseillais* in resisting the siren calls of extremists.[20] He also seized the opportunity to shift the focus onto the government, which he insisted bore a heavy responsibility for events in Marseille. As evidence that he had seen the problem coming and that his warnings had gone unheeded, Defferre pointed to an article he had published six months earlier in *Le Provençal*.

---

18   See V. de Rudder, '"Seuil de tolérance" et cohabitation pluriethnique', in P.-A. Taguieff (ed.), *Face au racisme: analyse, hypothèses, perspectives* (Paris: La Découverte, 1991).
19   *Le Provençal* (28 August 1973).
20   Reported in *Le Provençal* (31 August 1973).

The piece in question was provocatively entitled 'Immigrant workers: the new slavery'.[21] In it, he had begun by laying out familiar charges made on the left concerning the exploitation of foreign workers by unscrupulous employers. He also criticized the government's neglect of housing, the results of which were all too evident in the continuing proliferation of *bidonvilles* around Marseille. The article then went on to raise 'another problem about which very little has been said or written until now':

> The concentration of immigrant workers, and particularly North Africans, in certain areas of large cities creates a difficult situation which risks provoking a racist, xenophobic reaction in the near future. This phenomenon is aggravated by the fact that North Africans are not only more and more numerous, more and more concentrated in certain areas, but they now come over with their families, don't speak French, and live in France the way they did in their home villages.

The implied breach of a 'tolerance threshold' is again clear. Even more surprising than his public singling out of 'North Africans' was the emphasis he placed on family migration. In the aftermath of the bus killing, even the Front National-inspired Comité de Defense des Marseillais had not thought to mention families in its denunciation of 'l'immigration sauvage'. He concluded the piece by deploring the laxity of entry controls: 'When they stop working, they're allowed to stay in France. You only have to go to the place d'Aix to see, from morning till evening, bars full of North Africans playing cards and who clearly don't work.' By the end of the article, the reader could be forgiven for having forgotten the 'New Slavery' of the title. Yet this strange combination of denouncing the exploitation of immigrant workers in progressive, and at times almost *gauchiste* terms, with an essentialist discourse targeting 'North Africans', was a theme he would publicly develop and embellish during the autumn of 1973. Taking events in Marseille as vindication, he also began to put forward a number of policy proposals.

---

21 'Les travailleurs immigrés: le nouvel esclavagisme', published in *Le Provençal* (2 February 1973). Subsequent quotations are from this same article.

In a televised interview on 4 September, Defferre condemned far-right extremists 'who want to exploit the sentiments which, alas, exist in the hearts of men'.²² He went on to insist that 'it's not enough to talk about the suffering, the shameful, unacceptable exploitation of these workers by some business owners, with the complicity of the government. We also need to face the problem of the difficulties experienced by the population of Marseille when there is too large a concentration of foreigners in an area.' Returning to the parallel with slavery, he warned that France should not repeat the error of the United States, which 'based its economic development on foreign labour – subjected to the worst constraints – and which is now facing the consequences'. He then spelled out the challenge: 'will we have the foresight to tackle this problem, and to make the right choices, so as to avoid the racial conflict which threatens us?' The implication was increasingly clear. Ending the exploitation of immigrant workers and averting the dangers of 'racial conflict' necessitated a fundamental rethink of immigration policy.

Two weeks later, at the peak of the crisis and with reports of racist attacks widely relayed by the national press, the Algerian government made a shock announcement.²³ Pending guarantees concerning their security, the migration of Algerians to France was officially suspended. Most of the French media greeted the move as a reflection of the seriousness of the issue, and renewed criticism of the French authorities' reticence.²⁴ The Messmer Government, apparently wedded to the kind of analysis provided in the prefect's report, was thrown further onto the defensive. In the wake of the Algerian government's announcement, the Interior Minister, Raymond Marcellin, issued an extraordinarily ill-judged statement. Citing police statistics for the previous twelve months, he claimed these showed that more violent attacks had been committed by 'North Africans among

22   ORTF television news bulletin, 4 September 1973. INA.
23   19 September 1973.
24   One notable exception was *Le Méridional*'s editorial of 20 September. Apparently without irony, it criticized the Algerian president's decision under the headline 'Ni généraliser, ni grossir' ('Don't generalize, don't exaggerate').

themselves' than by 'Europeans against North Africans'.[25] Marcellin went on to dismiss the 'supposed anti-racist campaign' as the work of 'a certain number of extremist organizations and politicized trade unions'. Two days later at a presidential press conference, Pompidou was asked about racism and immigration. Seeking to play down recent events and ignoring Boumediene's announcement, he began by condemning racism in an abstract manner. He then fell back on the 'tolerance threshold' by pointing to the concentration of 'North Africans and particularly Algerians' in major French cities.[26] Elaborating, he suggested that 'it is obvious that in such a situation, when two communities find themselves living side by side with completely different habits, beliefs, and lifestyles, conflicts arise'. As a remedy to this problem, he offered the vague suggestion of 'a better distribution around the country'.

In contrast to the patrician reticence of the president, Marseille's mayor went out of his way to address the issue. On the television news Defferre argued that the Algerian government's announcement 'serves to clearly highlight the problem'.[27] In rhetorical fashion, he then proceeded to ask: 'do we want a French economy that depends on foreign labour in order to function? Do we import foreign labour into this country because French people refuse to take certain jobs or because some bosses refuse to pay French people a decent wage?' It seems unlikely that these were the questions that the Algerian President was seeking to raise. Like his emphasis on family migration, by questioning the utility of foreign labour to the French economy Defferre was once again breaking new ground. In his virulently racist polemic Gabriel Domenech had accused Algerians of everything from theft and rape to political subversion, but he had neglected to mention the taking away of jobs from the French. At the close of the interview Defferre made an explicit call for the negotiation of a new bilateral accord with Algeria. This, he insisted, should address both immigration policy

25 *Le Monde*, 26 September 1973.
26 Press conference held on 27 September 1973, transcript published in *Actualités-Documents: Comité Interministeriel pour l'Information*, 105, November 1973.
27 ORTF television news bulletin, 21 September 1973. INA.

and development aid, 'because it's quite clear that the two questions are linked'.

In fact, he had already elaborated on this linkage in an interview with *Les Dernières Nouvelles d'Alsace* published on the day of Boumediene's announcement.[28] The seemingly odd choice of forum reflected the relationship between regional newspaper groups. Intended for national syndication, his comments were duly reported in *Le Monde* and other national titles on 20 September.[29] The crux of his proposal was that henceforth foreign workers should only come to France 'alone ... and for a limited period'. As a counterpoint the French Government should expand its aid 'to under-developed countries, particularly for the construction of factories'. This assistance would ensure that 'in those countries industrialization can finally take off [...] and in a few years they will no longer need to export their labour force'.

The most comprehensive exposition of his ideas was published in *Le Provençal* in early November. By this time the national press and political agenda had moved on to other issues, not least the aftermath of the Yom Kippur War. Nonetheless, Defferre persevered in addressing a topic he presented as being of national importance.

The latent racialization that has characterized immigration debates in France over the last three decades was already emerging in this period.[30] The prefect's invocation of 'two communities' (accounting for around 3 per cent and 95 per cent of the population respectively) is one example. In public discourse, the interchangeable use of the terms 'immigrant worker', 'North African' and 'Maghrebi' by Government ministers was part of the same phenomenon. At the heart of this semantic Russian doll was the figure of the archetypal North African migrant, usually an Algerian. Such euphemisms were not confined to the French authorities. Even sympathetic press coverage and campaigning, by juxtaposing racism with immigrant

---

28 *Les Dernières Nouvelles d'Alsace* (19 September 1973).
29 *Le Monde* (20 September 1973).
30 See S. Bonnafous, *L'immigration prises aux mots: les immigrés dans la presse au tournant des années 80* (Paris: Editions Kimé, 1991).

workers, inadvertently contributed to the elision. Thus when *Le Provençal* ran a three-part investigative piece in November 1973 it was billed as an objective look at the situation of 'immigrant workers' in the region. Yet within the space of two paragraphs the journalist made clear that his reports would actually focus exclusively on migration from the Maghreb.[31] A conventional discussion of difficult housing and working conditions, the series was introduced and concluded with editorials penned by Gaston Defferre. He too oscillated between the terms 'immigrant worker', 'foreign worker' and 'Maghrebi', but specified that 'the most topical example is that of France and Algeria'.[32]

Defferre set out his stall with the observation that hitherto 'the problem posed by the presence of so many foreign workers in France has only been addressed in a limited way, either in terms of the exploitation of foreigners, or from the point of view of French people'. He then repeated his earlier rhetoric on the economy: 'Do we want a French economy that depends upon foreign labour in order to function and develop?' His follow-up question was addressed to businesses: 'Do French business owners employ immigrants because we lack workers, because the French refuse certain jobs, or because it's more profitable to employ foreigners than French workers?' Neither the government nor employers had ever dared to answer these questions, he declared, and 'their silence reveals their complicity'.

Turning to immigrants themselves, he identified family migration with vulnerability to exploitation 'from every point of view: housing, work etc.' In the case of Algerians, 'an almost tribal family structure facilitates the exploitation of immigrant workers. Living in families, they more readily accept being badly housed, badly paid, and badly treated.' Building on his earlier public statements, he reasserted the 'tolerance threshold' thesis: 'the concentration of foreigners in certain neighbourhoods [...] leads to

---

31 'La Provence des migrants', Jacques Bonnadier, published in *Le Provençal* on 5, 6 and 8 November 1973.
32 Editorials by Gaston Defferre which introduced and concluded a series of investigative articles by Jacques Bonnadier, 'La Provence des migrants', published in *Le Provençal* between 5 and 8 November 1973. Subsequent quotations are from Defferre's two-part editorial unless another source is specified.

dangerous racial tensions which can only get worse'. Exploitation and alterity then came together in his remarkable synthesis: 'There is both an economic and social conflict between the exploiters and the exploited, and a racial conflict between foreigners and the French.'

This then was the global vision Defferre had called for at the start of his article, bringing together the 'perspective of the exploitation of foreigners' and the 'perspective of the French'. Though not specified, the first of these perspectives was clearly also a French one. The undeniable novelty of this juxtaposition led to a striking assessment of 'the most topical example': 'Ten years after the evacuation of Algeria', he wrote, 'we find ourselves with a system which resembles both colonialism and the invasion of France.' Borrowing from the anti-colonial language of the far left and the alarmist vocabulary of the far right, Defferre's hybrid diagnosis led to a similarly eclectic prescription.

Turning to solutions, Marseille's mayor set out what he saw as the respective duties of the French government and those of labour-exporting countries. The French authorities must first ensure that sufficient professional and language training were provided to immigrants, adding that 'this could be done in the country of origin'. Secondly, he reiterated the call for a vast aid programme to be established, again stressing 'the construction of factories'. These factories would provide jobs for foreign workers 'who could then go back to their own country'. For their part, labour-exporting states, 'and notably the Algerian government', should undertake to allow only single or unaccompanied men to emigrate, and then for a limited period not exceeding three years. If these policies were introduced, he concluded, 'the grave difficulties and the risks of conflict stemming from the current situation, both for the French and for foreigners, would disappear'.

In effect, his position seemed to derive from a superficial paradox that had seen both the far-left Mouvement des Travailleurs Arabes and the far-right Comité de Défense des Marseillais publicly welcome Boumediene's suspension of Algerian migration to France. While the MTA had greeted the move as an acknowledgement of the racism and exploitation which the group had been seeking to highlight as well as a defiant anti-imperialist message to the former colonial power, the CDM had announced its satisfaction at the prospect of fewer Algerians arriving in the city's port. Clearly, this hardly represented a community of interests.

Defferre's proposed scheme to combine greater resources for professional training and an expansion of development aid to assist industrialization with an agreement to ensure that future immigration was both male and temporary was also somewhat original. Reminiscent of co-development, it would nonetheless be a full decade before the PS publicly floated a similar idea under this label.[33] By that time discussions of immigration policy were taking place in a very different political and economic environment in which the employment levels of the *Trente Glorieuses* seemed a distant memory. Even the attempts to introduce voluntary and forced repatriation during Valéry d'Estaing's presidency took place against a backdrop of sharply rising unemployment in which the equation of jobless Frenchmen with foreign workers was increasingly common. What stands out about Defferre's proposals in 1973 is the manner in which they challenged what remained a widespread political consensus on the utility of foreign labour to the French economy. Neither the centre-right government, nor *Le Méridional*, nor far-right campaigners, had publicly advanced such a claim. In a published response to Defferre's article, the president of the Union Patronale of the Bouches-du-Rhône, Henri Mercier, was unequivocal: 'Employers do lack workers – especially in our regions where the economy is growing rapidly – whatever the "partisan" claims some people may make.'[34] The originality of Defferre's proposals was thus in presenting a form of *co-développement* as a means of ending the exploita-

---

33 Jean Auroux, then ministre délégué in the ministry of social affairs, in an interview with *Paris-Match* in 1983. Cited in Gastaut, *L'immigration et l'opinion en France*, 400. Definitions of co-development have varied over time but as a strategy it has generally sought to link migration and development assistance within a mutually beneficial bilateral framework. In the formulation proposed by Auroux and in more recent European Union initiatives the notion of return and stemming future migration flows figure prominently.
34 *Le Provençal* (28 November 1973). A similar view on the importance of foreign labour to the French economy was expressed in the front page editorial of business monthly *Interprofessions: informations économiques et sociales de la region Provence – Côte d'Azur*, in October 1973.

tion of foreign workers (he made no reference to the equal rights clauses of the Programme Commun, nominally the Parti socialiste's solution to this problem) and as a way of averting 'racial conflict'.

Defferre vs. Defferre: Another *Exception Marseillaise*?

All of Gaston Defferre's public interventions between February and November 1973 were apparently made on his own initiative. Although he was a very senior figure within the party his statements were clearly not official policy, but rather the floating of ideas which ambiguously drew on his status as both local and national politician.

However, it is difficult to find any neat division between a more hardline message aimed at a local electorate and a more tolerant and consensual one for a national audience. In his interview with France-Inter he openly referred to the over-concentration of foreigners. The syndicated interview with *Les Dernières Nouvelles d'Alsace* (picked up by *Le Monde* and other national titles) saw him call for a halt to family migration and offer the first formulation of his proposal to end French dependency on foreign labour. The televised interviews he gave to the ORTF, in which he invoked the spectre of 'racial conflict', were also broadcast on a mixture of regional and national news bulletins. Similarly, his articles in *Le Provençal*, both in February and November, devoted as much space to attacks on the Government and 'certain employers' for their exploitation of immigrant workers as they did to calls for a more restrictive immigration policy.

While on at least two occasions he alluded somewhat vaguely to plans for a legislative proposal, there appears to be no trace of this either in his personal papers or in Parti socialiste documents conserved by the Fondation Jean Jaurès.[35] Indeed a statement issued by the party's Comité Directeur on 8 September condemned the unleashing of a 'racist campaign against immi-

---

35   Fonds 'Gaston Defferre'. Archives Municipales de Marseille 100II.

grant workers' and singled out *Minute* and *Le Méridional* for criticism, but made no reference to a 'tolerance threshold' or to immigration controls.[36] Instead it simply called for solidarity with immigrants and reiterated the proposals for equal rights contained in the Programme Commun. As has been noted, the latter did not merit a mention in any of Defferre's public statements. It may therefore be that his initiative was quietly dropped due to being overtaken by events (his long editorials in *Le Provençal* in November, which came after the Algerian government's unilateral suspension of emigration, repeated his earlier position but made no reference to a specific legislative proposal) or due to a lack of enthusiasm within his own party.

How then to account for Defferre's foray into immigration politics? As even the most sympathetic biographies acknowledge, the politician who managed to control the town hall of Marseille for thirty-three years without ever winning an election in the first round was an extremely sharp operator. Frequently accused of a cynical complacency towards the Front National when his re-election was threatened in the early 1980s, Defferre had in fact made pragmatic compromises with hard-right opinion long before Jean-Marie Le Pen's party rose to prominence. Unremarked at the time, the *Méridional* newspaper at the centre of the racist 'backlash' in 1973 had been effectively acquired a year earlier by the *Provençal* newspaper group that Defferre controlled. The two papers not only shared printing presses but also *faits divers* (news in brief) sections. Thus, somewhat surreally, during August and September 1973 both titles featured the same short pieces on robberies or attempted homicides, leaving *Le Méridional* free to elaborate at much greater length where witness descriptions implicated 'an Arab type'. The acquisition of Marseille's main right-wing newspaper seems to have been based on the logic of 'better the devil you own', and there is nothing to suggest Defferre exercised any direct editorial control of the kind so apparent with *Le Provençal*. Nonetheless, he was presumably grateful that the many voices condemning *Le Méridional* in 1973, including the leadership of his own party, failed to raise the issue of his ownership. Indeed one of the ironies of this episode is that, in the wake of Domenech's editorial,

---

36 'Communiqué du Comité Directeur', 8 September 1973, Fondation Jean Jaurès.

it was Defferre's newspaper group which was financially liable for one of the first fines ever imposed under the 1972 anti-racist legislation.

The eccentricities of Marseille politics seem relevant in other ways. The 'système Defferre', built up between the 1950s and 1970s, rested upon an electoral alliance between socialists, independents and the centre-right which was designed to keep the powerful local communist party out of power. Despite his support for Mitterrand's national Union de la Gauche strategy, Marseille's mayor maintained this coalition arrangement until the 1977 municipal elections.

Famously political arrangements in France's second city were also characterized by an entrenched *clientélisme*. In his thesis comparing practices in Marseille and Naples, Cesare Mattina provides a remarkable exposé of the ways in which public resources, in particular municipal employment and social housing, were diverted in order to consolidate an enduring electoral base in the *cité phocéenne*. During a period of unprecedented expansion in public-housing construction in response to a chronic shortage, and with perhaps 50,000 jobs indirectly controlled by the municipal administration, opportunities for patronage were adroitly exploited by local politicians.[37]

Through these material favours, which by their very nature could not satisfy everyone, politicians also sought to cultivate the support of symbolic, and to a large extent imagined, communities. Corsican, Italian, Armenian, Jewish, and after 1962 the *rapatriés* or *pieds noirs*, these notional groups conflated regional, national, ethno-linguistic and religious identity markers. Still, as various studies of the city's political history have shown, they maintained a very real hold on the imagination and political strategies of Marseille's *notables*.[38] The *pieds noirs* were problematic for Defferre, who by the standards of the mainstream political class had been an early advocate of negotiations with the FLN and had been on the receiving end

---

37 C. Mattina, 'La Régulation Clientélaire: relations de clientèle et gouvernement urbain à Naples et à Marseille (1970–1980)', unpublished PhD thesis (Université Pierre Mendès-France, Institut d'Etudes Politiques, Grenoble, 2003), 382–5.
38 See for example M. Peraldi and M. Samson, *Gouverner Marseille: enquête sur les mondes politiques marseillais* (Paris: La Découverte, 2005), 266–73.

of repeated death threats from the OAS.[39] However, as a constituency representing more than 6 per cent of the city's population they could scarcely be ignored, and room was found at the town hall for a moderate *rapatrié adjoint* elected on a centre-right list.[40]

Imagined or otherwise, not all communities were equal before the ballot box, and the most notable outsiders in this system, apart from perhaps communist voters, were Algerians (and, as they settled in growing numbers from the mid-1960s, Tunisians). Seen to carry little electoral weight, and lacking interlocutors integrated into the local political culture, the Algerian population tended to be viewed as a problem to be managed rather than a vein of support to be tapped. Jean Viard acidly summarizes how this *communautariste* logic could operate in 1970s Marseille: 'A new school in Menpenti satisfies all the Italians, the inauguration of an "avenue of the 1915 genocide" on the eve of an election satisfies all the Armenians, the widening of a road in Belsunce[41] that closes some of the local businesses satisfies those who think there are too many Arabs.'[42]

Such calculations were not confined to the ruling coalition. A proposal for HLM housing in St Henri, a communist stronghold, led to a bizarre exchange in the Municipal Council chamber in 1974. Introduced by a centre-right councillor, the project was opposed by the communist group after it failed to obtain assurances from the mayor that no more than 5–6 per cent of the apartments would be allocated to foreign families.[43] Surviving records of municipal council debates are also suggestive of how,

---

39   G. Marion, *Gaston Defferre* (Paris: Albin Michel, 1989), 205.
40   While often perceived as one of Marseille's 'communities' the *rapatriés* or Pieds Noirs were a heterogeneous group, and the growing radicalization of associations purporting to represent them also corresponded to a loss of membership during the 1960s.
41   An area of the city centre adjacent to the place d'Aix, which Defferre referred to in his February editorial as full of 'North Africans' drinking and gambling in bars.
42   J. Viard, *Marseille, une ville impossible* (Paris: Payot, 1995), 246–7.
43   *In extenso* minutes of a debate in the Marseille Municipal Council, session of 15 November 1974. Archives Municipales de Marseille, documents awaiting classification. It was this kind of calculation, and the perception among some local PCF sections that their electoral base was deliberately being undermined, that led *in extremis* to the 'bulldozer de Vitry' affair in 1980.

as early as 1973–4, problems in the *grands ensembles* and the confrontation with a new group of residents who did not correspond to the traditional *classe ouvrière* cultivated in the past, were increasingly viewed through an ethno-cultural lens. In discussions of social housing issues Defferre referred to the problem of residents not living 'in a European way'. In one (public) meeting the mayor regaled the assembled councillors with an anecdote about 'a donkey in a lift', an early forerunner of the 'sheep in the bathtub', which according to the clerk met with smiles on all sides.[44]

Can then Defferre's outspokenness on immigration simply be put down to electoral opportunism in a local political context where singling out Algerians seemed to be pushing at an open door? Besides the fact that he diffused his message on national as well as local platforms, there seem to be at least two other reasons for questioning this interpretation.

Although, during the summer and autumn of 1973, the eruption of far-right activism had received considerable attention in the press, and indirectly in the widespread condemnations by anti-racist opponents, as an electoral force it was still virtually non-existent. The Marseille section of the Front National had provided an early foretaste of future campaigns in seeking to get as much mileage as possible out of local tensions. The Comité de Défense des Marseillais claimed to have amassed around 3,000 signatures for its petition demanding a 'triple control: sanitary, judicial and professional' by the time it dissolved in the autumn. While not negligible, it was hardly a ringing endorsement in the context of such a heavily mediatized crisis. In fact it was almost exactly the same score that Jean-Marie Le Pen would achieve in Marseille during his first run at the presidency in 1974 (1.1 per cent). The city may have had a disproportionate number of voices eager to air *revanchard* prejudices tinged with a certain idea of *Algérie Française*, but the dynamics of the ruling coalition, which accounted for fifty-six of the sixty-three seats on the municipal council, insulated Defferre from a real challenge from the right. Two members of the ruling group, both from the Centre Democrate and one an *adjoint*, had joined in the lurid

---

44   *In extenso* minutes of a debate in the Marseille Municipal Council, session of 25 February 1974. Archives Municipales de Marseille, documents awaiting classification.

denunciations of 'North Africans' following the bus killing. However, the fact that they carefully avoided any criticism of Defferre's administration reflected just how closely tied into the system they were. As for the rump opposition of seven communists, whatever their private concerns about the changing demographics of their electoral fiefdoms, they were clearly in the anti-racist camp. So although – as the title of Charles Emile Loo's biographical homage suggests – for Gaston Defferre *c'était Marseille d'abord*, there seemed little reason to take up a position which was potentially embarrassing nationally, not least vis-à-vis his own party.[45]

That he continued to publicly develop his ideas after the question had faded from the headlines, and that he raised new issues – family settlement and the economic consensus on the utility of foreign labour – that even Domenech and the CDM had failed to mention, seems to suggest a conviction that went beyond dog-whistle politics.

The second reason to question a narrowly demagogic interpretation is the eclectic nature of Defferre's analysis and proposals. With repeated references to slavery and his striking phrase 'the conflict of the exploiters against the exploited', his evocation of the suffering of immigrant workers at the hands of employers and the Government was just as prominent as his Cassandra-like prophecies of racial conflict. It was certainly quite different from the *revanchard* outrage served up on the pages of *Le Méridional*, one of whose columnists described France as 'a milking-cow which continually turns the other cheek', or the local UJP's call to 'eliminate the North African, anti-French mob, who dare to break the laws of a country that offers them bread and work'.[46] The idea of expanding foreign aid was also ostensibly progressive, even if the repeated emphasis on factory-building suggested Defferre's enthusiasm for overseas development was not entirely disinterested. Instrumentalist or not, such calls were unlikely to find a sympathetic audience among *Algérie Française* nostalgists. Since the early 1960s

---

45   C. Loo and R. Colombani, *C'était Marseille d'abord. Les années Defferre* (Paris: Laffont, 1992).
46   Editorial in *Le Méridional* (29 August 1973) and a statement from the Bouches-du-Rhône section of the UJP published in *Le Méridional* (26 August 1973).

a major theme of *activiste* propaganda had been the 'scandal' of French financial assistance to independent Algeria. The Tixier-Vignancourt campaign of 1965, which scored 12 per cent in Marseille compared to five per cent nationally, featured the *cartieriste* slogan 'la Corrèze avant le Zambèze'.[47] For its part, *Le Méridional* took an interest in the economic plight of the Algerian Republic which bordered on the voyeuristic, and seldom failed to seize an opportunity to draw unfavourable comparisons with French colonial 'mise-en-valeur'.

Overall, the double-register of 'a system which resembles both colonialism and the invasion of France' seems so contrived, and so improbable as an appeal to radically opposed political tendencies, that it begs the question: was he perhaps sincere?

Taking Defferre's public statements seriously is not to divorce them from the political context in which they were made. Clearly he was first and foremost a politician. However, I would suggest that they may be read as an attempt to formulate a coherent position, rather than as opportunistic nods to seemingly polarized constituencies. Seen in this way his stance is suggestive of wider trends and tensions in the politics of immigration at this time.

What Zancarini-Fournel has described as a 'racialization of neighbourhood relations' in the mid-1970s, and the widely remarked tendency to employ an ethno-cultural lens in discussing socio-economic inequalities from the 1980s, are both pre-figured here.[48] Identification of Marseille's emerging social housing problems with residents not living 'in a European way' clearly stemmed from the same essentialist logic that saw Defferre publicly advocate the 'need to face the problem of the difficulties experienced by the population of Marseille when there is too large a concentration of foreigners in an area'. This apologia for racism in terms of a 'tolerance threshold' also characterized the private and public analysis of the French

---

47 A populist slogan, 'the Corrèze before Zambezi', encapsulated the arguments of Raymond Cartier for investment to be directed to metropolitan France rather than colonial possessions or development assistance to former colonies.

48 M. Zancarini-Fournel, 'Racisme et anti-racisme dans les "années 68"', in Boubeker and Hajjat (eds), *Histoire politique des immigrations (post)coloniales*, 122.

authorities. Unlike Pompidou, however, Defferre took this a stage further by openly calling for restrictions on immigration in the name of 'race relations'. He thereby broke with the prevailing consensus on the importance of foreign labour for the French economy, albeit by equating immigration with exploitation.

Yet his focus on Algerians and 'North Africans' – both in terms of exploitation and alterity – did not extend to the kinds of crude stereotypes about criminal and sexual proclivities relayed by *Le Méridional*. Rather, his assertions of difference and incompatibility, and his opposition to Algerian family migration (with the permanency it implied), more closely resemble the private views accorded to de Gaulle in 1959.[49] The link between these erstwhile political opponents may lie in Todd Shepard's notion of a 'tide of History thesis'. This narrative of decolonization 'allowed Algerian independence, which had been unthinkable for most French intellectuals and politicians until the late 1950s, to become obvious to almost all in the early 1960s.'[50] What the 'thesis' implied was that Algeria's independence was less the outcome of a political struggle than a natural, not to say inevitable, consequence of 'civilizational' difference. Defferre's reference to the 'evacuation' of Algeria and his emphasis on incompatible lifestyles seem to correspond to this vision.

When applied to postcolonial immigration, and especially family settlement, a 'common sense' distinction which appeared progressive in rejecting *Algérie Française* resembled something quite different: a crassly exclusionary essentialism. Clearly such attitudes, and a corresponding preoccupation with extra-European immigration, were not a monopoly of the far right. Nor were they necessarily derived from a *revanchard* reading of decolonization associated with *Algérie Française* nostalgists. Alongside scholarly interest over the last decade in the transfer of ideology and prac-

---

49 A. Peyrefitte, *C'était de Gaulle: La France redevient la France* (Paris: Editions de Fallois, 1994), 52.
50 T. Shepard, *The Invention of Decolonization: The Algerian War and the Remaking of France*, 2nd edn (Ithaca, NY: Cornell University Press, 2008), 271.

tice from colony to metropolis, the metropolitan roots of discrimination and racism should not be overlooked.[51]

In his unintentional *reductio ad absurdum* of common left-wing positions on immigration, Defferre also highlighted tensions between anti-colonialism and critiques of capitalist exploitation when bounded by national perspectives. Ironically, the position of Marseille's mayor had much in common with the official discourse of the Algerian government and its official Amicale des Algériens en France in the early 1970s, both of which emphasized the inevitable return of 'Algeria's emigration' as soon as the state-led modernization drive permitted.[52] Similarly Defferre's precocious (and instrumentalist) advocacy of a form of co-development underlines the ambiguity of the concept. Although today it is more likely to be couched in the language of empowerment and civil society, the term retains the connotations of a policy designed to stem migration flows from south to north.

The events of 1973 in Marseille and Gaston Defferre's public pronouncements underline the city's chequered history as a Mediterranean 'melting pot'. Yet the socialist mayor's interventions were arguably more than just another *exception marseillaise*. They may also offer certain insights into the emergence of a racialized 'immigration question' in the 1970s, and the complex relationship between narratives of decolonization and political discourse on postcolonial migration.

---

51   The most notable example is B. Stora, *Le Transfert d'une Mémoire: de 'l'Algérie Française' au racisme anti-arabe* (Paris: La Découverte, 1999). Other examples include: A. Spire, *Etrangers à la carte: l'administration de l'immigration en France (1945–1975)* (Paris: Grasset, 2005); J. House, 'Contrôle, encadrement, surveillance et répression des migrations coloniales: une décolonisation difficile (1956–1970)', *Bulletin de l'IHTP*, 83 (2005); and C. Hmed, '"Tenir ses hommes": la gestion des étrangers isolés dans les foyers Sonacotra après la guerre d'Algérie', *Politix*, 76 (2006), 11.

52   C. Wihtol de Wenden, *Les Immigrés et la Politique* (Paris: Presse de la Fondation Nationale des Sciences Politiques, 1988), 175.

Bibliography

Departmental Archives of the Bouches-du-Rhône, series 135 W.
Marseille Municipal Archives, 'Fonds Gaston Defferre', 100II.
Aissaoui, R., *Immigration and National Identity: North African Political Movements in Colonial and Postcolonial France* (London: Taurus, 2009).
Bonnafous, S., *L'immigration prises aux mots: les immigrés dans la presse au tournant des années 80* (Paris: Editions Kimé, 1991).
De Rudder, V., '"Seuil de tolérance" et cohabitation pluriethnique', in P.-A. Taguieff (ed.), *Face au racisme: analyse, hypothèses, perspectives* (Paris: La Découverte, 1991).
Gastaut, Y., *L'immigration et l'opinion en France sous la Vème République* (Paris: Seuil, 2000).
Hajjat, A., 'Des comités Palestine au mouvement des travailleurs arabes (1970–1976)', in A. Boubeker and A. Hajjat (eds), *Histoire politique des immigrations (post) coloniales: France 1920–2008* (Paris: Editions Amsterdam, 2008).
Hmed, C., '"Tenir ses hommes": la gestion des étrangers isolés dans les foyers Sonacotra après la guerre d'Algérie', *Politix*, 76 (2006).
House, J., 'Contrôle, encadrement, surveillance et répression des migrations coloniales: une décolonisation difficile (1956–1970)', *Bulletin de l'IHTP*, 83 (2005).
Laurens, S., '1974 et la fermeture des frontières. Analyse critique d'une décision érigée en *turning-point*', *Politix*, 82 (2008).
Loo, C., and Colombani, R., *C'était Marseille d'abord: Les années Defferre* (Paris: Laffont, 1992).
Marion, G., *Gaston Defferre* (Paris: Albin Michel, 1989).
Mattina, C., 'La Régulation Clientélaire: relations de clientèle et gouvernement urbain à Naples et à Marseille (1970–1980)', unpublished PhD thesis (Université Pierre Mendès-France, Institut d'Etudes Politiques, Grenoble, 2003).
Noiriel, G., *Immigration, antisémitisme et racisme en France (XIXème–XXème siècle): Discours publics, humiliations privées* (Paris: Fayard, 2007).
Peraldi, M., and Samson, M., *Gouverner Marseille: Enquête sur les mondes politiques marseillais* (Paris: La Découverte, 2005).
Peyrefitte, A., *C'était de Gaulle: La France redevient la France* (Paris: Editions de Fallois, 1994).
Shepard, T., *The Invention of Decolonization: The Algerian War and the Remaking of France*, 2nd edn (Ithaca, NY: Cornell University Press, 2008).

Silverman, M., *Deconstructing the Nation: Immigration, Racism and Citizenship in Modern France* (London: Routledge, 1992).
Spire, A., *Etrangers à la carte: l'administration de l'immigration en France (1945–1975)* (Paris: Grasset, 2005).
Stora, B., *Le Transfert d'Une Mémoire: de 'L'Algérie Française' au racisme anti-arabe* (Paris: La Découverte, 1999).
Viard, J., *Marseille, une ville impossible* (Paris: Payot, 1995).
Wihtol de Wenden, C., *Les Immigrés et la Politique* (Paris: Presse de la Fondation Nationale des Sciences Politiques, 1988).
Zancarini-Fournel, M., 'Racisme et anti-racisme dans les "années 68"', in A. Boubeker and A. Hajjat (eds), *Histoire politique des immigrations (post)coloniales: France 1920–2008* (Paris: Editions Amsterdam, 2008).

ANDREA BRAZZODURO

Postcolonial Memories of the Algerian War of Independence, 1955–2010: French Veterans and Contemporary France[1]

> Cette guerre coloniale est originale même dans la pathologie qu'elle sécrète.[2]

> Dans l'Armée, nous avons fait notre devoir, pour rien finalement, mais nous avions trahi l'Algérie française, et depuis, personne ne nous entend, sauf vous.[3]

Elie was born in Rennes, in Bretagne, on April 1939, no more than a couple of months before the beginning of the Second World War and the Nazi occupation of France. In 1960, aged twenty-one, Elie was called up by his country to serve in the army in Algeria. There, France is conducting something called a 'public order operation' (*opérations de maintien de l'ordre*) against an internal '*rébellion*' – as it was called at the time – which has, during the previous six years, enflamed the French *départements* of Algeria.

---

[1] I would like to thank Peter Lang's anonymous reviewers for helpful critical suggestions. All translations are mine unless otherwise indicated.
[2] 'This colonial war is singular even in the pathology that it gives rise to.' F. Fanon, *Les Damnés de la terre* (Paris: Maspero, 1961), 191.
[3] 'In the army we did our duty, in the end for nothing, but we betrayed French Algeria, and since then no one listens to us, except you.' Jacques (1936, 60–1/A), dentist, 4th section d'infirmiers militaires, second lieutenant. Letter to the author, 17 January 2008. ('60–1' gives the year and semester of conscription; the final letter (which can be either A, B, or C) refers to a division of the semester into thirds, which was implemented after August 1954. So 60–1/A means that Jacques was drafted in 1960, in the first semester, and in the first two months of that semester. Conscription was automatic at age twenty, except in special cases such as Jacques': born in 1936, he was supposed to leave in 1956 but as a university student he was able to put off the call-up until he had completed his studies.)

Elie will spend two years – as a simple, second-class, infantry man in the ranks of the 156th infantry regiment – patrolling the region of Constantine, moving from one operation to another as far as the Tunisian border (and beyond) and as far as Khenchela in the Aurès-Nememcha.

Forty years later, in January 2000, Elie decided to recount his experiences by writing a book. In the absence of a publisher his memoir takes the form of forty-seven A4 pages, with maps, pictures and drawings. The chosen title is 'Drafted in Algeria: The Retrieved Voice' (*Appelé en Algérie. La parole retrouvée*). The foreword begins with these words, which demand close attention from the reader:

> Je viens d'achever la lecture d'un livre acheté par Maïwenn, ma fille, qui voulait en savoir plus sur la question de l'Algérie.
>
> En premier lieu, lorsque j'ai vu l'ouvrage, j'ai d'abord été surpris par le choix de Maïwenn, puis je me suis laissé dire qu'elle avait besoin d'être un peu plus éclairée. Je n'avais pas fait mon devoir d'information envers mes enfants et je me sentais frustré. C'est précisément parce que ma parole aussi avait été *confisquée*.
>
> [I have just finished reading a book bought by Maïwenn, my daughter, who wanted to know more about the issue of Algeria.
>
> Initially, when I saw the work, I was taken by surprise by Maïwenn's choice, but then I brought myself to understand that she needed to be a bit more informed. I had not done my duty to inform my children, and I felt frustrated. This was precisely because my voice too had been *confiscated*.]

This text, which superficially might seem anodyne and anecdotal, is in fact full of information useful to the historian of the present day who is concerned with memorial processes – i.e. with the way in which groups and individuals remember the past in the present.[4]

First let us note how the recollection in Elie's case begins as a reaction to a feeling of frustration, which in psychoanalysis may be associated with the onset of a removed desire. Second, let us note also that this feeling is

---

4   Memory is the presence of the past, according to the acute definition offered by Augustine of Hippo, *Confessiones*, XI: 20, 26; for more on this, see the commentary by P. Ricœur, *Temps et Récit III, Le Temps Raconté* (Paris: Seuil, 1985), 19–53.

taking form within an intergenerational relationship. In response to his daughter's (implicit) question, Elie engages in memory-work on a period of his life which he had not so much forgotten as repressed or removed.[5] Since forgetting and silence are not the same, Elie comes to encounter a new framework in which someone pays attention to his voice, listening to his combat (and homecoming) stories, and thereby allowing him to start this work which ends in speaking out (*prise de parole*).[6] He finds once again the words to tell his experience and therefore to tell himself. The philosopher Paul Ricœur showed us, with the notion of narrative identity, the extent to which we are what we tell ourselves (*on est ce qu'on se raconte*):[7] recounting life events – particularly combat ones – allows one to create a coherence, to 'make the events understandable to [one]self too in order to provide an interpretation of them to an outsider'.[8] As in the case of Elie – writing to explain himself to his daughter – we need this outsider to such an extent that in order to think we have to invent a fictive other to talk with. If veterans remain silent for a long time, this is primarily because French society does not want to listen to them. We don't remember alone, as Halbwachs was the first to notice.[9]

5   On memory transmission, see M. Bloch, 'Mémoire collective, Tradition et coutume. A propos d'un livre récent', in A. Becker (ed.), *L'Histoire, la Guerre, la Résistance* (Paris: Gallimard, 2006 [1925]), 335–46; see also F. Dosse, *Les Héritiers du silence. Enfants d'appelés en Algérie* (Paris: Stock, 2012).
6   This was the title chosen by M. de Certeau, *La Prise de parole. Pour une nouvelle culture* (Paris: Desclée De Brouwer, 1968). See the useful distinction between amnesia and aphasia suggested by A.L. Stoler, 'L'aphasie colonial française: l'histoire mutilée', in N. Bancel et al. (eds), *Ruptures postcoloniales. Les nouveaux visages de la société française* (Paris: La Découverte, 2010), 71–2.
7   See P. Ricœur, *Temps et Récit III, Le Temps Raconté*, 347–92.
8   F.H. Allison, 'Remembering a Vietnam War Firefight: Changing Perspectives Over Time', in R. Perks and A. Thomson (eds), *The Oral History Reader* (London and New York: Routledge, 2006 [2004]), 226.
9   'En résumé, il n'y a pas de mémoire possible en dehors des cadres dont les hommes vivant en société se servent pour fixer et retrouver leurs souvenirs': M. Halbwachs, *Les Cadres sociaux de la mémoire* (Paris: Albin Michel, 1994 [1925]), 79.

I would like to stress this point – in regard to Elie's and other combat veterans' difficulties in verbalizing their own war experiences – by taking as an example one of the most ancient stories of a soldier's homecoming, namely Homer's 'Odyssey'. If we look at the structure of the poem, the 'Odyssey' appears as the narration *à rebours* that Odysseus, coming back from war, is giving to the Phaeacian court, solicited by the blind bard Demodokos. One cannot help but be struck by how thoroughly the Greek poem depicts the archetype of the veteran's condition. Odysseus returns from war to a thriving and peaceful society: there, he is listening to the blind bard (a kind of historian, who did not directly see the things he is relating) who is singing the history of the Trojan War to the assembly of Phaeacians (i.e. the civilians).[10] Hearing his combat experience objectified as an epic, Odysseus – who actually fought in that war – bursts into tears and is moved to react, thus beginning to speak himself.[11] Returning to Elie, we also find a blind bard, an historian, standing between him and his daughter. Thus it is important to notice the way in which Elie identifies the book bought by his daughter Maïwenn ('who wanted to know more' (*qui voulait en savoir plus*), as he phrases it) as the origin of the process leading him to voice his war experience. The book is *Appelés en Algérie: La parole confisquée* ('Draftees in Algeria: The Confiscated Voice'), published the previous year, 1999, by the historian Claire Mauss-Copeaux.[12]

This book made such an impression on Elie that he used the title, *détourné*, for his own testimony: in reaction, as it were, to that confiscated

10   See F. Hartog, *Régimes d'Historicité: Présentisme et Expérience du Temps* (Paris: Seuil, 2003), 59–65. H. Arendt wrote that, 'poetically speaking', we can see the beginning of the category of history in the meeting of Odysseus and the bard Demodokos, 'when Ulysses, at the court of the king of the Phaeacians, listened to the story of his own deeds and sufferings, to the story of his life': H. Arendt, 'The Concept of History: Ancient and Modern', in *Between Past and Future* (London: Penguin, 1993 [1958]), 29.
11   See J. Shay, *Odysseus in America: Combat Trauma and the Trials of Homecoming* (New York: Scribner, 2002), 11–18.
12   C. Mauss-Copeaux, *Appelés en Algérie: La Parole confisquée* (Paris: Hachette, 1999). I am deeply grateful to Claire Mauss-Copeaux for letting me see Elie's unpublished manuscript and discussing various points of the present article with me; responsibility for any errors in my account is obviously mine.

or seized voice ('parole confisquée') of the drafted servicemen analysed by Mauss-Copeaux during the 1990s, Elie's own voice is retrieved ('retrouvée'). As in the case of Odysseus (the veteran), Demodokos (the historian), and the Phaeacians (the civilians), we are here witnessing the productive meeting between one of the scholarly vectors of memory (namely historiography) and the social claim to knowledge.[13]

In what follows, I want to explore some aspects of the way in which veterans compose their memories of the war within present-day social frames of memory. How have veterans developed their stories faced with a modern audience that has not experienced war? And how do they cope with the ongoing historical research, now being taken up by a third generation of scholars? Are war veterans still locked into silence and shame, as has long been suggested? (*Le silence et la honte* was the title both of a book by the psychiatrist Bernard W. Sigg and of one by the journalist Andrew Orr, published in 1989 and 1990 respectively.)[14] In recollecting their memories of the last (colonial) French war, to what extent are veterans influenced by contemporary (postcolonial) social and political events – if indeed they are?

* * *

My analysis, in this chapter, is based on interviews conducted with thirty-one French veterans between 2007 and 2010.[15] The sample size has been kept reasonably low in order to allow long and articulated interviews and, where possible, repeated interviews in successive meetings. In the first

---

13   In his path-breaking book on the memory of the Vichy regime, Henry Rousso underlines the role of *interface* which characterizes the works of historians concerning collective memory. In the 'interactive chain of transmission' contributing to the 'shaping of the representations of the past', historians are placed between the available sources of information and the social claim to knowledge. H. Rousso, *Le Syndrome de Vichy: De 1944 à Nos Jours* (Paris: Seuil, 1990), 297.

14   See B.W. Sigg, *Le Silence et la honte: Névroses de la guerre d'Algérie* (Paris: Messidor, 1989); A. Orr, *Ceux d'Algérie: Le silence et la honte* (Paris: Payot, 1990).

15   Records and transcriptions of the interviews are stored at the Bibliothèque de Documentation Internationale Contemporaine, in Nanterre, France.

encounter with each veteran I used a chronological life story approach, and then went on to focus in subsequent meetings on specific events or relevant themes.

Combat memories are not monolithic or homogeneous: they demand that we acknowledge a plurality of perspectives. The memories depend on the zone and the period of service in the long, yet fast-moving, duration of the war. Thus, because *homo bellicus algerianus* exists, but in 'fragmented pieces' (*en pièces détachées*), I chose to analyse and compare the memories of men whose experience shared a spatial and temporal unity.[16] I therefore looked for veterans who had fought in Wilaya I: the Aurès-Nememcha area, corresponding, in the terminology of French military bureaucracy, to the 'Zone du Sud Constantinois', identified by the south-eastern towns of Batna, Tébessa, Négrine, and Biskra. As epitomized by the military daily *Aurès-Nemencha*, published for the area troops, the motto of the French army was 'Pacify, Protect, Pursue' (*Pacifier, Protéger, Poursuivre*).[17]

Besides the spatial unity, I wanted the interviewees to be not career but drafted servicemen. This was because I wanted to avoid the *esprit de corps* that characterizes enlisted soldiers and officers. Thus, I hoped to avoid simplistic stances of accusation and defence, and sought instead to highlight the depth of social impact of a war which, through the resort to the draft, concerned a whole generation.[18] For instance, compared to America's war in Vietnam, the Algerian War of Independence – which lasted the same length of time and demanded a similar level of troops in proportion to population – hit French families, measured by deaths of servicemen,

---

16 For the quotation, see J.-C. Jauffret, *Soldats en Algérie 1954–1962: Expériences Contrastées des Hommes du Contingent* (Paris: Autrement, 2000), 9.
17 See *Aurès Nemencha*, 5 (juin 1956), in SHD 1H 4600/1 (Annexes).
18 Between 1954 and 1962, 2.5 million Frenchmen crossed the Mediterranean to combat Algerian independence fighters. Among them, the *appelés*, the draftees, drawn from all over France and from all social classes, numbered 1,179,523 – according to calculations by Jean-Charles Jauffret, 'Une armée à deux vitesse en Algérie (1954–1962): Réserves générales et troupes de secteur', in J.-C. Jauffret and M. Vaïsse (eds), *Militaires et Guérilla dans la Guerre d'Algérie* (Brussels: Complexe, 2001), 21–38.

25 per cent harder than American families (24,000 French soldiers died in Algeria; and at least 250,000 Algerians).[19]

The majority of those interviewed were born between 1938 and 1940.[20] In 1958–60, when they joined the 'pacification' in Algeria as draftees, the war had been going on for at least four years. In May 1958, de Gaulle returned to power in order to seek an end to the conflict, which was tearing the country apart; yet it was to last four more years. Between 1959 and 1960, the Challe and Constantine Plans hit the Algerian independence fighters hard on both the military and psychological fronts, combining war actions with a strong propaganda campaign devoted to gaining the support of the population. Moreover, the 'Battle of Algiers', militarily won by the paratroops in 1957, had stirred up a harsh debate about the methods used by the French army, and had shocked metropolitan and international public opinion.[21] Even if not declared, and therefore 'without a name', the war was a major issue for French society – to such an extent that it triggered a regime change.

Given this historical framing, what at first is striking is that most of the interviewees, fifty years after the war, still remember themselves as young and careless, unaware of the issues at stake. They depict themselves as 'kids' – as is stressed by one interviewee, Yves, while showing me his

---

19 58,000 American soldiers died in Vietnam. See M. Alexander, M. Evans, and J. Keiger, 'The "War Without a Name": The French Army and the Algerians: Recovering Experiences, Images and Testimonies', in Alexander et al. (eds), *The Algerian War and the French Army, 1954–62: Experiences, Images, Testimonies* (Basingstoke and New York: Palgrave Macmillan, 2002), 32. See also D.L. Schalk, *War and the Ivory Tower: Algeria and Vietnam* (Lincoln and London: University of Nebraska Press, 2005), 29. For figures of deaths, see G. Pervillé, *La guerre d'Algérie: combien de morts?*, in M. Harbi and B. Stora (eds), *La Guerre d'Algérie. 1954–2004, la fin de l'amnésie* (Paris: Robert Laffont, 2004), 483.

20 A few, coming from upper classes, are older (as is Michel, discussed below, born in 1931). As university students they had the opportunity to leave later (they were *sursitaires*): therefore the majority of those interviewed were in Algeria in the second half of the war, after 1958.

21 See R. Branche, *La Torture et l'armée pendant la guerre d'Algérie, 1954–1962* (Paris: Gallimard, 2001), 147–75.

military record: '*Ça c'est mon livret militaire ... vous voyez que j'étais un petit garçon, hein? La première réflexion qu'un des sergents m'a fait c'est: 'Tiens, on les prend dans les maternités maintenant?'* ('That is my military record ... you can see that I was a little boy, eh? The first remark one of the sergeants made to me was: "What, we're taking them out of the maternity wards now?"').[22] In the interviews, after having explained the reasons for the research, I try to collect a story which is as chronologically coherent as possible by asking the following question: 'Please talk about yourself at the moment of departure, when you receive the call-up card'. Most often veterans answer like Jean-Claude, drafted in April 1958:

> JEAN-CLAUDE: Moi, je suis jeune, déjà ... puisque ... j'ai été incorporé dans l'armée à vingt ans.
> Je suis né en trente-neuf et donc je pars au service militaire j'ai vingt ans, j'ai tout juste vingt ans ... vingt ans et un mois. Vous voyez, c'est ... Je suis politiquement mpfff... immature, quoi. A vingt ans, bon, je me préoccupe plus de mes copains, un petit peu des mes copines, un petit peu de savoir si mon scooter va bien marcher ... Et j'ai déjà commencé à travailler. Je travaille depuis deux ans. Je travaille depuis deux ans. Je suis ... J'ai arrêté mes études en seconde, donc j'ai pas le bac. [...] là à dix-huit ans, j'étais un peu désœuvré ... donc je rentre dans le monde du travail. [...] Et alors je travaille, je travaille sans me soucier trop de quoi que ce soit.[23]
>
> [Me, I was young, already ... then ... I was drafted into the army aged twenty.
> I was born in thirty-nine and so when I leave for military service I'm twenty, I'm only just twenty years old, twenty years and one month. You see, it's ... I'm politically mpfff ... immature, you know? At twenty years, well, I'm more preoccupied with my friends, a little bit with my girlfriends, a little bit about whether my scooter is going to work okay ... And I had already started to work. I've been working for two years.

---

22 Yves (1939, enlisted in 1958), worker, 457th groupe d'artillerie antiaérien légère, sergeant. Interviewed at Villers-Saint-Sépulcre on 10 June 2009 and 18 January 2010. Yves enlisted (entering a military engineering school) to avoid the war and acquire a trade. He did acquire a trade, but, as for heterogenesis of intents, had to go to Algeria. I decided to keep him in the sample because he is closer to the conscripts than to the army *esprit de corps*.

23 Jean-Claude (1939, 59–1/B), employee, 457th groupe d'artillerie antiaérien légère, second lieutenant. Interviewed in Bretignolles on 7 July and in Nantes on 26 November 2008.

> I've been working for two years. I'm ... I quit my studies in second grade, so I don't have the bac. [...] from there until I was eighteen, I was a bit idle ... so I entered into the world of work. [...] And so I'm working, I'm working without bothering too much about the things that I should.]

If it is given sufficient room, the characterization by the narrator of himself as 'young', namely carefree and politically inexpert, reveals a broad arc of meaning to be investigated. With the distance of age (today's narrator is no longer the young man of those days past), isn't there also the trace of a judgement, which is at the same time a note of condemnation (I shall not do again what I did) and a suggestion of exculpation (but I was young...)?

Another interviewee, Michel, was not so very young when he left for Algeria. Belonging to the small group who then attended university, he could put off the call-up until he was twenty-eight. A reader of the left-wing Catholic journal *Esprit*, he was studying Latin and Greek literature and became fond of contemporary Algerian writers: 'Je lisais pas mal des romans des algériens ... Mouloud Feraoun, Mouloud Mammeri, Kateb Yacine, Driss Chraïbi ... et j'étais très intéressé par cette question et d'un point de vue intellectuel ... Ça c'est dans les années cinquante-quatre, cinquante-cinq, cinquante-six que je me suis intéressé à ces livres-là ... Avant de partir en Algérie' ('I was reading quite a few Algerian novels ... Mouloud Feraoun, Mouloud Mammeri, Kateb Yacine, Driss Chraïbi ... and I was very concerned with this question and from an intellectual point of view ... It was in the years fifty-four, fifty-five, fifty-six that I was interested in those books ... before leaving for Algeria').

> AB: Vous connaissiez donc les livres comme *Des rappelés témoignent*, les manifestations de rappelés?
> MICHEL: Pratiquement pas. Très peu. Sauf par la presse un peu ... Je suis pas un militant, je suis un, je suis un ... je suis un rigolo, je suis un ... comment on dit en italien? ... un *amateur* ... J'ai jamais été très sérieux.[24]

---

24 Michel (1931, 58–2/A), high-school teacher, 26th régiment d'infanterie, second lieutenant. Interviewed in Paris on 20 May and 14 June 2010.

[AB: So you knew about books like *Des rappelés témoignent* [The testimony of the returned soldiers], the demonstrations by the returnees?
MICHEL: Practically not. Very little. Except from the press a little ... I'm not an activist, I'm a, I'm a ... I'm a joker, I'm a ... how do you say it in Italian? ... an *amateur* ... I was never very committed.]

The theme of youth, running like a bass-line through the memories of the majority of the interviewees, and rooted in their (actual, biological) youth, recurs in the retrospective self-representation of Michel as an easy-going, existential youth, of a joker ('je suis un rigolo ... un amateur ... j'ai jamais été très sérieux ...'). As an aside, it is also worth noting the sly wink, the *clin d'œil*, with which the interviewee then looks to the interviewer for empathy, referring to the proverbial Italian character of levity and nonchalance expressed by Michel with the French expression *amateur* (unpaid). Being tall and lanky, one can easily imagine him in Monsieur Hulot's shoes. But for both, for the younger interviewees as well as for Michel, the memory-work seems to be the same. Almost all of them are looking for shelter in the image of the childish existential condition – which Roland Barthes identified as the epitome of the lower-middle-class culture which became dominant as mass culture – and which answers, if not precisely to a search for forbearance, to the difficulty of expressing with modern words a transparent judgement about one's behaviour in the past or taking up one's responsibilities.[25] The 'Djebel generation', lacking accomplishments through which to recognize itself, crushed by the giants of 1914–18 and 1939–45 on the one hand, and by the disrespectful dissenters of May 1968 on the other, seems to find in the fragile myth of lost youth the only feature that allows them to think of themselves as a collectivity. Youth, then, seems to function negatively, as a shield that prevents one from looking at the ugly side of the war.

Nonetheless, it is worth noting that the informants' insistence on that 'juvenile' existential dimension, related to themselves as they were and expressed at the threshold of their narrations, does not mean that they want to give me a sanitized version of the war. On the contrary, it

25   See R. Barthes, *Mythologies* (Paris: Seuil, 1957), 143.

seems that it is precisely that introduction which releases their voice and makes possible – after having defused the ethical and political question of responsibility – accounts of war violence, torture, and others' atrocities. Moreover, this lack of self-consciousness – whether or not it reflects the intellectual horizon of the informants at the time of the events (and this is an inescapable zone of indeterminacy given the absence of adequate sources) – is, on the contrary, a faithful reflection of the time in which the interview is taking place, and in which the veterans are verbalizing their memories. And if the present time characterizes itself as 'the victim's time' – rather than 'the witness's one' – it is easy to see how the interviewees, by laying emphasis on their past youth, are trying to build an image of themselves by an analogy with the moral figure that has emerged as so successful at the turn of the century: namely the victim, the sacrificial lamb that has no knowledge or guilt.[26] In this context it is clear how the voice of the former fighters is released (or retrieved) only after the borders between victims and perpetrators become uncertain, to the point that agents appointed by the state to exercise violence can passively describe themselves as innocent casualties of the events.[27] Alongside those social frames of memory, which draw on tendencies that seem common on a European if not a worldwide scale, the 'juvenilistic' theme also echoes cultural frames more properly linked to the recent French political mood.

As a matter of fact, one of the issues which characterized Nicolas Sarkozy's 2007 presidential campaign was the rejection of 'repentance', an issue that was conjured up in almost all his speeches during the campaign – that is, at least twenty-seven times between 1 March and 3 May.[28] 'A tous

26 See H. Rousso, 'Vers une mondialisation de la mémoire', *Vingtième Siècle. Revue d'histoire*, 23/94 (2007), 5–6. Rousso is referring to the classic study by A. Wieviorka, *L'ère du témoin* (Paris: Seuil, 1998). See also J.-M. Chaumont, *La Concurrence de victimes: Génocide, identité, reconnaissance* (Paris: La Découverte, 2002).

27 On this topic, see the issue of *Memory Studies* edited by J. Dunnage, 'Perpetrator Memory and Memories about Perpetrators', *Memory Studies*, 3/2 (2010).

28 As a sample one can quote: 'Je déteste cette mode de la repentance qui exprime la détestation de la France et de son histoire' (Lyon, 5 April 2007); 'Nous devons aujourd'hui construire ensemble l'avenir, sans repentance, sans réécrire notre histoire

les peuples de la Méditerranée qui passent leur temps à ressasser le passé et les vieilles haines de jadis', Sarkozy declares in Montpellier, on the last day of the campaign, 'je veux dire ce soir que le temps est venu de regarder vers l'avenir' ('To all the people of the Mediterranean who spend their time dwelling on the past and on former hatreds, I want to say this evening that the time has come to look to the future'). Thus, behind the appearances of a 'futurist' attitude, this rupture, this clean break for which Sarkozy calls, functioned as a blueprint for selective reinterpretation of national history. This reinterpretation was aimed at avoiding the link between the splendour of the colonial empire which is receding into an imaginary past, and the complex social reality of contemporary France which is becoming run through by a colour line, a legacy of that colonialism.[29] His heart of darkness, the Algerian War of Independence, stands, by synecdoche, for the colonial relationship in its entirety.

* * *

I began to quote Elie's written memoirs to my other interviewees, because his case is representative of that of many other veterans at the beginning of the new century. Indeed, a major difference with the 1990s is that veterans, when asked to tell of their war experiences within the framework of academic research, actually speak. That very fact leads us to consider the word

---

avec l'Algérie' (16 April 2007, letter to Denis Fadda); 'Ici on n'aime pas la repentance qui est un dénigrement systématique de la France et de son histoire. Je déteste cette repentance qui est une forme de la détestation de soi parce que l'on n'a qu'un pays. Le détester c'est se détester soi-même. Je déteste cette repentance qui est une falsification de l'histoire de France. Car la France n'a pas à avoir honte de son histoire' (Metz, 17 April 2007). The French Heads of State's speeches have been collected, since 1974, on the website 'documentation française' (www.vie-publique.fr/discours/). See S. Lefranc, 'Repentance', in L. De Cock, F. Madeline, N. Offenstadt, and S. Wahnich (eds), *Comment Nicolas Sarkozy écrit l'histoire de France* (Marseille: Agone, 2008), 156.

29 See C. Coquery-Vidrovitch, 'Un faux concept: la repentance', in *Enjeux politiques de l'histoire coloniale* (Marseille: Agone, 2009), 139–41; V. Martigny, 'Le débat autour de l'identité nationale dans la campagne présidentielle 2007: quelle rupture?', *French Politics, Culture and Society*, 27/1 (2009), 23–42.

as an historical event, meaning – drawing from Arlette Farges's remarks – that we are trying to show that, starting from it, we can perform a *récit d'histoire*: if 'hearing a testimony is comparable to the reading of archives, one is supposed to manage an effort to grasp the historical meaning and insert it into the community from which it is coming'.[30] The reason why the veterans are now speaking can be explained by changes in the modalities, both individual and collective, which govern the organization of the memory.

The key moment for this shift was the Papon double trial in 1997–9, which marked 'the transition from an obsessive concern with the "Vichy syndrome" to what might be termed an "Algerian syndrome"'.[31] That shift is already well known to historians: in a very short period of time, between 1999 and 2003, one saw – among other events – state recognition that in Algeria there had been a war (1999), the speech by the Algerian president to the Assemblée nationale (2000), a large public debate on the conflict and in particular on the use of torture (2000–1), the installation in Paris of a commemorative plaque to honour the Algerians killed on 17 October 1961 (2001), of a monument to the memory of the soldiers who died in North Africa (2002), and also the visit of the President of the French Republic to Algeria (2003).[32] The opening of the military archives in 1992 allowed historical research to arrive at that rendezvous on time: thus, a decade later, the publication of research by a new generation of historians came to join with these symbolic acts, and thereby encountered a powerful social claim to a 'specific knowledge', very far from the kind of knowledge proper to the

---

30   A. Farge, *Quel bruit ferons-nous?* (Paris: Les Prairies ordinaires, 2005), 93, 170. See also A. Portelli, 'L'inter-vista nella storia orale', in M. Pistacchi (ed.), *Vive voci. L'intervista come fonte di documentazione* (Rome: Donzelli, 2010), 8.
31   N. MacMaster, 'The Torture Controversy (1998–2002): Towards a "New History" of the Algerian War?', *Modern & Contemporary France*, 10/4 (2002), 450.
32   For more details, see: the issue of *Modern & Contemporary France* edited by A. Hargreaves, 'France and Algeria, 1962–2002: Turning the Page?', *Modern & Contemporary France*, 10/4 (2002); A. Brazzoduro, 'La guerra d'Algeria nel discorso pubblico francese. Quaranta anni dopo', *Mondo Contemporaneo. Rivista di storia*, 4/1 (2008), 67–93.

'universal intellectual' of previous times – the major exemplar of which was Jean-Paul Sartre, signing, in 1961, the vitriolic foreword to Frantz Fanon's *Les Damnés de la terre*.[33]

In this frame, for some veterans the interview takes on the features of a cathartic experience. To Yves, a blue-collar chemical worker in the capital's hinterland, it is a last chance to get it off his chest (*'Et enfin j'étais content de vider mon sac...'*):[34] and as a matter of fact, he speaks to me in a tone of voice which sometimes is barely audible, as if with a secret to be protected from indiscreet ears.

Amand, a carpenter in Le Havre shipyards, feels the moment of dizziness always associated with the start of an interview and encourages me not to have reticence of any sort, warmly expressing his trust:

> AB: Alors ... moi je voudrais ... euh ... je sais pas on peut ...
> AMAND: Vas-y, vas-y! Mène ... mène l'affaire!
>
> [AB: Well ... I would like ... uh ... I don't know if we can ...
> AMAND: Go for it! Bring ... bring it on!][35]

Even if it is hidden behind the façade of irresponsible youth, the question of why, of the meaning of the war and of the motivations of the parties involved, lies in the background of almost all these narrations. Whether the issue is explicitly tackled or not, the strategic and cumbersome presence of the microphone constructs a situation in which the dialogical

---

33   For the opposition between 'specific' and 'universal knowledge', see Michel Foucault, 'Intervista a Michel Foucault', in A. Fontana and P. Pasquino (eds), *Microfisica del Potere. Interventi Politici* (Turin: Einaudi, 1977), 3–28. I take this point from H. Rousso, 'Intellectuals and the Law', *Modern & Contemporary France*, 17/2 (2009), 153–61. On Sartre, see P. Arthur, *Unfinished Projects: Decolonization and the Philosophy of Jean-Paul Sartre* (London and New York: Verso, 2010).

34   Yves (1939, enlisted in 1958), worker, 457th groupe d'artillerie antiaérien légère, sergeant. Interviewed in Villers-Saint-Sépulcre on 10 June 2009 and on 18 January 2010.

35   Amand (1938, 58–1/C), carpenter, 18th bataillon de transmissions, then 'disciplinary regiment' of the Bérets noirs, private. Interviewed in Harfleur on 27 June 2008 and on 14 February 2009.

relationship between interviewer and narrator is taking place in the present, starting from the past but striving towards a future for the sake of which the conversation is consciously being recorded. In this connection it is worth noting that none of the informants refused to be recorded – on the contrary almost everyone considered this to be the primary reason for our meeting. The consent to the research project was given at the outset: 'si vous êtes arrivé jusque-là', Jean Pierre tells me, 'c'est bien que je suis d'accord, hein?'[36] Equally, nobody asked to protect himself by remaining anonymous. As recently as 1991, Anne Roche – author of the first studies of oral history on *pieds-noirs* and the draft contingents – had highlighted how the collection of soldiers' testimonies had been problematic: those who refused to speak were still 'many', while 'those who agreed [to speak] did so, in over a third of cases, anonymously'.[37]

Thus, once the microphone is switched on, the meeting always starts with a formal question about the interviewee's authorization for the recording. However, still earlier there is a scene, which constantly recurs, in which the sharply defined and reassuring borders dividing 'interviewee' from 'interviewer' mingle dangerously. The agreement which authorizes the interview – which is also an exercise of power – is called into question precisely by the fact that, in the preliminary stages, that agreement is suspended, thus establishing its artificiality and, hence, revocability. Such a scene can be rather puzzling for the neophyte interviewer. But veterans want to know who is talking to them, what he is looking for and why: the interviewer chooses whom to interview, and in the same way the interviewee chooses the interviewer.[38]

---

36 Jean-Pierre (1939, 59–1/A), florist, 457th groupe d'artillerie antiaérien légère, sergeant. Interviewed in La Chapelle Saint Sépulcre on 20 January 2010.
37 A. Roche, '"Je vous le raconte volontiers, parce qu'on ne me l'a jamais demandé". Autobiographies d'appelés en Algérie', in L. Gervereau, J.-P. Rioux, and B. Stora (eds), *La France en guerre d'Algérie* (Paris: BDIC, 1992), 264.
38 For a similar remark, see M. Pollak, *L'Expérience concentrationnaire. Essai sur le maintien de l'identité sociale* (Paris: Métailié, 2000 [1990]), 77; and A. Portelli, 'History-Telling and Time: An Example from Kentucky', *The Oral History Review*, 20/1–2 (1992), 51–66.

Thus, the generational gap (I am the same age as their children, if not their grandchildren), and, on the whole, the cultural one (which expresses itself as a feeling of my non-involvement with Franco-French narratives), generate in most cases an unprecedented no-man's-land in which the possibilities for speech seem to be different than is customary. There is also a 'cognitive break' in play here, represented not only by the interviewer's absence of direct experience of war, but also by an (illusory?) departure from the horizon of possibility in the Western world of the very notion of a war fought through conscription.[39] Moreover, the veterans perceive that, through myself, the university itself is addressing them; and thus the veterans see this as an opportunity to reintegrate themselves into History, from which they feel they have been excluded.

Except for a very few, all my informants were offering their testimony for the first time – indeed, it was noticeable that their wives often stayed with us, sitting in a corner, or barely hidden in the kitchen with the door wide open: they themselves had never listened to these stories. The veterans accepted that presence, and indeed sometimes urged their wives to stay: their memory, this exercise of oblivion, works nowadays with them. Women can have an inhibitory effect or, on the contrary, can press their husband to tell more: they can urge the telling of stories of violence, but they tacitly stop every word concerning sexuality and sexual violence. Moreover, the wives constitute the privileged connective tissue between the veterans' private memory-work and the public space wherein the antagonistic Algerian memories that traverse French society fight for hegemony. Often younger than their husbands, it is they who move more deftly through the new media. For instance, it is interesting to note on what subjects, and in what manner, Elisabeth – out of sight in the kitchen, but listening – interrupts her husband's account, as he relates to me his painful feelings about the burden of war memory in his present life.

---

39 As stressed by S. Audoin-Rouzeau and A. Becker, 'La guerre, qui a marqué de sa présence récurrente, jusqu'à la fin du conflit algérien, les générations successives du XXᵉ siècle, est désormais en position de complète extériorité', *14–18, Retrouver la Guerre* (Paris: Gallimard, 2000), 18.

AB: Est-ce que vous y pensez, là, quarante ans après?
YVES: Ah oui oui ... Ça m'arrive très souvent. Mais en premier, les gens que j'ai connu ... les gens aussi qui sont morts ... et je trouve que ça c'est ce qu'il me reste un peu à travers de la gorge ... ils sont morts ... je veux pas dire pour rien ... mais pratiquement si, hein? [...] Puis surtout, il faut quand même le dire, les pieds-noirs ... Je suis pas pied-noir, mais je me mets à leur place ... d'un seul coup vous partez avec ... comme on dit vos deux pantoufles, un pantalon et ... Ça, ça a été dramatique quand même, hein? [...] C'est un peu ça le sentiment que j'ai ...[40]
ELISABETH: T'oublies quelque chose ... T'oublies quelqu'un ... Et les harkis?

[AB: Do you think of it, now, forty years afterwards?
YVES: Oh yes, yes ... it often comes to mind. But mostly the people that I knew ... the people too who are dead ... and I find that that is what somehow sticks in my throat ... they are dead ... I don't want to say for nothing ... but practically so, eh? [...] Then of course one has to mention the *pieds-noirs* ... I'm not a *pied-noir*, but I put myself in their place ... in one fell swoop you leave with ... as we say with your slippers, your trousers and ... That, that was dramatic all the same, eh? [...] That's kind of the feeling that I have ...
ELISABETH: You're forgetting something ... You're forgetting someone ... And the *harkis*?]

Elisabeth was not only listening but also carefully watching over her husband's narration. Yves remembers with deep compassion his comrades in combat, particularly the fallen ones, and then the French settlers, the *pieds-noirs*, in a hierarchical order ranging from the inner to the outermost circle, manifesting a perspective on the events which is typical of those who lived a military life on the ground. His memory is definitively a Franco-French narrative, wherein the Others, the 'French Muslims' (not yet 'Algerians'), are invisible. Meanwhile, drawing attention to the *harkis* – Algerians who fought for the French as auxiliaries – Elisabeth is clearly verbalizing the common sense which had recently become widespread in French society and which, starting from the 1990s, accommodated the *harkis* – paternalistically identified as 'victims' of an anonymous historical force – into the *roman*

---

40 Yves (1939, enlisted in 1958), worker, 457th groupe d'artillerie antiaérien légère, sergeant. Interviewed in Villers-Saint-Sépulcre on 10 June 2009 and on 18 January 2010.

national.[41] But the key moment of this shift was the presidency of Jacques Chirac (1995–2007), that lead to the institution, on 25 September 2001, of a day of national homage to the *harkis* (Journée d'hommage nationale aux harkis). The former president – who fought as a conscript in Algeria in the ranks of the 6th Régiment de chasseurs d'Afrique – was always very sensitive to the fate of the *pieds-noirs* and the *harkis*.[42]

After Elisabeth's interruption, Yves goes on:

> YVES: Ah oui! Ça, ça a toujours était ... c'est la honte de l'armée française ... La honte de l'armée française. Ces gens se battaient ... vous pouvez pas savoir ... Ils croyaient en nous et tout ... Et quand je pense qu'on les désarmait avec les 'fells' qui attendaient sur la colline ... pour les passer au ... au barbecue, pour les ébouillanter, pour ... [...] Parce que vous savez que les harkis ... ça a été chaud, hein? ... on a fait les pires saloperies. Je peux pas vous raconter ... [...] C'est ... ils savent très bien qu'ils ont fait des cochonneries, des saloperies ... C'était pas la peine de faire tout ça. Ils ont massacré ces gens là ... femmes, enfants, tout le bazar y est passé, hein?

> [YVES: Ah, yes! That, that has always been ... it's the shame of the French army ... the shame of the French army. Those people fought ... you cannot know ... they believed in us entirely ... And when I think that they were disarmed with the 'fells' waiting in the hills ... to carry them off to ... to the barbeque, to boil them, for ... [...] Because you know that the *harkis* ... that was hot, eh? ... there were some dirty tricks. I can't tell you ... [...] It's ... they knew very well that they had done some beastly, dirty tricks ... There was no point in doing all that. They massacred those men there ... women, children, the whole market went that way, eh?]

41 The first law which aimed at providing compensation for the *harkis* was the Romani law, passed on 11 June 1994 (Loi n. 94-488, 'Loi relative aux rapatriés anciens membres des formations supplétives et assimilés ou victimes de la captivité en Algérie'). For a helpful genealogical insight into the public emergence of the *harkis'* memories, see C. Eldrige, '"We've Never Had a Voice": Memory Construction and the Children of the Harkis (1962–1991)', *French History*, 23/1 (2009), 88–107.

42 'Les anciens des forces supplétives, les harkis et leurs familles, ont été les victimes d'une terrible tragédie. Les massacres commis en 1962, frappant les militaires comme les civils, les femmes comme les enfants, laisseront pour toujours l'empreinte irréparable de la barbarie. Ils doivent être reconnus. ... Les harkis ne sauraient demeurer les oubliés d'une histoire enfouie. Ils doivent désormais prendre toute leur place dans notre mémoire'. J. Chirac, *Mon Combat pour la France: Textes et interventions* (Paris: Odile Jacob, 2007), 46.

But Yves was in Algeria until October 1961. Therefore he had not seen *harkis* disarmed by the French army nor 'cooked' by FLN fighters 'waiting on the hill' (something that could have happened only in the aftermath of summer 1962). On this point, although his memory is completely mediated it is nonetheless interesting if we see it as an echo of widespread common sense. Thus, it is worth noting that Yves's first statement is that the destiny of *harkis* was 'the shame of the French army': even if – as Yves will later tell me – contemptuous statements like 't'es pire qu'un harki' ('you are worse than a *harki*') were common in the army. Moreover, it is revealing that once the initial statements have been made, as if by way of an overture, the *harkis* become an excuse to talk of 'them' (*ils*) and their '*saloperies*'.[43] But who exactly are these mysterious 'fells', 'waiting on the hill' to cook *harkis*, and of course slaughter women and children?

* * *

Algerians rarely appear in the veterans' pictures. All interviewees keep pictures of the war, taken by themselves or comrades, stored, in order, in an album or a box.[44] What effect does looking at these photos have on them today? 'C'est difficile à exprimer', Claude says to me, 'car d'une part c'est un retour dans une autre existence [...], mais c'est aussi un retour sur soi-même [...]. Enfin, malgré tout c'est aussi le souvenir d'une aventure exotique, un peu comme je le dis pour expliquer ce ressenti, comme si j'avais été acteur dans un film comportant de multiples figurants'.[45] ('It is difficult to express, because in part it is a return to a different existence [...], but it is also a return on itself [...] In the end, despite everything, it is also a memory of an

---

43 On the political use of the *harkis*' memory by the powerful *pied-noir* community, see C. Eldrige, 'Blurring the Boundaries between Perpetrators and Victims: Pied-noir Memories and the Harki Community', *Memory Studies*, 3/2 (2010), 123–36.

44 On veterans' photographs, see C. Mauss-Copeaux, *À travers le viseur. Images d'appelés en Algérie 1955–1962* (Lyon: Ædelsa 2003); and M. Chominot, 'Guerre des images, guerre sans image? Pratiques et usages de la photographie pendant la guerre d'indépendance algérienne 1954–1962', unpublished PhD thesis (Université de Paris 8, 2008).

45 Claude (1939, 59–2/B), employee, 457th groupe d'artillerie antiaérienne légère, private. Letter to the author, 4 May 2010.

exotic adventure, a bit like I say to explain this feeling, as if I had been an actor in a film composed of many characters.') The main subjects in these pictures – the 'extras' in the film – can be roughly divided in three groups: the soldier and his comrades; the landscape and military equipment; and, more rarely, pictures portraying Algerians. Even more difficult to find are the pictures of war actions, where the three groups mingle and sometimes even independence fighters appear, as captives.

The most numerous group is the first, who impress because of the juvenile features of the soldiers and their *bon enfant* mood – as if they were on a school tour, and not in the least warlike. Those pictures were taken close-up and the relationship between the photographer-soldier and his subjects – comrades whose pictures he is taking during pauses in military life – is of complete agreement if not complicity. As for landscapes and military equipment, on the contrary, the field expands to the point of boundlessness. Halfway between a close portrait and the infinite landscape, the rare French Muslims occupy an indefinite area in the pictures of French soldiers, neither close nor far, revealing the ambiguous reality of these citizens *entièrement à part*. The distance between the two groups is also a matter of complying with security rules. Michel tells me: 'Quand on allait voir la population d'abord on disait, je me rappelle, "Matrafouch! Matrafouch!" (c'est-à-dire: "n'ait pas peur!") ça signifie quelque chose, non? Vous arrivez dans un … moi quand je me balade en France dans un village je dis pas aux gens "n'ait pas peur!"'[46] ('When we would go and see the population we would first say, I recall, "Matrafouch! Matrafouch!" (which means "don't be frightened") that means something, no? You arrive in a … me when I go for a stroll in a village in France I don't say to people "don't be frightened!"') As a matter of fact the relationship between the soldiers and the Algerian population is one of war, no matter how deeply this is hidden beneath the litotes of 'pacification'. As a consequence, the few pictures taken close-up give the impression of being stolen or extorted: the Algerians portrayed look to the lens with surprise and diffidence; others

---

46 Michel (1931, 58–2/A), high-school teacher, 26th régiment d'infanterie, second lieutenant. Interviewed in Paris on 20 May and 14 June 2010.

pretend an unconcern under which proud opposition seems barely hidden. One can easily imagine the situations in which these pictures came to be taken: most frequently just after a patrol of heavily armed (and overexcited or fearful) soldiers has intruded into the everyday life of a civil population fixed on their own activities.

One of Yves's pictures portrays two women under a tent, likely a mother and her daughter. The youngest is staring at the ground while the woman on her left, spinning wool, is vexedly looking straight ahead, avoiding the lens.[47] Yves has written on the back: 'une zouze [sic] qui file la laine. La fille pudique baisse la tête car on la regarde. Chéria 1961' [An old woman spinning wool. The modest daughter lowers her head because we are looking at her]. In this manner the soldiers read an unknown reality, which they do not understand, through stereotypes drawn from colonial times. The woman spinning wool is the perfect image of a pre-modern past looked at with a condescending air as folklore: the woman is an *azouze* condemned by age (biological, but also historical). The young one, sitting beside her and avoiding the soldiers' look, is on the contrary the object of a unreciprocated desire. The appeal to the virginal topos – the girl is 'modest' (*pudique*) – reveals the sexualizing colonial imagery deployed by soldiers who are excited by the (lack of) interaction, and thereby conveniently misunderstand the political implication of the girl's refusal to look at their lenses and guns.

Rather than focus on violence – a key issue within the colonial relationship that changes political and military conflict into a 'clash of civilizations' ('primitive violence' being an essentialist character of 'savages') – I will end by focusing on two elements which recur in the narrations of the veterans concerning the Algerians: the backwardness of Algerian society and the image of their women.[48]

---

47 Similar acts of resistance faced with the photographers' violence are described by C. Mauss-Copeaux, *A travers le viseur. Images d'appelés en Algérie 1955–1962*, 68–9.
48 On violence in this specific colonial war, see R. Branche, *L'Embuscade de Palestro: Algérie 1956* (Paris: Armand Colin, 2010), in particular ch. 3, 'L'Algérie sauvage' (77–104).

The soldiers were indeed struck by the poverty and backwardness of the *indigène* – and the accompanying dirtiness. The soldiers, who watch the Algerians – and photograph them – while always keeping their distance and often without getting down from their jeeps or trucks, all retain a similar (indeed, stereotyped) image: that of a people mired in the past, outside of history. Veterans often refer to them as dwelling in the 'Middle Ages' or the 'Stone Age'. The description is sometimes nuanced, even to the extent that it becomes associated with a certain form of deference: for Michel, the inhabitants of the region were 'dans un mot, nobles' ('in a word, noble'); mais 'd'une noblesse médiévale, cornélienne. Des misérables: loques, sales, déchirés' ('but of a Medieval nobility, Cornelian. Wretched: their rags dirty and ripped'). But this theme of misery, perceived in the first instance as a lack of hygiene – and thus read as an imperviousness to the course of progress of civilization of manners – contributes powerfully to the construction of a mental image characterized by radical alterity. However, having been constructed through sedimentation of successive layers over the last fifty years, this image no longer belongs just to this single period of war, but also to the postcolonial presents of France and Algeria. It is clear, incidentally, that France in the 1950s – and most notably the rural areas from where the vast majority of the conscripts came – had only just begun to impose on itself the disciplines of hygiene, which today have triumphed to the point of obsession.

> AB: Est-ce que c'était une guerre coloniale, selon vous?
> JEAN-CLAUDE: Moi je crois qu'on avait tous, on avait tous une attitude quand même un peu colonialiste. Par exemple, l'exemple type, quand sur une piste on croisait un vieux, un gars qui été monté sur son, sur son âne et que la femme marchait derrière avec des gosses, on disait 'non, toi tu descends', c'était à la femme de monter. Je crois que tous on avait cette idée de leur apporter l'hygiène, je crois qu'on avait tous cette idée là, de leur apporter la civilisation. Donc, je sais pas ce que t'appelles une attitude colonialiste, mais en tout cas une attitude de dire c'est nous qui avons la civilisation, et vous, regardez comment vous vivez dans la crasse, dans le déni de femmes, dans ...[49]

---

49 Jean-Claude (1939, 59–1/B), employee, 457th groupe d'artillerie antiaérien légère, second lieutenant. Interviewed in Bretignolles on 7 July and in Nantes on 26 November 2008.

[AB: Was it a colonial war, in your opinion?
JEAN-CLAUDE: Me I think that all of us had, we all had an attitude that was a bit colonialist. For example, a classic example, when you're on a track and you come across an old man, some guy who is riding his, his donkey and the woman is walking behind with her kids, you say 'no, you get down', it's the woman's place to ride. I think we all had this idea that we could bring them hygiene, I think we all had that kind of idea, to bring them civilization. So, I don't know what you call a colonialist attitude, but in any case an attitude of saying that it's us who have civilization, and you, look how you live in filth, in the denial of your women, in ...]

In this manner, gender – the 'women' topic, of 'their women', whom one should defend, protect, rescue[50] – becomes a strategically central theme, shaping the memories that swing between past and present and are epitomized, one might say, in Gayatri Spivak's ironic definition of colonialism: 'white men saving brown women from brown men'.[51] It is in this sense that the centrality of women in the veterans' discourse acquires a particular heuristic value: during the war, the frontier between 'us' and the 'other' was defined with respect to the role of women, and in a certain fashion this process of framing continues today.

'Les femmes', says Marc, 'pour eux c'est rien du tout, hein? ... c'est prrrr ... d'ailleurs ils les cachent, ils les voilent, ils les machinent ... heureusement qu'elles ne sont pas très belles ... c'est un avantage, comme ça on les voit pas! [laughs] Non non, je regrette rien, je regrette pas cette période de guerre d'Algérie. Au contraire je suis bien content qu'on s'en soit débarrassé ...' ('The women, for them it's nothing at all, eh? ... it's prrrr ... anyway they hide them, they veil them, they control them ... luckily they aren't very pretty ... it's an advantage, that's how we don't see them! No, no, I regret nothing, I don't regret that time of war in Algeria. On the contrary I was

50 See R. Branche, 'Des viols pendant la guerre d'Algérie', Vingtième Siècle. Revue d'histoire, 18/75 (2002), 124–5.
51 See his groundbreaking study on the British abolition of suttee in nineteenth-century India: G.C. Spivak, 'Can the Subaltern Speak?', in C. Nelson and L. Grossberg (eds), Marxism and the Interpretation of Culture (Urbana: University of Illinois Press, 1988), 271–313.

quite happy that we got rid of it ...').⁵² It is rare to hear positions expressed as stridently as this; most views are more nuanced. Nevertheless, both strident and nuanced opinions echo the ideological offensive pursued in France, still ongoing after twenty years, against female Muslims, who are for the most part French citizens.⁵³ Since the 'headscarf affair' in 1989, up until President Sarkozy's peremptory declarations in June 2009 ('La burqa ne sera pas la bienvenue sur le territoire de la République'), women have reappeared as a region of confrontation between civilizations.⁵⁴ It has not been 'got rid of', as Marc wanted to say: on the contrary, the ambiguities of colonial Algeria reappear as nightmarish monsters in postcolonial France. The campaigns for *dévoilement* [unveiling] have a long colonial history, as Christine Delphy has pointed out.⁵⁵

Conclusion

Nonetheless, 'such historically dense vocabularies raise basic problems of interpretative licence. To what degree [are veterans'] descriptions of the colonial past to be read as commentaries on the postcolonial present? And given that they must be to some extent 'about' both, how are we to calibrate the relative weighting of the two?'⁵⁶ Veterans' memories swing between past

---

52   Marc (1940, 60–1/A), garage worker, 457th groupe d'artillerie antiaérien légère, sergeant. Interviewed in Les Pavillons sous Bois on 13 July and 11 December 2008.
53   For a concise account of the debates on this issue, see J. Baubérot, 'La Commission Stasi: Entre Laïcité Républicaine et Multiculturelle', *Historical Reflections*, 34/3 (2008), 7–20.
54   See E. Dorlin, 'Le grand strip-tease: féminisme, nationalisme et *burqa* en France', in N. Bancel et al. (eds), *Ruptures postcoloniales: Les nouveaux visages de la société française*, 429–42. Nicolas Sarkozy's speech was delivered in Versailles, 22 June 2009.
55   C. Delphy, *Classer, dominer. Qui sont les 'autres'?* (Paris: La fabrique, 2008), 144.
56   A.L. Stoler, *Carnal Knowledge and Imperial Power: Race and the Intimate in Colonial Rule* (Berkeley: University of California Press, 2002), 273.

and present: even if their memories of the past are profoundly influenced by the present realities of French and Algerian society, at the same time they are still one of the privileged vectors of transmission of past colonial imagery within French society. Nonetheless, in that overused term 'postcolonialism', one must not disregard the heuristic potential rooted in the prefix 'post' – one should not, that is, give way to an implacable logic of continuity or ahistorical permanence of the 'yet'.[57] Thereby, the postcolonial time 'is the one in which colonial experience appears, simultaneously, to be consigned to the past and, precisely due to the modalities with which its "overcoming" comes about, to be installed at the centre of contemporary social experience'.[58]

Given that witnesses tell stories in different ways, we are supposed to deal with the veterans' memories in the plural, without doing them violence by looking for an authoritative synthesis imposed from outside the conflict environment and military culture: moreover, veterans differ because of their social condition, education, and political stances. Nevertheless, they attempt retrospectively to give a meaning to their war experience through the contemporary socio-cultural frames to which they belong – as does everybody in looking back at their lives.

If we return to the 'Odyssey', and consider Odysseus as the archetype of the veteran narrating himself, we can, perhaps, better understand the manner in which he is *polytropos*, as the first line of the 'Odyssey' says:[59] *polytropos*, certainly as in a man with many experiences ('skilled' in the Fitzgerald translation), but also polyvalent, concealable. Haunted by the uneasy task of composing the past and the present, the war time and its aftermath.

---

57 See Chakrabarty's discussion of historicism – as the Modern European idea of history – through the concept of the 'not yet', in D. Chakrabarty, *Provincializing Europe: Postcolonial Thought and Historical Difference* (Princeton: Princeton University Press, 2000).

58 S. Mezzadra, *La Condizione Postcoloniale. Storia e Politica nel Presente Globale* (Verona: ombre corte, 2008), 25.

59 'Sing in me, Muse, and through me tell the story of that man skilled in all ways of contending' ('Ἄνδρα μοι ἔννεπε, Μοῦσα, πολύτροπον), *Odyssey*, I, vol. 1, trans. Robert Fitzgerald (New York: Farrar, Strauss, and Giroux, 1998).

Bibliography

Alexander, M., Evans, M., and Keiger, J., 'The "War Without a Name", the French Army and the Algerians: Recovering Experiences, Images and Testimonies', in M. Alexander, M. Evans, and J. Keiger (eds), *The Algerian War and the French Army, 1954–62: Experiences, Images, Testimonies* (Basingstoke and New York: Palgrave Macmillan, 2002), 1–39.

Allison, F.H., 'Remembering a Vietnam War Firefight: Changing Perspectives Over Time', in R. Perks and A. Thomson (eds), *The Oral History Reader* (London and New York: Routledge, 2006 [2004]), 221–9.

Arendt, H., *Between Past and Future* (London: Penguin, 1993 [1958]).

Arthur, P., *Unfinished Projects: Decolonization and the Philosophy of Jean-Paul Sartre* (London and New York: Verso, 2010).

Audoin-Rouzeau, S., and Becker, A., *14–18 Retrouver la Guerre* (Paris: Gallimard, 2000).

Augustine of Hippo, Book XI, *Confessions* (London: Signet Classics, 2001), 251–80.

Barthes, R., *Mythologies* (Paris: Seuil, 1957).

Baubérot, J., 'La Commission Stasi: Entre Laïcité Républicaine et Multiculturelle', *Historical Reflections* 34/3 (2008), 7–20.

Bloch, M., 'Mémoire collective, Tradition et coutume. A propos d'un livre récent', in A. Becker (ed.), *L'Histoire, la Guerre, la Résistance* (Paris: Gallimard, 2006 [1925]), 335–46.

Branche, R. 'Des viols pendant la guerre d'Algérie', *Vingtième Siècle. Revue d'histoire*, 75/3 (2002), 124–32.

Branche, R., *L'Embuscade de Palestro: Algérie 1956* (Paris: Armand Colin, 2010).

Branche, R., *La Torture et l'armée pendant la guerre d'Algérie, 1954–1962* (Paris: Gallimard, 2001).

Brazzoduro, A., 'La guerra d'Algeria nel discorso pubblico francese. Quaranta anni dopo', *Mondo Contemporaneo. Rivista di storia*, 4/1 (2008), 67–93.

Certeau, M. de, *La Prise de parole: Pour une nouvelle culture* (Paris: Desclée De Brouwer, 1968).

Chakrabarty, D., *Provincializing Europe: Postcolonial Thought and Historical Difference* (Princeton: Princeton University Press, 2000).

Chaumont, J.-M., *La Concurrence de victimes: Génocide, identité, reconnaissance* (Paris: La Découverte, 2002).

Chirac, J., *Mon Combat pour la France. Textes et interventions* (Paris: Odile Jacob, 2007).

Chominot, M., 'Guerre des images, guerre sans image? Pratiques et usages de la photographie pendant la guerre d'indépendance algérienne 1954–1962', unpublished PhD thesis (Université de Paris 8, 2008).

Coquery-Vidrovitch, C, 'Un faux concept: la repentance', *Enjeux politiques de l'histoire coloniale* (Marseille: Agone, 2009), 139–41.

Delphy, C., *Classer, dominer: Qui sont les 'autres'?* (Paris: La fabrique, 2008).

Dorlin, E., 'Le grand strip-tease: féminisme, nationalisme et burqa en France', in N. Bancel et al. (eds), *Ruptures postcoloniales. Les nouveaux visages de la société française* (Paris: La Découverte, 2010), 429–42.

Dosse, F., *Les Héritiers du silence. Enfants d'appelés en Algérie* (Paris: Stock, 2012).

Dunnage, J., 'Perpetrator Memory and Memories about Perpetrators', *Memory Studies*, 3/2 (2010), 91–4.

Eldrige, C. '"We've Never Had a Voice": Memory Construction and the Children of the Harkis (1962–1991)', *French History* 23/1 (2009), 88–107.

Eldrige, C., 'Blurring the Boundaries between Perpetrators and Victims: Pied-noir Memories and the Harki Community', *Memory Studies*, 3/2 (2010), 123–36.

Fanon, F., *Les Damnés de la terre* (Paris: Maspero, 1961).

Farge, A., *Quel bruit ferons-nous?* (Paris: Les Prairies ordinaires, 2005).

Foucault, M., 'Intervista a Michel Foucault', in A. Fontana and P. Pasquino (eds), *Microfisica del Potere. Interventi Politici* (Turin: Einaudi, 1977), 3–28.

Halbwachs, M., *Les Cadres sociaux de la mémoire* (Paris: Albin Michel, 1994 [1925]).

Hargreaves, A., 'France and Algeria, 1962–2002: Turning the Page?', *Modern & Contemporary France*, 10/4 (2002), 445–7.

Hartog, F., *Régimes d'Historicité: Présentisme et Expérience du Temps* (Paris: Seuil, 2003).

Homer, *Odyssey*, trans. Robert Fitzgerald (New York: Farrar, Strauss, and Giroux, 1998).

Jauffret, J.-C., 'Une armée à deux vitesse en Algérie (1954–1962): Réserves générales et troupes de secteur', in J.-C. Jauffret and M. Vaïsse (eds), *Militaires et Guérilla dans la Guerre d'Algérie* (Brussels: Complexe, 2001), 21–37.

Jauffret, J.-C., *Soldats en Algérie 1954–1962. Expériences Contrastées des Hommes du Contingent* (Paris: Autrement, 2000).

Lefranc, S.,'Repentance', in L. De Cock, F. Madeline, N. Offenstadt, and S.Wahnich (eds), *Comment Nicolas Sarkozy écrit l'histoire de France* (Marseille: Agone, 2008).

MacMaster, N., 'The Torture Controversy (1998–2002): Towards a "New History" of the Algerian War?', *Modern & Contemporary France*, 10/4 (2002), 449–59.

Martigny, V., 'Le débat autour de l'identité nationale dans la campagne présidentielle 2007: quelle rupture?', *French Politics, Culture and Society*, 27/1 (2009), 23–42.

Mauss-Copeaux, C., *Appelés en Algérie: La Parole confisquée* (Paris: Hachette, 1999).

Mauss-Copeaux, C., *A travers le viseur: Images d'appelés en Algérie 1955–1962* (Lyon: Ædelsa 2003).

Mezzadra, S., *La Condizione Postcoloniale. Storia e Politica nel Presente Globale* (Verona: Ombre corte, 2008).

Orr, A., *Ceux d'Algérie: Le silence et la honte* (Paris: Payot, 1990).

Pervillé, G., 'La guerre d'Algérie: combien de morts?', in M. Harbi and B. Stora (eds), *La Guerre d'Algérie. 1954–2004, la fin de l'amnésie* (Paris: Robert Laffont, 2004), 477–93.

Pollak, M.L., *Expérience concentrationnaire. Essai sur le maintien de l'identité sociale* (Paris: Métailié, 2000 [1990]).

Portelli, A., 'L'inter-vista nella storia orale', in M. Pistacchi (ed.), *Vive voci. L'intervista come fonte di documentazione* (Rome: Donzelli, 2010).

Portelli, A., 'History-Telling and Time: An Example from Kentucky', *The Oral History Review*, 20 (1992), 51–66.

Ricœur, P., *Temps et Récit III, Le Temps Raconté* (Paris: Seuil, 1985).

Roche, A., '"Je vous le raconte volontiers, parce qu'on ne me l'a jamais demandé": Autobiographies d'appelés en Algérie', in L. Gervereau, J.-P. Rioux and B. Stora (eds), *La France en guerre d'Algérie* (Paris: BDIC, 1992), 264–72.

Rousso, H., 'Intellectuals and the Law', *Modern & Contemporary France*, 17/2 (2009), 153–61.

Rousso, H., 'Vers une mondialisation de la mémoire', *Vingtième Siècle. Revue d'histoire*, 94/2 (2007), 3–10.

Rousso, H., *Le Syndrome de Vichy: De 1944 à nos Jours* (Paris: Seuil, 1990).

Schalk, D.L., *War and the Ivory Tower: Algeria and Vietnam* (Lincoln and London: University of Nebraska Press, 2005).

Shay, J., *Odysseus in America: Combat Trauma and the Trials of Homecoming* (New York: Scribner, 2002).

Sigg, B.W, *Le Silence et la honte. Névroses de la guerre d'Algérie* (Paris: Messidor, 1989).

Spivak, G.C., 'Can the Subaltern Speak?', in C. Nelson and L. Grossberg (eds), *Marxism and the Interpretation of Culture* (Urbana: University of Illinois Press, 1988), 271–313.

Stoler, A.L., 'L'aphasie colonial française: l'histoire mutilée', in N. Bancel et al. (eds), *Ruptures postcoloniales. Les nouveaux visages de la société française* (Paris: La Découverte, 2010), 62–78.

Stoler, A.L., *Carnal Knowledge and Imperial Power: Race and the Intimate in Colonial Rule* (Berkeley: University of California Press, 2002).

Wieviorka, A., *L'ère du témoin* (Paris: Seuil, 1998).

NATALYA VINCE

Questioning the Colonial Fracture: The Algerian War as a 'Useful Past' in Contemporary France and Algeria

On 21 May 2010, under intense police surveillance and the angry shouts of protestors, Franco-Algerian director Rachid Bouchareb presented his latest film, *Hors-la-loi* (Outside of the Law), at the Cannes film festival. Bouchareb's previous film, *Indigènes* (Days of Glory, 2006), had been a largely consensual affair, recovering the forgotten role of soldiers from France's North African colonies in the fight to liberate Nazi-occupied Europe and reinserting them into the national historical narrative as loyal, albeit marginalized, subjects. The four main characters were collectively rewarded with the Cannes prize for best male actor and President Jacques Chirac increased the pensions of former soldiers from the colonies following a screening of the film. Pitched as a form of sequel to *Indigènes*, *Hors-la-loi* enters into a highly contentious period of history in which 'goodies' and 'baddies' remain a matter of virulently defended opinion: the build-up to, and the unfolding of, the Algerian War of Independence (1954–62). The film opens with a representation of the Sétif massacre of 8 May 1945. Following the liberation of Europe, Algerians in the eastern regions of Sétif and Guelma flooded the streets demanding the liberation of imprisoned nationalist Messali Hadj and the right to self-determination. The sight of forbidden nationalist flags and pro-independence banners in this pacifist protest triggered the wrath of the French authorities. Violence broke out, and in the repression which followed, thousands of Algerians were killed by the army and settlers; 103 Europeans also died.¹

The polemical reaction to *Hors-la-loi* had begun in the weeks and months before its Cannes screening, with the vast majority of those participating in the debate yet to see it. Leading the attack were activist groups representing *pieds noirs* – settlers of European origin who were 'repatriated' to France after independence in 1962. *Pied-noir* website Bab el Oued Story described the film as an 'anti-French propaganda prop peppered with crude historical errors.'² Alpes-Maritimes UMP deputy Lionnel Luca accused the film of being 'anti-French', 'a hemiplegic view of history', 'negationist' by not explicitly representing the killing of Europeans in Sétif and encouraging 'a permanent repentance.'³ Luca had not actually seen the film, but had read an 'analysis of its accuracy' produced by the Service historique de la Défense (the French Army Archives) on the orders of another critic, Hubert Falco, minister for veterans and mayor of Toulon (Var) where, as in Alpes-Maritimes, many *pieds noirs* live. In Algeria, Abdelaziz Belkhadem, general secretary of the party of the Front de Libération nationale (National Liberation Front, FLN), praised the film for bringing the crimes of colonization to the big screen whilst somewhat conveniently ignoring that the film's depiction of the wartime FLN is often less than flattering.⁴

---

1   The numbers of Algerians killed in Sétif and Guelma is subject to intense debate. The French government recognizes 1,165 deaths, whilst Algerian nationalists put forward the figure of 45,000. French and British civilian archives suggest 6,000–15,000 were killed. J.-P. Peyroulou, 'Le rapport Tubert sur les événements de mai 1945 dans le Constantinois' (2005), http://www.ldh-toulon.net/spip.php?article600 (accessed 1 October 2011).
2   'Hors la loi: un très mauvais film', http://babelouedstory.com/thema_les/desinformation/8300/8300.html (accessed 1 October 2011).
3   *Le Point* (7 May 2010); *Libération* (7 May 2010); *Bakchich* (22 May 2010).
4   *Algérie 360* (3 May 2010); M. Rahal '*Hors la loi*, un film dans l'histoire', *Mediapart* (13 October 2010).

The Insights and Limitations of a Postcolonial Framework

The enflamed responses generated by *Hors-la-loi* can be seen as yet another example of the intense politicization of memory surrounding the Algerian War. One of the most solicited historians called upon to comment upon the release of the film was Benjamin Stora, author of numerous works which argue that the unhealed wounds of the Franco-Algerian past poison present relations:

> Franco-Algerian relations were forged in violence, by the imposition of the colonial system and by a seven-year war which enabled Algeria to acquire independence. This is why, thirty years later, time has not appeased passions. [...] From 1962 onwards, the Mediterranean, whose name in Arabic, *al-bahr al-abyad al-mutawassat* means 'the white sea in the middle', became a fracture line, an imaginary blue 'wall'. The violent divorce has unceasingly fed tensions, obsessions and fantasies from one shore to the other.[5]

Stora's work on the enduring impact of the Algerian War – this *passé qui ne passe pas*[6] – on both Franco-Algerian relations and attitudes in France towards North African immigrants fits in with the recent engagement by sections of French academia with postcolonial theory. Born amongst Anglophone intellectuals, postcolonial theory was originally applied to tracing the legacies of and ruptures with colonialism in literary works produced by writers from former colonies.[7] In French thought, postcolonial theory has been used as a framework to analyse a range of contemporary

---

5   B. Stora, *La Gangrène et l'oubli: la mémoire de la guerre d'Algérie* (Paris: La Découverte, 1991), 317. The translations are mine, unless otherwise stated.
6   Title of Eric Conan and Henry Rousso's work on France's difficulty coming to terms with the Vichy regime 1940–4, translated as 'an ever-present past' in the English-language edition. E. Conan and H. Rousso, *Vichy, un passé qui ne passe pas* (Paris: Folio, 1996).
7   B. Ashcroft, G. Griffiths and H. Tiffin's *The Empire Writes Back: Theory and Practice in Post-colonial Literature* (London and New York: Routledge, 1989, 2nd edn, 2002) is a seminal work.

social and political issues, including debates about citizenship and belonging, *laïcité* and racism. Questions about the relationship between history, memory and politics have been very much to the fore.

France in recent years has certainly had many colonial 'echoes'. Arguments claiming that the 2004 law banning the wearing of ostentatious religious symbols, notably the Muslim headscarf, in French schools promoted women's equality can be seen as having a colonial precedent in attempts to win the hearts and minds of Muslim women during the Algerian War by 'emancipating' them from the oppression of tradition – including (largely staged) unveiling ceremonies.[8] During the November 2005 riots, many commentators noted that the curfew imposed in certain *banlieues* was reminiscent of that imposed on North Africans living in France during the Algerian War.[9] However, it is clearly simplistic to draw a straight line between these past and present events, and argue that France is still 'colonial' in its treatment of immigrants from the former colonies. As Pascal Blanchard, Nicholas Bancel and Sandrine Lemaire underline in *La Fracture coloniale*, 'postcoloniality' is about analysing the multiple realities of heterogeneous present-day situations which can *in part* be explained by looking at history in the *longue durée*, including colonial history: 'We must not therefore seek out a systematic coherence in the effects of this fracture: it has different effects on diverse fields, which are not necessarily linked.'[10]

---

8   For more on the widespread use of the stereotype of the violent, oppressive Arab/Black male versus the assimilated *beurette* (young woman of North African origin) see N. Guénif-Souilamas, 'The Other French Exception: Virtuous Racism and the War of the Sexes in Postcolonial France', *French Politics, Culture and Society*, 24/3 (2006), 23–41. For the colonial period see N. MacMaster, *Burning the Veil: The Algerian War and the 'Emancipation' of Muslim Women, 1954–1962* (Manchester: Manchester University Press, 2009).

9   The Mouvement des Indigènes de la République drew a direct parallel between the colonial period and the contemporary situation, publishing a press statement on 9 November 2005 declaring: 'No to the colonial curfew! Revolt is not a crime! The true fire starters are those in power!'

10 Quoted in P. Blanchard and N. Bancel, *Culture post-colonial 1961–2006* (Paris: Autrement, 2005), 13.

We cannot understand contemporary French Republicanism without understanding the simultaneously symbiotic and contradictory relationship which developed in the nineteenth century between 'universal' Republicanism, the 'assimilatory' 'civilizing mission' and colonial practices of racialization and differential treatment.[11] At the same time, 'French Republicanism' is not fixed in time, but is instead an ideology which has proved to be constantly adaptable to present concerns. Moreover, there are fundamental structural differences between colonial and contemporary France: the legal basis for discrimination between citizen and subject which was the cornerstone of colonial rule no longer exists. Furthermore, we need to engage critically with what the concept of 'colonial' means, rather than assume that everything associated with the period of colonial rule has some kind of specificity, and, by extension, that we can relocate this specificity in the present. In Algeria, the nature of Algerian nationalism – and notably the emphasis on Islam and Arabic as markers of identity – was in many ways forged through conflict with French colonial rule.[12] Yet by concen-

11   See: A.L. Conklin, *A Mission to Civilise: The Republican Idea of Empire in France and West Africa* (Stanford: Stanford University Press, 1997); P. Lorcin, *Imperial Identities: Stereotyping, Prejudice and Race in Colonial Algeria* (London and New York: I.B. Tauris, 1995); G. Manceron, *Marianne et les colonies: une introduction à l'histoire colonial de la France* (Paris: La Découverte, 2005); T. Shephard, *The Invention of Decolonisation: The Algerian War and the Remaking of France* (Ithaca, NY: Cornell University Press, 2006); T. Stovall and G. Van Den Abbeele (eds), *French Civilisation and its Discontents* (Lanham, MD: Lexington, 2003); T. Todorov, *On Human Diversity: Nationalism, Racism and Exoticism in French Thought* (Cambridge, MA: Harvard University Press, 1993); P. Weil, *Qu'est-ce qu'un Français?* (Paris: Grasset, 2004).

12   See O. Carlier, *Entre nation et jihad: histoire des radicalismes algériens* (Paris: Presses de Sciences Po, 1995); J. Clancy-Smith, 'Islam, Gender and Identities in the Making of French Algeria, 1830–1962', in *Domesticating the Empire: Race, Gender and Family Life in French and Dutch Colonialisms*, ed. Clancy-Smith and F. Gouda (Charlottesville and London: University of Virginia Press, 1998), 154–74; M. Gadant, *Islam et nationalisme en Algérie: d'après El Moudjahid, organe central du FLN de 1956 à 1962* (Paris: L'Harmattan, 1988); M. Harbi, *Le FLN, mirage et réalité: des origines à la prise du pouvoir (1945–1962)* (Paris: Editions Jeune Afrique, 1980); M. Kaddache, *Histoire du nationalisme algérien*, vols 1 & 2 (Algiers: Entreprise nationale du livre, 1993);

trating on the narrow Franco-Algerian sphere and the period 1830–1962 there is not only the danger of obscuring all other sources of influence, we also risk locking formerly colonized peoples in one particular aspect of their past, involuntarily promoting 'the absurd idea that domination made them enter into history.'[13]

Using the example of the memory of the Algerian War, this chapter provides a case study which illustrates emerging theoretical critiques of the limitations of postcolonial theory. The chapter seeks to demonstrate that recent debates, propaganda and laws raising issues about French colonialism and the Algerian War reflect and promote internal political causes in both countries rather than being symptomatic of Franco-Algerian conflict, and that, moreover, these debates are only partly the product of the colonial past. Meanings are ascribed to the past rather than the unfiltered past providing a straightforward frame of reference for understanding current discourse. This point may seem obvious, but is one that is often obscured in the Franco-Algerian case by a dominant rhetoric of colonial nostalgia and anti-colonialism.

Relations between the French and Algerian States since 1962: Conflictual Discourse and Close Cooperation

A brief outline of foreign relations between the French and Algerian states since 1962 reveals two key features. Firstly, the weight of conflictual history is more evident in discourse than in the practice of state interactions.

G. Manceron and H. Remaoun, *D'une rive à une autre: la guerre d'Algérie de la mémoire à l'histoire* (Paris: Syros, 1993); J. McDougall, *History and Culture of Nationalism in Algeria* (Cambridge: Cambridge University Press, 2006); G. Meynier, *Histoire Intérieure du FLN 1954–1962* (Paris: Fayard, 2002).

13   E. Sibeud, 'Post-colonial et Colonial Studies: enjeux et débats', *Revue d'histoire moderne et contemporaine*, 51/4 (2004), 87–95 (90).

Secondly, since 1962, French and Algerian political elites have often closely cooperated, and, particularly in recent years, this has often been at the expense of the wider Algerian population.

Since independence, France and Algeria have cultivated intense economic ties, most obviously based on Algeria's status as a major source of gas and petrol. At the same time, both France and Algeria have a specific vision of their place in the world and for neither country can foreign policy decisions simply be reduced to economic pragmatism. The evolving manifestations of France's *politique arabe* – a nebulous and catch-all concept – have been discussed elsewhere in this volume. Upon independence, the Algerian state was determined to break away from the dependency evident in former French sub-Saharan colonies and position itself as a leader of the Third Worldist movement. In the 1960s and 1970s, Presidents Ahmed Ben Bella (1962–5) and Houari Boumediene (1965–78), key politicians and foreign policy actors were not career diplomats or civil servants but war veterans with 'revolutionary experience': Amine Ait Chaalal characterizes this period as one of fixed goals and stubborn determination.[14] From the 1980s onwards, the mechanisms of foreign policy moved more into the hands of younger generations with professional training. Nevertheless, since 1999, having largely recovered from the internal crisis and diplomatic inertia of the previous two decades, ending Algeria's 1990s reputation as an international pariah and promoting an assertive foreign policy has been used by President Abdelaziz Bouteflika – Boumediene's former foreign minister – as a tool for domestic unity and internal legitimacy. Algeria has remained reluctant to participate in multilateral frameworks, from the Organisation internationale de la Francophonie (OIF), which Algeria has never fully integrated despite being one of the most francophone countries

---

14 A. Ait Chaalal, 'La politique étrangère de l'Algérie: entre héritage et originalité' in *La politique étrangère: le modèle classique à l'épreuve* (Brussels: PIE-Peter Lang, 2004), 203–15 (206). For more on post-independence Algerian foreign policy see: N. Grimaud, *La Politique extérieure de l'Algérie (1962–1978)* (Paris: Karthala, 1984).

in the world, to EU frameworks such as the ENP.[15] The reception in Algiers of Sarkozy's Union for the Mediterranean was decidedly lukewarm, with Bouteflika maintaining the suspense surrounding whether or not Algeria would participate in the July 2008 founding talks, before deciding to attend at the last minute. The Algerian state has preferred to cultivate bilateral diplomatic and trade relations with a range of partners, including France, and more increasingly the United States, China, Russia and Turkey. Algeria is able to maintain an assertive foreign policy because it has a significant economic power as a major hydrocarbons producer. New discoveries of oil and natural gas in the mid-1990s only reinforced this status.

Neither Algerian nor French foreign policy are simply forged in the shadow of the colonial past and in opposition to the colonial/ anti-colonial 'other' – foreign policy by its very nature has to take into account a range of factors and constantly shifts and adapts to present needs. De Gaulle took a very selective view of France's historical role in the world to suit foreign policy needs after 1962, reinventing himself as a decolonizer and seeking to position France as a friend of the Third World, and in particular the Arab world.[16] In Algeria, the memory of anti-colonial struggle has been used instrumentally to justify foreign policy decisions to domestic audiences and legitimize the regime in power. Yet closer analysis of the different contexts in which this discourse has been used – and its very flexible adaptation – reveals that it is not necessarily revealing of a 'specificity' in Franco-Algerian relations. In February 1971, Boumediene unilaterally nationalized Algerian petrol exploitation, a decision which had a major impact on French economic interests in the region. In the middle of the ensuing crisis, Boumediene called upon the heritage of 1 November 1954, the date which marks the outbreak of the liberation struggle, arguing that controlling one's petrol was part of 'the

---

15 H. Darbouche and Y. Zoubir 'The Algerian Crisis in European and US Foreign Policies: A Hindsight Analysis', *The Journal of North African Studies*, 14/1 (2009), 33–55 (48).
16 J.-F. Daguzan, 'Les Relations franco-algériennes ou la poursuite des amicales incompréhensions', *Annuaire français des relations internationales*, 2 (2001), 438–50; see also D. Sieffert, *Israël Palestine: une passion française* (Paris: La Découverte, 2004), 123.

historical logic of the Revolution.'[17] Yet the previous year, Boumediene had used the fifth anniversary of his coup, or 'redressement', on 19 June 1970, to announce the nationalization of four (non-French) petrol companies, including Anglo-Dutch Shell and American Phillips as well as German and Italian interests.[18] So it is not just French companies nor just dates related to the anti-colonial struggle which are mobilized in politico-economic projects, and the audience for these speeches is primarily an Algerian one. Indeed, the same wartime revolutionary discourse was also instrumentalized to justify very different foreign and domestic decisions made in the 1980s. The political and economic crisis in Algeria under President Chadli Bendjedid (1979–92) meant that Algiers was more susceptible to outside economic influence. In return for financial assistance, Algeria began to yield to demands for economic liberalization and Algerian debt was rescheduled within the IMF's structural readjustment programme. In the state-owned press, this liberalization – the desire to 'make profitable one's economic potential and extract the maximum gain' was justified to its readership as being 'in the spirit and the letter of the November revolution' The decade previously, the same state-owned press had argued that 1 November represented socialism, nationalization and *autogestion*.[19]

Despite an often hostile Algerian anti-colonial rhetoric since 1962, the French and Algerian states have cooperated closely, notably in areas relating to economics, politics, migration and security. Algerians remain France's biggest foreign population: in 2007, 702,324 Algerians were recorded as living in France. This figure does not include French nationals of Algerian origin.[20] For much of the 1960s, 1970s and 1980s, the Algerian govern-

---

17 V.-B. Rosoux, 'Poids et usage du passé dans les relations franco-algériennes', *Annuaire français de relations internationales*, 2 (2001), 451–65 (457).
18 ORTF, Le Pétrole en Algérie, 24 July 1970, http://www.ina.fr/economie-et-societe/vie-economique/video/CAF86015723/le-petrole-en-algerie.fr.html (accessed 1 October 2011).
19 *El Moudjahid* (31 October 1970) and (3 November 1987).
20 Institut national des etudes démographiques, 2007, http://www.ined.fr/fr/pop_chiffres/france/immigres_etrangers/pays_naissance_1999/ (accessed 1 October 2011).

ment was considered by the French government to be the main actor in the organization and control of the Algerian immigrant community in France. Massive Algerian migration followed independence, as economic migrants sought to provide for their families in an Algeria ravaged by war. Both states viewed immigration as a temporary phenomenon. Until 1973, when in the wake of a series of racist murders Boumediene unilaterally declared the end of Algerian migration to France, it was the Algerian authorities which issued exit visas allowing its citizens to leave for France. The origins of the Amicale des Algériens en Europe (≈ Friendly Society for Algerians in Europe, AAE) dates back to the War of Independence, and the networks in mainland France which collected dues from Algerian immigrants to fund the FLN. After 1962, the Amicale took the form of 'a quasi Ministry of Emigration'[21] collecting subscriptions from Algerians for repatriation insurance in case of death on French soil, bringing imams to France, organizing Arabic lessons and concerts and training youth workers to work in areas with high immigrant populations. Given that Algeria was a one-party state, the Amicale represented the interests of the FLN and sought to extend their control over Algerian immigrants, conscious that France, both during and after the war, was a space for Algerian political movements challenging the hegemony of the FLN. Indeed, in 1965, the French authorities connived in suppressing Amicale dissent in France in the wake of Boumediene's coup.[22]

Another institution linking Paris and Algiers is the Grande Mosquée de Paris (Grand Paris Mosque in the fifth arrondissement). Since 1982, this mosque has been under the joint control and finance of the French and Algerian states with the position of rector collectively agreed. The Mosquée de Paris maintained close ties with the AAE, until the demise of the latter in the late 1980s following the introduction of multipartyism in Algeria. Despite the French Republican model's rejection of the use

---

21 J.-C. Scagnetti, 'Pays d'origine et encadrement des pratiques religieuses: l'Algérie et ses émigrés (1962–1988)', *Cahiers de la Méditerranée*, 78 (2009), 177–202.
22 E. Naylor, 'The Politics of a Presence: Algerians in Marseille from Independence to "Immigration Sauvage" (1962–1974)', unpublished PhD thesis (University of London, 2011), 190–3.

of 'intermediaries' to negotiate community relationships, and despite the proclaimed desire since the 1990s to create a 'French Islam' rather than just passively witness the development 'Islam in France', the French state accepted – albeit implicitly – that the AAE and the Mosquée de Paris had the legitimacy to collectively represent immigrants, and indirectly benefited from the AAE and the Mosquée de Paris's provision of social and cultural structures. It is only more recently that such an arrangement has begun to be questioned by the French political class: Algerian immigration (and immigration in general) has proved not to be a transitory phenomenon, new challenges have emerged and the effectiveness of this delegation of responsibilities to engage with and/or control second and third generations of immigrant origin has been undermined. Indeed, having shared a mutual interest in managing immigration to France, the Algerian state now emphatically argues that problems of socio-economic deprivation and racial and cultural discrimination faced by French citizens of Algerian origin are France's problem.[23]

Today, the Algerian state plays an important role in preventing new migrants reaching France – both Algerians and those of other African nationalities – whilst the elite enjoys privileged ties with France. Stora describes the Mediterranean as a 'wall' between France and Algeria since 1962. Yet this wall does not apply to all. Although unsurprisingly not officially confirmed, it is widely understood in Algeria that a number of past and present government ministers have dual nationality, in most cases French. Indeed, despite Bouteflika's diatribe against dual nationals

---

23 This can be demonstrated by Algerian reactions to the France–Algeria football match at the Stade de France on 6 October 2001. The match got off to a tense start when the Marseillaise was booed, and ended early with a pitch invasion by fans of Algeria – most of whom were French-born dual nationals. For the Algerian press, and indeed the Algerian ambassador to France, this was a Franco-French problem. The *Jeune Indépendant* argued that it was the French authorities who had created these '"beurs" socialement inassimilables' and warned France 'Break down your ghettos before it is too late, [the Algerian football team] did not arrive [in France] with 80,000 spectators.' A. Baghzouz, 'France-Algérie: rejouer le match?' *Outre Terre*, 8 (2004), 191–4; See also *L'Express* (14 February 2002).

in 2006,[24] on 28 June 2009 Algerian newspaper *Echorouk* reported that according to 'well-informed sources', 60 per cent of Algerian government officials were dual nationals. In late 2005, Bouteflika himself came to France for medical treatment for a mysterious illness. For those Algerians without wealth or connections, the reinforcement of Europe's frontiers is a joint European–North African enterprise, with the Moroccan, Algerian – and until the revolts of 2011 – Tunisian and Libyan states all participating in the policing of the Mediterranean. On 25 February 2009, the Algerian National Assembly passed a law criminalizing *harraga* ('those who burn [frontiers]'), making attempts to illegally leave Algeria punishable by prison or a fine.

For European states, controlling migration is closely tied to security issues. With the Algerian economic and political crisis of the 1980s, and the subsequent rise of the Front islamique du salut (Islamic Salvation Front, FIS) and its armed offshoots in the aftermath of the attempt to introduce multipartyism in 1989, Algeria was one of the first countries to face the global challenge of political Islam. The position of France during the civil violence which ravaged Algeria in the 1990s has come under much scrutiny. Internal disagreements were rife within the French élite and political parties over how to respond to the Algerian crisis. On the one hand, there was the need for the 'motherland of the Rights of Man' to be seen to promote respect for democratic mechanisms, including the FIS's right to participate in elections and eventually win them, even if this led to the creation of a theocratic state. On the other hand, there were deep concerns about growing Islamism and the threat it could present on French shores – fears which were realized by the Saint Michel metro bombing in 1995. Ultimately, France chose to support the Algerian regime against the Islamist threat, whatever its democratic and human rights failings, shielding Algeria to a certain degree from international and European interference by positioning itself as a privileged intermediary.[25] Paris's hesitations nevertheless riled Algiers and in a speech in the French National Assembly during his 2000 state visit,

---

24   RFI, 26 June 2006, http://www.rfi.fr/actufr/articles/078/article_44733.asp (accessed 1 October 2011).
25   Daguzan, 'Les Relations franco-algériennes', 444.

Bouteflika attacked France for 'non-assistance to a person in danger' and 'neo-colonialism' hidden behind the 'pretext of non-interference'.²⁶ A year previously, Bouteflika had criticized French doubts about the legitimacy of his election by declaring that 'this form of protectorate, this form of limited sovereignty is totally unacceptable to me.'²⁷ Yet whilst France may have had a vacillating public discourse on the Algerian crisis, in practical affairs there was significant Franco-Algerian cooperation. The French and Algerian secret services, often with the connivance of senior politicians, worked closely together in the 1990s to crush the Islamists, often using dirty tricks. Franco-Algerian relations had shifted a gear: 'Paris's relations with Algiers degenerated into the French government's involvement with the *personnel* of the Algerian regime', argues Hugh Roberts, generating a 'complex and unsavoury system of patronage, reciprocal back-scratching and corruption' which both contributed to delegitimizing the Algerian regime in the eyes of its citizens and compromising members of the French political class.²⁸

Franco-Algerian relations since 1962 have been shaped according to each country's *certaine idée de soi*, taking into account the balance of economic and commercial needs, the domestic political context and external pressures – including actors and issues beyond the Franco-Algerian relationship. Key areas of tension today include the Arab-Israeli conflict and the Western Sahara. Of course, past experiences shape frames of reference and how the world is seen, but foreign policy decisions have to take into account a host of contemporary considerations. The weight of colonial history in determining positions has been only one of many factors, often employed as a post-hoc justificatory discourse. Without romanticizing the

26   Bouteflika's speech to the National Assembly, 14 June 2000, www.assemblee-nationale.fr/international/reception-algerie-cr.asp (accessed 1 October 2011).
27   *France 3* (16 April 1999).
28   H. Roberts, *The Battlefield Algeria 1988–2002: Studies in a Broken Polity* (London and New York: Verso, 2003), 307. See also L. Aggoun and J.-B. Rivoire, *Françalgérie: Crimes et mensonges d'Etat* (Paris: La Découverte, 2004); M. Evans and J. Phillips, *Algeria: Anger of the Dispossessed* (New Haven and London: Yale University Press, 2007).

authoritarian and repressive Boumediene regime, the discourse of national economic liberation in the 1960s and 1970s had a certain sincerity. Today the nationalist discourse is often used as a smokescreen for the ideological paucity of the current regime and the disjunction between a self-serving elite and the wider population which reaps few rewards from Algeria's petrol wealth. Pronouncements aimed at domestic audiences can of course have repercussions on external relations, but as we will see in the next section, both the French and Algerian states have selective hearing.

Actors, Activists and Reluctant Travellers in Memory Conflicts Surrounding the Algerian War

In the past decade, memory conflicts have frequently sprung up across the Mediterranean. In analysing the genesis and the unfolding of these debates, it is important to pinpoint the different roles of the French and Algerian states, pressure groups and individual initiatives. This section seeks to do this by looking at the torture scandals of 2000 following the publication the account of a former FLN activist, Louisette Ighilahriz, in *Le Monde*; the French law of 23 February 2005, notably its article 4 stating that the school curriculum should 'recognize the positive role of the French overseas presence'; and the proposal in Algeria to create a law criminalizing colonization in 2010.

On 20 June 2000, just a few days after Bouteflika's state visit to France, an interview by Florence Beaugé with FLN activist Louisette Ighilahriz made front-page news in *Le Monde*. Ighilahriz's account of the prolonged and sadistic torture she suffered in the hands of the French army during the 1957 'Battle of Algiers' created a media storm and provoked a 'retour de mémoire' (return of memory) from a number of actors across the Mediterranean. Long-retired army generals nearing the end of their lives agreed to be interviewed. Jacques Massu confessed to and regretted the use of torture, Paul Aussaresses admitted to and justified torture as an interrogation technique, Marcel Bigeard and Maurice Schmitt vigorously denied all accusations.[29] Other Algerian victims of torture also spoke out and were interviewed in the French and Algerian press. On 31 October 2000, Communist newspaper *Humanité* published an 'Appel des douze', signed by twelve French men and women, including Henri Alleg, Germaine Tillion, Pierre Vidal-Naquet, Josette Audin and Gisèle Halimi, who had sympathized with or been active in the Algerian struggle for independence, calling on the memory of the war to be collectively addressed so that healing could take place.[30]

In Algeria, Ighilahriz's story was not welcomed by all of her fellow veterans, revealing fissures within the war veteran community. Amongst female veterans, some seemed to resent the foregrounding of her personal story in what had largely been an anonymous collective history. One veteran went as far as claiming that Ighilahriz's limp and walking stick are the result of a car crash, not an injury sustained in the *maquis*. Others found her revelations about rape too hard to admit because of social taboos.[31] When Ighilahriz was invited to participate in a French *Envoyé spécial* documentary on torture and rape during the war,[32] she asked a number of other female veterans to add their testimony to hers. Ighilahriz says that after initially agreeing, women backed out under pressure from their families: 'I found myself alone. They let me drop. It was very hard. They let me be taken for

29   *Le Monde* (22 June 2000); *Le Monde* (23 November 2000); P. Aussaresses, *Services spéciaux, Algérie 1955–1957: Mon témoignage sur la torture* (Paris: Perrin, 2001); F. Beaugé, *Algérie: une guerre sans gloire: histoire d'une enquête* (Paris: Calmann-Lévy, 2005).
30   Note that Germaine Tillion's position during the war and in relation to this *Appel* was rather more ambiguous than that of the other signatories. Whilst she spoke out against torture, she did not support the methods of the FLN and to a certain extent, through her work with her *Centre sociaux*, established in 1955, embodied a kind of 'benevolent colonialism'.
31   N. Vince, 'To be a Moudjahida in Independent Algeria: Itineraries and Memories of Women Veterans of the Algerian War', unpublished PhD thesis (University of London, 2008), 208–9.
32   P. Jasselin, V. Gaget and A.-C. Bequet, 'Les viols pendant la Guerre d'Algérie', *Envoyé special*, France 2, 7 February 2002.

a liar.' Within her own family, Ighilahriz says her husband encouraged her to speak out, but her children had greater difficulty accepting her stance – her son was initially unsupportive and the shock of her mother's revelations led to her daughter going on long-term sick leave.[33] Whereas in 1959 the FLN seized upon the torture and rape of another of its militants, Djamila Boupacha, to discredit the French colonial war internationally, in 2000 official Algerian support for Ighilahriz (either from the state or the party of the FLN) was at best discreet. The Algerian state, usually quick to capitalize on populist anti-colonialism, refrained from entering into the debate – its own 'dirty war' against the Islamist challenge, during which it too used torture, imprisonment without trial and assassination, was all too recent.

In France, the *pied-noir* association Cercle Algérianiste (established 1973) condemned the *Envoyé spécial* programme as an invention, describing it as a 'pseudo-documentaire'. Henri Pouillot, an ex-French soldier who witnessed rape during the war and was interviewed for the programme also came under personal attack.[34] The Association soutien à l'armée française (Association to support the French army, ASAF), created in 1983 with the aim of defending 'the honour of the army and its soldiers', unsuccessfully sought to lodge a complaint for defamation in relation to Ighilhariz's accusations about Captain Graziani.[35] The Cercle pour la défense des combattants d'Afrique française du Nord (Society to defend veterans of French North Africa, CDCAFN, created in 1997, bringing together a number of associations) and the more mainstream Union Nationale des Combattants (National union of veterans, UNC, established 1918) also supported General Schmitt in his judicial proceedings against Ighilahriz.[36] The perspective of many senior generals and Algeria veterans was reproduced in *Livre blanc de*

---

33   Interview with Ighilahriz, Algiers, 8 June 2005.
34   Henri Pouillot, 'C'était un reportage réalisé par France 2', 27 May 2010, http://henri-pouillot.fr/spip.php?article11 (accessed 1 October 2011).
35   Reported on the website of the Collectif des rapatriés internautes http://www.lecri.net/desinformation/Lessons/Lesson_Files/ASAF.htm (accessed 1 October 2011).
36   Letter from Maurice Schmitt, reproduced on the website Chemin de mémoire des parachutistes http://www.chemin-de-memoire-parachutistes.org/algerie-1954–

l'armée française en Algérie (Paris: Contretemps, 2001) which attacked the research of Raphaëlle Branche, whose thesis on torture during the Algerian War was examined in December 2000, as well as Aussaresses, who was seen as shooting the French army in the foot.[37]

President Chirac and his socialist party prime minister, Lionel Jospin, were very hesitant to enter into the debates, both repeating that attributing responsibility for torture during the Algerian war should remain the work of historians.[38] Chirac did not build upon his famous Vel d'Hiv speech in 1995, when he recognized the responsibility of the French state in the Holocaust, nor indeed go beyond the 1999 law adopting 'Algerian War' as the official title for what had previously been designated 'operations to maintain order in North Africa'. The silence of the French state during this media storm was perhaps hardly surprising: official recognition would ignite the controversial question of the extent to which politicians sanctioned or encouraged torture in Algeria, generate tensions between politicians and the army, and from a financial perspective could potentially lead to compensation claims.

1962-f66/general-m-schmitt-fin-des-peripeties-judiciaires-t186.htm (accessed 1 October 2011).

37   Amongst army associations there are venomous disputes about the interpretation of the Algerian War. A recent dispute between the FNACA and the CDCAFN focuses on when the end of the Algerian War should be commemorated. The FNACA argues for 19 March 1962 – the day the Evian Accords came into force. CDCAFN and Jeune Pied Noir (JPN), the Collectif des Rapatriés Internautes (CRI) condemn this choice, arguing that fighting continued and that many civilians, *harkis*, and soldiers who had defected to the OAS were killed after this date. Their choice of commemorative date is 5 December 1962, chosen as it was a number of months after the ceasefire and on 5 December 2002 the National Memorial to the Algerian War and combats in Morocco and Tunisia was inaugurated by Chirac on the Quai Branly. In turn, the FNACA condemns *pied-noir* and CDCAFN anti-Gaullism and glorification of the OAS, a subversive organization. www.ldh-toulon.net/spip.php?article727 (accessed 1 October 2011).

38   AIDH, La torture pendant la guerre d'Algérie, http://www.aidh.org/faits_documents/algerie/verite.html (accessed 1 October 2011).

The tensions created by the reopening of the torture debate were thus Franco-French and Algero-Algerian. Of course, the Algerian War was also a Franco-French and Algero-Algerian conflict. Yet the influence of new dynamics must be taken into account. We are now looking at the interactions of two nation states, both of which refrained from entering into the debate. Generational change has also taken place, as the unsympathetic reaction to Ighilahriz by younger generations of Algerians suggests. Furthermore, Ighilahriz's initial motivation for collaborating in the publication of Beaugé's article was very personal – she wanted to find a 'Doctor Richaud' whom she described as having saved her life by removing her from the hands of French army torturers. Indeed, Ighilahriz dedicated her 2001 memoirs (*Algérienne*, Paris: Fayard and Calmann Lévy) to him as well as her own family. Ighilhariz also uses the final pages of *Algérienne* to attack the 'Islamization' of Algerian society, unfavourably comparing contemporary society with an idealized view of the liberal cultural practices of the Islam of her parents' generation. She fustigates an Algerian education system which has only transmitted an asceptized and truncated version of the war to young people, who, disillusioned, she argues, reach out to 'foreign' forms of identification from the Middle East. Reading *Algérienne*, it becomes clear that this is a memoir written in the light of the 1980s rise of Islamism in Algeria and the civil violence of the 1990s. It is in the 2006 (Algerian) Casbah Editions version of *Algérienne* where a preface to the Algerian publication explicitly foregrounds the anti-colonial reading of Ighilahriz's story. Refering to the 23 February 2005 law (passed after the memoirs were written), Ahcène Madjoudj declares that Ighilahriz's account 'also aims to counter a version of colonialism, its methods and its misdemeanours which certain groups have sought to assert and diffuse. This is the idea of an occupation which brought civilizational progress.' When an anti-colonial 'reading' of Ighilahriz's memoirs is made, it is by a third party for a specific, contemporary political use.

The 23 February 2005 law (2005–158) followed the form of another 'memory' law in 2001, recognizing slavery as a crime against humanity. Law 2005–158 stated in article 4 that the school curriculum should recognize 'the positive role of the French overseas presence, notably in North Africa', proposed a state-sponsored foundation for the memory of the Algerian war, and provided financial compensation for *harkis* (Algerians who fought in the French army) and those repatriated, including members of the extreme-right Organisation de l'Armée secrète (OAS).³⁹ Pushed for by certain *pied-noir* and *harki* associations, with the text prepared by deputies on the right of the UMP keen to attract this Midi-based electorate, the principle of a law to honour and promote the history of those 'repatriated' aroused little controversy in the National Assembly and Senate. Debates that did take place largely focused on the detail rather than the principle of the law, involved a small number of deputies and often thwarted party lines.⁴⁰

Law 2005–158 was seized upon by historians and teachers in France who denounced the imposition of an official history, and moreover, an official history which ignored the inherent violence and racism of the colonial relationship. A campaign followed, led by French and Algerian historians working in France. Other critics of the law included rival *pied-noir* and *harki* associations, albeit for often differing reasons. A statement issued by Harkis et droits de l'homme (Harkis and Human Rights) on behalf of 'children of harkis' welcomed certain aspects of the law, such as those protecting *harki* from insults and discrimination, but rejected colonialism as 'indefensible' and declared 'we refuse to remain hostages of the defenders of *Algérie française*.'⁴¹ Jeune Pied-Noir (Young Pied Noir, JPN)

---

39   Organization of the Secret Army which led a terrorist campaign in 1961–2 to try to maintain 'French Algeria'.

40   For example, Kléber Mesquida (PS, Hérault) François Liberti (PCF, Hérault) and Rudi Salles (UMP, Alpes Martimes) fought – ultimately unsuccessfully – to get those who were killed on 26 March 1962 in Algiers, during the Rue d'Isly confrontations between the French army and pro-*Algérie française* demonstrators, recognized as 'morts pour la France' – the attribution given to soldiers killed fighting for the nation. See: http://www.ldh-toulon.net/spip.php?article1189 (accessed 1 October 2011). For a detailed analysis of the origins and evolution of the law, notably article 4, see C. Liauzu and G. Manceron (eds), *La colonisation, la loi et l'histoire* (Paris: Syllepse, 2006).

41   Harkis et Droits de l'Homme, Appel d'enfants de harkis contre les articles 4 et 13 de la loi du 23 février 2005. http://harki.net/article.php?id=50 (accessed 1 October 2011). In the past, many *harkis* have been linked to *pied-noir* and/or right-wing organizations. Organizations such as Harkis et Droits de l'Homme have sought to distance

– which is an unfailing defender of 'French Algeria' – criticized the failure of an amendment (number 35), presented by members of the socialist group which would have recognized the responsibility of the French state and nation in the deaths of *pieds noirs* and *harkis* following the 19 March 1962 ceasefire.[42] In justifying the rejection of the amendment, the *rapporteur* for the law, Alain Gournac (Yvelines, UMP), declared – seemingly without a trace of irony – that 'it is the role of historians and not us, senators, to determine the responsibilities in the future.'[43]

The Algerian press began to report on the 23 February law and Algerian associations such as the Fondation du 8 mai 1945, presided by historian Mohamed El Korso, also joined the charge. The Fondation's critiques focused less on the imposition of an official history, and more on the promotion of an 'official lie', paying homage to '*harkis*; and to all those who were at the origin of the martyrization which Algerians suffered for seven-and-half years of a pitiless war' and occulting colonial crimes and crimes against humanity.[44] The Fondation was created in 1990 by Bachir Boumaza, war veteran and, in 1998, president of the Conseil de la Nation (the upper house), to campaign for recognition of crimes committed in

---

themselves from the glorification of 'French Algeria', rejecting a simplistic demonization of the FLN, and recently lobbying Sarkozy to get him to fufil a pre-electoral promise to recognize the responsibility of the French state in the abandonment and massacre of the *harkis* in 1962, see www.harki.net/article.php?id-485 (accessed 1 October 2011).

42 B. Coll and T. Titraoui, 'Lettre ouverte', 18 December 2004. Note that in this open letter they do not state that amendment 35 was a socialist initiative. There was a similar amendment proposed (amendment 11) by Centre-Right Senator Jean-Paul Alduy, who did not come to the Senate to defend the amendment as he did not want to see it mixed up with the socialist proposal. In this example we see the tensions between differing memories of the Algerian War, often driven by differing local electoral allegiances, and present-day party loyalties.

43 Sénat, Séance du 16 décembre 2004 (compte rendu intégral des débats), http://www.senat.fr/seances/s200412/s20041216/s20041216004.html#SOM7 (accessed 1 October 2011).

44 Fondation du 8 mai 1945, Mémorandum contre la falsification de l'histoire de l'Algérie par l'ancienne puissance coloniale, 20 May 2005.

Sétif. Its founder is described as having one foot inside the FLN and the other on the outside.⁴⁵ The Fondation's version of history is not necessarily opposed to the state narrative – it can go in the same direction – but is also a lobby for events that are sometimes more convenient for the government to forget. Indeed, the event chosen to provide the name for the association is not one easily assimilated into official history – Sétif was nine years before the creation of the FLN and key players included Messali Hadj, an alternative and long-marginalized nationalist leader.

In both France and Algeria, the response of the heads of state and their governments was delayed. In his 2006 new year message to the press, Chirac announced that article 4 of the 23 February law would be rewritten because 'The current text divides the French people'. This, alongside a greater emphasis on equal opportunities, was part of Chirac's response to the crisis in the French *banlieues* in November 2005.⁴⁶ This link between the 23 February law and the intensive riots later that year is curious given that this was not a 'demand' of the rioters. Indeed, one of the features of the riots was the lack of representation of the perspectives of rioters themselves; in the mainstream media they were alternatively described as mindless vandals, angry young men suffering socio-economic exclusion, Muslim extremists leading an *infitada*, Frenchmen acting in a revolutionary tradition, the post-colonial dispossessed. The 2005-158 law was followed by a flurry of publications on 'postcolonial' France – in the sense of 'the forms of continuity with and rupture from the colonial past'⁴⁷ – from the French academic community, which had thus far been reticent to engage with this 'Anglophone' concept. A measure of the extent to which the mainstream media was drawn towards the postcolonial explanation can be demonstrated by debate programmes on themes such as 'Colonisation,

---

45   A. Cheurfi, *La Classe politique algérienne de 1900 à nos jours: dictionnaire biographique* (Algiers: Casbah Editions, 2001), 124.
46   RFI, Voeux présidentiels, 4 January 2006, http://www.rfi.fr/actufr/articles/073/article_40852.asp (accessed 1 October 2011).
47   C. Forsdick and D. Murphy, *Postcolonial Thought in the French-Speaking World* (Liverpool: Liverpool University Press, 2009), 16.

banlieue: is France guilty?' (*France 3*, 14 December 2005), which included the participation of Algerian War specialist Benjamin Stora.

The most pronounced use of this postcolonial framework was by the Movement des indigènes de la République (Movement for the indigenous of the Republic, MIR). MIR was formed in January 2005, just before the passing of the 23 February law, as a new political movement consisting of various anti-discrimination movements, immigrant and Muslim groups and pro-Palestinian networks, issuing a statement which declared that 'Nous sommes les indigènes de la République' ('We are the indigenous of the Republic'). Using the historically loaded term 'indigène', MIR made an explicit link between colonial and 'postcolonial' France, arguing that contemporary xenophobia and racism take specific and particularly discriminatory forms for people from former and current colonies (the DOM-TOM are classed as the latter). MIR has no relations with any mainstream political party, nor with movements such as SOS Racism. MIR is regularly accused of anti-Republicanism, misogyny, overly focusing on the Algerian War and being disconnected from the socio-economic realities of the *banlieue*, an accusation it denies, instead stating that 'social exclusion which results from class relations is further entrenched by imperial prejudice.'[48]

The Algerian state was again slow to enter into the debate about the law. Indeed, the Algerian state has very few lessons to give on the subject of the independence of history as a discipline. When a group of French and Algerian historians gave a press conference and launched a petition, the Algerian state did not react. Historian and opponent of the law Mohamed

---

48   R. Laffitte, 'Banlieues des villes, banlieues du monde', 31 July 2006, http://www.indigenes-republique.org/spip.php?article388 (accessed 1 October 2011). Looking at the website of MIR, its message seems less simplistic about the links between the colonial past and present than impassioned criticisms of the organization suggest. Abdellali Hajjat takes a more sympathetic view of the MIR in his exploration of the tensions between the roles of researcher and political activist. He argues that the MIR consciously uses postcolonial discourse to rewrite defunct revolutionary far left themes: another example of a selective reading of the past providing a 'useful' narrative to structure present-day concerns. 'Les usages politiques de l'héritage colonial' in B. Stora and E. Témime (eds), *Immigrances* (Paris: Hachette, 2007), 195–210.

Harbi underlines the very selective way in which the Algerian state waves the old colonial scarecrow, when he asks why the Algerian government did not enter into the debate at the time: 'I would really like to know why? Algerians need to know. They should have access to information. They should not be treated like puppets that are mobilized by a click of the fingers.'[49] Nevertheless the much-talked about *Traité d'amitié* (Friendship Treaty to consolidate the success of the 2003 Year of Algeria in France) which was meant to be signed by France and Algeria did not take place, with the common assumption being that this was a result of the 2005-158 law. Yet whilst in a speech on 8 May 2005 in Sétif, Bouteflika violently fustigated colonialism, he also sent a letter on Bastille Day to Chirac making positive noises about the signing of the Treaty.[50] In an article in French weekly *L'Express* on 23 August 2005, a 'source close to de Villepin' described the law as a 'colossal error' – no doubt hoping that the Algerian authorities would understand this message and negotiations get back on track; so far they have not, and Sarkozy does not seem particularly interested in pushing forward the issue.

The purported circumstances surrounding the failure of the *Traité d'amitié* are interesting for two reasons. Firstly, as Harbi underlines, and as we will see in the example of the 2010 proposition to criminalize colonialism, Algiers can sideline anti-colonial repentance when it suits, so one might well suspect that the 23 February law provided a useful 'get out' clause. Secondly, there is a sense that the message Paris sends to Algiers through diplomatic channels is often different to that which dominates political discourse in France. It is no coincidence that the closest France has got to meeting Algerian demands for a formal apology for Sétif or recognition of crimes against humanity was through its ambassadors: Hubert Colin de Verdière described the 'massacres of 8 May' as an 'inexcusable tragedy' in 2005; in 2008 Bernard Bajolet stated that 'the period of negational-

---

49 *El Watan* (20 March 2010).
50 *Quotidien d'Oran* (23 August 2005).

ism' of colonial crimes 'is over'[51] Bouteflika's diplomatic team have also had to smooth over some of the outburts of politicians such as Veterans Minister Mohamed Chérif Abbas, who attacked the 'Jewish lobby' just before Sarkozy's visit to Algiers in December 2007. Whilst the French Parti socialiste demanded that Sarkozy cancel his trip, the Quai d'Orsay simply asked for 'clarifications' whilst Bouteflika underlined that foreign affairs were his 'reserved domain'.[52]

The *décalage* between the Algerian executive and the legislative – even in a system where there is no genuine separation of powers – and within the French executive, legislative and diplomatic corps on questions about memory is demonstrated in the next example. In February 2010, 125 Algerian deputies, of various parties, including the FLN, signed a proposition for a law criminalizing colonization and envisaging the creation of special tribunals to judge those accused of colonial crimes. The bill was proposed during a period of social conflict in Algeria, just before the FLN congress, and just before the (ultimately postponed) visit of French foreign minister Bernard Kouchner, where controversial topics such as compensation for victims of French nuclear testing in the Sahara and the visas and status of Algerian immigrants were to be discussed. There was also a lot of popular anger that Algerians had been placed on aviation security 'blacklists' and that Bouteflika had seemed powerless to prevent this.[53] Pied-noir organizations such as the Cercle Algérianiste reacted to the suggested bill with predicable anger, whilst some members of the French government, including Veterans Minister Falco and Eric Besson, of the now defunct Ministry for Immigration, Integration and National Identity, expressed their concerns over the project.

Kouchner downplayed the situation: 'the Algerian executive has not taken any position on this proposal, and its inclusion in the agenda is not

---

51 Ambassade de France en Algérie: http://www.ambafrance-dz.org/ambassade/spip.php?article755 and http://www.ambafrance-dz.org/ambassade/spip.php?article2000 (accessed 1 October 2011).
52 *20 minutes*, 29 November, 2007.
53 *Liberté*, 8 April 2010.

guaranteed.'[54] Indeed, by May 2010, associations in favour of the law, such as the Fondation du 8 Mai 1945, were already concerned that the government lacked the will to follow the proposition through. Former deputy and historian Mohammed Arezki Ferrad declared: 'Those who block the adoption of this law are people more concerned with preserving their personal interests to the detriment of those of the nation.'[55] In September 2010 the president of the Algerian National Assembly and FLN deputy Abdelaziz Ziari announced that the bill would not make it onto the agenda, vaguely referring to 'diplomatic, international and judicial' reasons. Algerian newspaper *Liberté* speculated that these reasons are most likely linked to close Franco-Algerian cooperation in anti-terrorism activities and dealing with hostage-taking of French nationals in the south of Algeria.[56] At this point, strengthening Franco-Algerian ties was seen as more important than domestic point-scoring.

We see that these controversies are linked to internal not external politics, and indeed, in many cases, local and regional politics. Debates about the past reveal Franco-French and Algero-Algerian tensions and divisions and the impact on external relations neither immediate nor inevitable. As far as possible, both French and Algerian states are very selective about what memory wars they tune into or the polemics they choose to fuel, where possible – clearly governments also at times struggle to keep memorial lobbies under control. These 'memory wars' are driven by an activist minority, split into a wide range of political allegiances with specific memorial axes to grind, who manage to get their voices heard through the media or through political networks. To what extent do these memory lobbies represent wider fractures in Algerian and French society which are the product of the colonial period? Stora argues that debates about the Algerian War are not a minority issue: 'This is a very broad question which touches upon French identity as Algeria was France.'[57] Yet when we scratch below the

54   *Le Nouvel Observateur* (9 February 2010).
55   *Liberté* (6 May 2010).
56   *Liberté* (26 September 2010).
57   *Quotidien d'Oran* (3 November 2008).

surface, these 'memory wars' reveal less about the impact of the colonial period on the post-independence relationship between two nation states and much more about internal Franco-French and Algero-Algerian tensions and divisions *which are only partly linked to colonial history*. It is to exploring some of these tensions that we shall now turn.

The Algerian War: Colonial Fracture or 'Useful Past'?

Discourse on Immigration in France

In his monograph on the teaching – or rather the occultation of – the Algerian War in French schools, Jo McCormack argues that 'the resentment and the bitterness of defeat in the Algerian War continue to feed into racism in France and to a general ignorance of this period of French history that is inhibiting Algerian French [sic] being able to find a home in France.'[58] In *Le Transfert d'une mémoire: de 'l'Algérie française' au racisme anti-Arabe* (Paris: La Découverte, 1999), Stora puts the case for the existence of a 'sudisme' *à la française* (pioneer spirit, economic exploitation, destruction and form of segregation). But is the extreme right in France – or extreme-right sympathies – a product of colonization and the bitterness of decolonization, or has it just re-appropriated the Algerian War as a 'useful past'? Christopher Flood and Hugo Frey underline the 'instrumental value of the link to decolonization' in the way in which immigration is used in extreme-right discourse, notably that of the Front National (FN, established 1972), demonstrated by expressions such as 'colonisation à rebours' and 'colonisation inversée' (literally: inverse colonialism). As the 'immigrant menace' became a 'Muslim immigrant menace' in the course of the 1990s, the past could also be adapted to reflect this new threat – the wartime FLN

---

58 J. McCormack, *Collective Memory: France and the Algerian War (1954–1962)* (Plymouth: Lexington, 2007), 3.

became 'a disguised extension of the eternal holy war against the infidel.'[59] The vocabulary of the Crusades has been revived, linked to the perceived need to protect France from a borderless Europe. The virulence of the anti-immigrant/ anti-colonial repentance discourse can partly be understood as the product of the background and experiences of older generations on the extreme right – who served as soldiers in Algeria (such as Jean-Marie Le Pen, first accused of torture in the *Canard enchaîné* in the 1980s) and former members or sympathizers of the OAS. The FN also taps into some of the *pied-noir* vote, by praising the achievements of settlers in the colonies whilst nevertheless denouncing the folly of the 'civilizing mission' and emphasizing the impossibility of 'assimilation.'

Yet meanings are ascribed to the past rather than the unfiltered past providing a clear frame of reference for understanding current extreme-right discourse, itself often confused and contradictory. Discussing the broader appeal – avowed or not – of extreme-right rhetoric, Ann Laura Stoler, a specialist of the history of French and Dutch colonialisms, emphasizes the 'polyvalent mobility' of racialized discourses. She warns against seeing racism in France today as either a replica of the old or a new kind of racism based on culture rather than skin colour, concluding that: 'Geneologies of racisms must reckon with racism's power to rupture with the past and selectively and strategically recuperate it at the same time.'[60]

Stoler warns us of the complexities of trying to assess the impact of the kaleidoscope of memories and vocabulary surrounding the Algerian War on the formulation of contemporary attitudes and policy on immigration across the political spectrum. There are a whole range of other, post-independence factors to take into account when seeking to understand how discourses and frames of reference develop: the 1979 Iranian revolution, the end of the Cold War, European integration, globalization

---

59 C. Flood and H. Frey, 'Questions of Decolonisation and Post-Colonialism in the Ideology of the French Extreme Right', in *The Decolonisation Reader*, ed. J. Le Sueur (London: Routledge, 2003), 399–413 (404, 406).

60 A.L. Stoler, 'Racist Visions for the Twentieth Century: On the Cultural Politics of the French Radical Right', in D.T. Goldberg and A. Quayson (eds), *Relocating Postcolonialism* (Oxford: Blackwell, 2002), 103–21 (106, 119).

and September 11, 2001 to name but a few. Indeed, a key criticism which can be made of Stora's *Le Transfert d'une mémoire* is its failure to account for the 'wilderness years' of both the extreme right and its key figures in the 1960s and 1970s, instead juxtaposing *Algérie française*, the OAS and the tensions between the army and politicians with extreme-right rhetoric in the 1980s and 1990s. Moreover, there is not a convincing explanation for how right-wing sections of the *Algérie française* generation manage to transmit their *sudiste* attitudes – which are the product of a specific historical context – to younger generations. Is this memory hereditary? Contagious? Broader questions are thus raised about the reception of extreme-right discourse: how can we assess whether 'colonial' frames of reference are implicitly assimilated by the majority of the French population whom – as is constantly deplored by memory activists of all political tendencies – have no knowledge of the Algerian War?

In *Identities, Discourses and Experiences*, Nadia Kiwan seeks to place the subjectivity of French youth of North African origin (Algerian, Tunisian and Moroccan) at the heart of her study. She highlights the disjuncture between often inflammatory public discourses and the frames of reference which young people use to understand their lives – and in her analysis of the semi-structured interviews she carried out with youths, the Algerian War does not emerge as a category. One of Kiwan's key conclusions is that one of the reasons why a collective French–North African voice has not emerged in France is because of the 'success' of the French Republican model – not in the sense that it has successfully challenged socio-economic, racial and cultural discrimination, but in that it has convinced a generation of young people that linking culture or cultural origins with a political or civic engagement is either irrelevant or stigmatizing.[61]

It is worth pointing out that new generations of Franco-Algerians are dual nationals in the true sense – i.e. they hold two passports – rather than just being Algerians living in France, which was the case for most immigrants living in France in the 1960s and 1970s. The stigma attached

---

61   N. Kiwan, *Identities, Discourses and Experiences: Young People of North African Origin in France* (Manchester: Manchester University Press, 2009), 213–14.

to taking a French nationality has passed. Instead, it is seen as a document which makes life easier. In terms of Franco-Algerian history, young Franco-Algerians have Algerian cultural references (as well as French ones), but they do not have the same historical references as young Algerians – which is hardly surprising given the lack of transmission in either the family or French school programmes.[62] When Algerian manager Rabah Saadane screened the iconic 1966 Gillo Pontecorvo film *The Battle of Algiers* to motivate his team during the 2010 World Cup, many of his players, notably dual nationals who had grown up in France, had not seen it.[63] It is impossible to imagine the same being true of Algerians who grew up in Algeria where the film is regularly screened on national television. To return to our opening example, despite the publicity which it generated, *Hors la loi* attracted low audience numbers in France. Released on 22 September 2010, after a week in cinemas, only 165,000 tickets had been sold. This compared to an audience of 350,000 for Xavier Beauvois's *Des hommes et des dieux* (2010) in the same week, giving a total of 1.3 million tickets sold for the latter film in the space of three weeks.[64] *Des hommes et des dieux* tells the story of French monks living in Tibhirine in Algeria, murdered in 1996 during the civil violence ravaging the country. However, the focus of the film is not Franco-Algerian relations, nor does it engage with the highly politicized question of who killed the monks. Instead, the film explored the nature of spirituality, a universal theme which seemed to attract audiences living in a materialist, consumerist and credit-crunched world.

If the French state were to heed the appeals to 'come to terms' with the French colonial past, and the Algerian state were to strive to create a more inclusive and less glorified collective memory, such a 'travail de mémoire' across both sides of the Mediterranean would represent an important moral and intellectual endeavour. However, it is erroneous to suggest that this would then resolve a host of contemporary issues linked to socio-economic inequalities and structural problems in the Republican model

62 McCormack, *Collective Memory*.
63 *The Guardian* (18 June 2010).
64 *Jeune Afrique* (7 October 2010).

of integration. Indeed contemporary anxieties in France about the place of immigrants from former colonies – notably North Africa – reflect a wider range of anxieties across the Western world about immigration, and in particular Muslim immigrants. In studying the origins and developments of these debates in a French context, we could learn much from a comparative approach which has so far been lacking in the existing literature, notably with countries such as Switzerland, Denmark, Austria and Sweden, which have neither France's colonial history nor a seminal event such as the Algerian War as a frame of reference.

Algerian Discourses on Legitimacy and the Nature of the 'Revolution'

As in France, Algerian organizations which get involved in cross-Mediterranean conflicts about the nature of colonialism and the Algerian War are a small minority. In addition to the Fondation du 8 mai, the Organisation nationale des mujahidin (National organization for veterans, ONM) and the Organisation nationale des enfants de chouhada (National organization for war orphans, ONEC) represent economic as well as commemorative interests. Both are very closely tied to the FLN but are not just the instrument of the FLN party, as they also lobby and apply political pressure around issues such as financial benefits and commemoration.[65] The main priorities for the vast majority of Algerians – particularly the under-thirties who make up around 70 per cent of the population – are an often difficult search for housing and employment. Yet history, and in particular the history of the Algerian War, does occupy an important place in the public sphere because it is linked to debates about political legitimacy. In the context of this chapter on Franco-Algerian relations, we will discuss two

---

65 T. De Georges argues that through promoting local and regional memory, the mujahidin version of history 'subtly differs' from the national narrative. 'The Shifting Sands of Revolutionary Legitimacy: the role of the former mujahidin in the shaping of Algeria's collective memory', *Journal of North African Studies*, 14/2 (2009), 273–88.

aspects of the relationship with the past in Algeria – the notion of 'traitor' and perceptions of the Algerian War as an 'anti-colonial' war.

The debates which occupy the French memorial space – whether colonialism was 'positive' or 'negative' or what should be done about *harkis*, are non-debates in Algeria. There is broad consensus that colonialism was profoundly exploitative and destructive, and the figure of the *harki* is the epitome of treachery, from the statements of the Fondation du 8 mai to the playground insult, where it has become a synonym for traitor and has perhaps lost some of its historical specificity. In oral testimony, notably in rural areas, *harkis* tend to be associated with extreme violence and rape. As Fatima B. from the village of Agraradj in Kabylia euphemistically puts it: 'When it was the French [soldiers], it was OK. But as soon as they brought in the *harkis*, they knew the population and the problems began.'[66] The extent to which the factual basis of this reputation has been conflated with the image of traitor is certainly open to debate. What one can say is that *harkis* presented much more of a threat to rural communities helping the FLN because of their linguistic and local knowledge. Somewhat disingenuously, given *harki* self-identification with Algerian land and culture, the Algerian state insists on underlining that the *harkis* are a French problem, with Algerian Foreign Minister Salah Eddine Belabas declaring in response to a question about *harkis* obtaining visas to return to visit Algeria that: 'Harkis are French citizens who possess French passports and are subject to the same legal requirements as all French citizens.'[67]

Debates in Algeria are not about *how harkis* should be seen and treated but *who* the 'traitors' and the 'patriots' are. This question is of central importance to political life in Algeria as since 1962 political legitimacy has been based not on democratic mechanisms but on the real or proclaimed participation of the 'revolutionary family' in the War of Independence. It is difficult to imagine Bouteflika coming to power in 1999 without the weight of his past as a *mujahid* (veteran) in the Algerian War. In addition to their political power, recognized *mujahidin* are entitled to a whole

66 Interview with Fatima B., Agraradj, Kabylie, 16 June 2005.
67 *El Watan* (17 July 2004).

range of financial benefits which have included not only a pension but also taxi and café permits, free airline tickets, and the possibility to import a car without paying tax. Veteran status is thus highly attractive, can be bought and sold, and the phenomenon of 'false *mujahidin*' is recognized in Algeria as a widespread political and social problem.[68] Injustice in independent Algeria is often expressed by arguing that 'the true liberators' did not take power upon independence, instead Algeria was confiscated by opportunistic last-minute resisters, often Algerians who were French army officers, who capitalized on the decimation of the real *mujahidin* to seize power for their own benefit. Whatever the historical merits of this interpretation, and it certainly has a historical foundation, it is a powerful strand of popular discourse.

Since the liberalization of the press in 1989, there are regular controversies over wartime heroes whose legacies are subject to revisionism and rehabilitation. These debates are usually linked to contemporary conflicts opposing different groups associated with particular 'clans', often associated with specific regions. One recent polemic was around the publication of Said Sadi's *Amirouche: une vie, deux morts, un testament* (Paris: L'Harmattan, 2010), the biography of an FLN commander in Kabylie, who fell out of favour during the war and has subsequently been accused of paranoid purges and anti-intellectualism. Amirouche was killed in 1959 on his way to Tunisia, by a French army seemingly well-informed of his movements. Sadi, president of the opposition party Rassemblement pour la culture et la démocratie (Rally for culture and democracy, RCD) is from the same region as Amirouche, and argues that he was betrayed by the FLN clan of Boussouf and Boumediene – the faction who successfully seized power after independence, and Bouteflika (FLN) was of course a key figure in the Boumediene regime. A virulent debate ran for weeks in the independent press.

Conflicts about 'true' and 'false' *mujahidin* on one level demonstrate the historical origins of contemporary divisions surrounding political legitimacy. Yet one might more persuasively interpret today's debates as not

---

68 Meynier, *Histoire intérieure du FLN*, 288–9.

simply a teleological continuation of past tensions, but instead a set of easily accessible metaphors to discuss a broader range of political and social issues not necessarily linked to either the colonial past or anti-colonial conflict. The pejorative and widely used Algerian term *hogra*, used to describe the state's contempt for 'ordinary' Algerians, is not just aimed at those perceived to be historically illegitimate, or those *mujahidin* or (now adult) war orphans with a wide range of advantages, but all those who through corruption work only to benefit themselves, politically and financially. The widespread use of the term *hogra*, and its comprehensive denunciation of the political/financial/military elite, emphasizes that Algerians today do not simplistically see their country's problems through the lens of colonial after-effects.

In Algeria, as in France, the past is ascribed multiple meanings to suit present political needs. The term used most commonly to refer to the Algerian war is *al-thawra al-jazā'iriyya*, which is generally, although not entirely accurately, translated as 'the Algerian revolution.'[69] The version of the 'Algerian revolution' promoted by the state and taught in schools in Algeria is glorified and asceptized. Key elements include the sacrifice of the 'martyrs', the exploitation and violence of the homogenous colonizer, a strong emphasis on the role of the *'ulama'* (Muslim theologian) movement in the development of Algerian nationalism and the unified people behind the FLN. Other nationalist strands are at best marginalized: this is why the commemoration of Sétif is potentially problematic, as is the commemoration of the events in France of 17 October 1961.[70] The way in

---

69 On this point see G. Meynier, 'La "Révolution" du FLN (1954–1962)', *Insaniyat*, 25–6 (2004), 7–25.

70 Demonstration against the curfew imposed on North Africans, violently repressed by the Parisian police and resulting in a number of Algerian deaths. The demonstration was organized by the French FLN Federation and the FF-FLN ruptured with those who seized power in 1962 making commemoration difficult. Boumediene made 17 October a 'national day for emigration' in 1968, emphasizing the heroic role played by immigrants in the liberation struggle, as part of a broader attempt at reconciliation. See J. House and N. MacMaster, *Paris 1961: Algerians, State Terror and Memory* (Oxford: Oxford University Press, 2006).

which the state has chosen to teach history since 1962 – in a similar way to religious education – and the international context from the late 1970s onwards and the rise in Muslim identity politics across North Africa and the Middle East mean that young people today often have a perception of the war which is far removed from its anti-colonial context. In a study which I carried out in 2007 amongst university students in Algiers on perceptions of female veterans (*mujahidat*), many students had a tendency to 'rewrite' the Algerian War as a primarily religious, rather than anti-colonial struggle. In interviews which I carried out with female veterans, it was very clear that the enemy was the colonial oppressor. In many of my interviews with students today, women's stories of concrete human suffering had become heroic tales of religious martyrdom, with outspoken, feisty and often unconventional women been rewritten as saintly grandmothers[71] – another example of how our present view of the world defines the historical narrative which we choose to weave.

Conclusion

The purpose of this chapter has not been to deny the influence of the colonial period or the Algerian War on contemporary France and Algeria in an ahistorical or reactionary way but to emphasize the internal dynamics not necessarily related to the colonial period which create what are often hastily classed as Franco-Algerian memory wars. Whilst foregrounding these national contexts, we also need to highlight the international currency of 'memory' and its vocabulary of pardon, forgetting, reconciliation and repentance: 'In Africa and in Europe, the dialogue on the colonial past is not limited to the hermetic space of relations between the ex-colonized and the ex-colonizers, it also borrows from this universal language and

---

71   Vince, 'To be a *mujahida*'.

motifs and arguments.'[72] This 'dialogue' on the colonial past needs to be approached with caution. Charles Forsdick emphasizes that: 'In talking of sharing – of histories, of narratives, of memories – there is a need to reflect vigilantly on the nature of the common ownership of the past this implies, to ask whether it is latent or engineered, existing or fantasized.'[73] Not only can the 'common ownership' of the past be imagined, so can conflicting versions of the past be engineered and fantasized for present needs. René Gallissot puts it pithily: we need to avoid an analytically lazy obsession with 'dramatic outbreaks of "bursts of memory" or "bleeding memory".'[74]

On both sides of the Mediterranean there are extremely well-organized memorial lobbies which link into local and national networks of politicians, and reactive and proactive independent presses interested in memory debates. The opportunism of many politicians who engage in the politics of memory is perhaps best embodied by President Nicolas Sarkozy who has engaged with both the politics of politically correct repentance and unashamed colonial nostalgia and revisionism, alternately satisfying and enraging a range of memory lobbies and civil society actors. To caricature the Algerian state as bringing out the old colonial scarecrow every time they need an injection of popular legitimacy is to take a static and simplistic view of Algerian society. As Frederick Cooper warns, we need to be aware of the dangers of the field of postcolonial study becoming 'unmoored from analysis of processes unfolding over time.'[75]

The Mediterranean no longer crosses through 'France' as the Seine crosses through Paris. Nor is the Mediterranean a 'wall' dividing two nations torn apart by a violent divorce. Indeed, both of these powerful images are based on questionable stereotypes. The first ignores that colonial rule was

---

72   C. Deslaurier and A. Roger, 'Mémoires grises. Pratiques politiques du passé colonial entre Europe et Afrique', *Politique africaine*, 120 (2006), 5–27 (19).

73   C. Forsdick, 'Colonialism, Postcolonialism and the Cultures of Commemoration' in *Postcolonial Thought*, ed. Forsdick and Murphy, 271–84; 281–2.

74   R. Gallissot, 'Libérez l'histoire de la guerre d'indépendance algérienne des allégeances nationales', *L'Année du Maghreb*, 1 (2004), 501–17.

75   F. Cooper, *Colonialism in Question: Theory, Knowledge, History* (London: University of California Press, 2005), 52.

based on the legal exclusion of the majority from citizenship. The second distracts us from the fact that the most virulent memory wars are Franco-French and Algero-Algerian. In doing so, dynamics within and beyond the Franco-Algerian relationship which shape present attitudes and policy are obscured and the Algerian War is used as a convenient metaphor, a 'useable past', rather than providing a convincing explanation for contemporary political and social fractures.

Bibliography

Aggoun, L., and Rivoire, J.-B., *Françalgérie: Crimes et mensonges d'Etat* (Paris: La Découverte, 2004).
Ait Chaalal, A., 'La politique étrangère de l'Algérie: entre héritage et originalité', in *La politique étrangère: le modèle classique à l'épreuve* (Brussels: PIE–Peter Lang, 2004), 203–15.
Ashcroft, B., Griffiths, G., and Tiffin, H., *The Empire Writes Back: Theory and Practice in Post-colonial Literature* (London and New York: Routledge, 1989, 2nd edn, 2002).
Baghzouz, A., 'France-Algérie: rejouer le match?', *Outre Terre*, 8 (2004), 191–4.
Beaugé, F., *Algérie: une guerre sans gloire: histoire d'une enquête* (Paris: Calmann-Lévy, 2005).
Blanchard, P., and Bancel, N., *Culture post-colonial 1961–2006* (Paris: Autrement, 2005).
Carlier, O., *Entre nation et jihad: histoire des radicalismes algériens* (Paris: Presses de Sciences Po, 1995).
Cheurfi, A., *La Classe politique algérienne de 1900 à nos jours: dictionnaire biographique* (Algiers: Casbah Editions, 2001).
Clancy-Smith, J., 'Islam, Gender and Identities in the Making of French Algeria, 1830–1962', in J. Clancy-Smith and F. Gouda (eds), *Domesticating the Empire: Race, Gender and Family Life in French and Dutch colonialisms* (Charlottesville and London: University of Virginia Press, 1998), 154–74.
Conan, E., and Rousso, H., *Vichy, un passé qui ne passe pas* (Paris: Folio, 1996).
Conklin, A.L., *A Mission to Civilise: The Republican Idea of Empire in France and West Africa* (Stanford: Stanford University Press, 1997).

Cooper, F., *Colonialism in Question: Theory, Knowledge, History* (London: University of California Press, 2005).
Daguzan, J.-F., 'Les Relations franco-algériennes ou la poursuite des amicales incompréhensions', *Annuaire français des relations internationales*, 2 (2001), 438–50.
Darbouche, H., and Zoubir, Y., 'The Algerian Crisis in European and US Foreign Policies: A Hindsight Analysis', *The Journal of North African Studies*, 14/1 (2009), 33–55.
De Georges, T., 'The Shifting Sands of Revolutionary Legitimacy: The Role of the Former Mujahidin in the Shaping of Algeria's Collective Memory', *Journal of North African Studies*, 14/2 (2009), 273–88.
Deslaurier, C., and Roger, A., 'Mémoires grises: pratiques politiques du passé colonial entre Europe et Afrique', *Politique africaine*, 120 (2006), 5–27.
Evans, M., and Phillips, J., *Algeria: Anger of the Dispossessed* (New Haven and London: Yale University Press, 2007).
Flood, C., and Frey, H., 'Questions of Decolonisation and Post-Colonialism in the Ideology of the French Extreme Right', in *The Decolonisation Reader*, ed. J. Le Sueur (London: Routledge, 2003), 399–413.
Forsdick, C., and Murphy, D., *Postcolonial Thought in the French-Speaking World* (Liverpool: Liverpool University Press, 2009).
Gadant, M., *Islam et nationalisme en Algérie: d'après El Moudjahid, organe central du FLN de 1956 à 1962* (Paris: L'Harmattan, 1988).
Gallissot, R., 'Libérez l'histoire de la guerre d'indépendance algérienne des allégeances nationales', *L'Année du Maghreb*, 1 (2004), 501–17.
Grimaud, N., *La Politique extérieure de l'Algérie (1962–1978)* (Paris: Karthala; 1984).
Guénif-Souilamas, N., 'The Other French Exception: Virtuous Racism and the War of the Sexes in Postcolonial France', *French Politics, Culture and Society*, 24/3 (2006), 23–41.
Hajjat, A., 'Les usages politiques de l'héritage colonial', in B. Stora and E. Témime (eds), *Immigrances* (Paris: Hachette, 2007), 195–210.
Harbi, M., *Le FLN, mirage et réalité: des origines à la prise du pouvoir (1945–1962)* (Paris: Editions Jeune Afrique, 1980).
House, J., and MacMaster, N., *Paris 1961: Algerians, State Terror and Memory* (Oxford: Oxford University Press, 2006).
Ighilahriz, L., with Nivat, A., *Algérienne* (Paris: Fayard and Calmann Lévy, 2001).
Kaddache, M., *Histoire du nationalisme algérien*, vols 1–2 (Algiers: Entreprise nationale du livre, 1993).
Kiwan, N., *Identities, Discourses and Experiences: Young People of North African Origin in France* (Manchester: Manchester University Press, 2009).

Liauzu, C., and Manceron, G. (eds), *La colonisation, la loi et l'histoire* (Paris: Syllepse, 2006).
Lorcin, P. *Imperial Identities: Stereotyping, Prejudice and Race in Colonial Algeria* (London and New York: I.B. Tauris, 1995).
MacMaster, N., *Burning the Veil: The Algerian War and the 'Emancipation' of Muslim Women, 1954–1962* (Manchester: Manchester University Press, 2009).
Manceron, G., *Marianne et les colonies: une introduction à l'histoire colonial de la France* (Paris: La Découverte, 2005).
Manceron, G., and Remaoun, H., *D'une rive à une autre: la guerre d'Algérie de la mémoire à l'histoire* (Paris: Syros, 1993).
McCormack, J., *Collective Memory: France and the Algerian War (1954–1962)* (Plymouth: Lexington, 2007).
McDougall, J., *History and Culture of Nationalism in Algeria* (Cambridge: Cambridge University Press, 2006).
Meynier, G., 'La "Révolution" du FLN (1954–1962)', *Insaniyat*, 25–6 (2004), 7–25.
Meynier, G., *Histoire Intérieure du FLN 1954–1962* (Paris: Fayard, 2002).
Naylor, E., 'The Politics of a Presence: Algerians in Marseille from Independence to "Immigration Sauvage" (1962–1974)', unpublished PhD thesis (University of London, 2011).
Roberts, H., *The Battlefield Algeria 1988–2002: Studies in a Broken Polity* (London and New York: Verso, 2003).
Rosoux, V.B., 'Poids et usage du passé dans les relations franco-algériennes', *Annuaire français de relations internationales*, 2 (2001), 451–65.
Sadi, S., *Amirouche: une vie, deux morts, un testament* (Paris: L'Harmattan, 2010).
Scagnetti, J.-C., 'Pays d'origine et encadrement des pratiques religieuses: l'Algérie et ses émigrés (1962–1988)', *Cahiers de la Méditerranée*, 78 (2009), 177–202.
Sibeud, E., 'Post-colonial et Colonial Studies: enjeux et débats', *Revue d'histoire moderne et contemporaine*, 51/4 (2004), 87–95.
Sieffert, D., *Israël Palestine: une passion française* (Paris: La Découverte, 2004).
Shephard, T., *The Invention of Decolonisation: The Algerian War and the Remaking of France* (Ithaca, NY: Cornell University Press, 2006).
Stoler, A.L., 'Racist Visions for the Twentieth Century: On the Cultural Politics of the French Radical Right', in D.T. Goldberg and A. Quayson (eds), *Relocating Postcolonialism* (Oxford: Blackwell, 2002), 103–21.
Stora, B., *Le Transfert d'une mémoire: de 'l'Algérie française' au racisme anti-Arabe* (Paris: La Découverte, 1999).
Stora, B., *La Gangrène et l'oubli: la mémoire de la guerre d'Algérie* (Paris: La Découverte, 1991).

Stovall, T., and Van Den Abbeele, G. (eds), *French Civilisation and its Discontents* (Lanham, MD: Lexington, 2003).

Todorov, T., *On Human Diversity: Nationalism, Racism and Exoticism in French Thought* (Cambridge, MA: Harvard University Press, 1993).

Vince, N., 'To be a Moudjahida in Independent Algeria: Itineraries and Memories of Women Veterans of the Algerian War', unpublished PhD thesis (University of London, 2008).

Weil, P., *Qu'est-ce qu'un Français?* (Paris: Grasset, 2004).

Notes on Contributors

GÉRALD ARBOIT is research director of the Centre Français de Recherche sur la Renseignement. He holds a PhD in contemporary history from the University of Strasbourg. He is a consultant to France's Ministry of Foreign Affairs, as part of the Mediterranean Cultural Workshop (2008–11), and to the Council of Europe in Strasbourg. The author of several books and articles, he also teaches at various universities in France.

MANAR EZZAT BAYOUMI was awarded her PhD in economics by Cairo University in 2009. Her research centred on climate change in the Mediterranean, with a focus on actors, strategy and representation. She is currently a researcher at the Institute of Agricultural Economics, Cairo.

MARY BEN BONHAM is an assistant professor in the Department of Architecture and Interior Design at Miami University, Ohio. In her teaching and research, she investigates contemporary buildings as manifestations of culture that incorporate evolving building technologies. Having worked as an architectural intern in Paris, she has a special interest in the relationship between building technology and culture in France.

ANDREA BRAZZODURO holds a PhD from 'La Sapienza' University of Rome and the University of Paris X (2011). His book on French Veterans of the Algerian War of Independence is to be published in 2012 (*Soldati senza causa in una guerra senza nome*, Roma-Bari: Laterza). Beyond Francophone studies, his research interests include the history of colonialism, the anthropology of war and violence, the history of memory, oral history and post-colonial theory. He is a member of the editorial board of *Zapruder: Rivista di storia della conflittualità sociale*.

IVANO BRUNO is a PhD candidate at the University of Portsmouth. He is currently finalizing his doctoral thesis under the working title of 'The European Union and Norm Transfer: The Case of the ENP in Egypt'. He is currently a programme manager for the EU Delegation in Cairo. Any views or opinions presented in Ivano Bruno's chapter are solely those of the author and do not represent those of the European Union Delegation in the Arab Republic of Egypt.

PATRICIA CAILLÉ is maîtresse de conférences in the Information-Communication Department at the Université de Strasbourg, where she works on images, films and, more particularly, Maghrebi cinemas. She is the coordinator of the research project 'Maghreb et cinémas: circulation des films, production des savoirs et constitution d'un patrimoine' (2010–12), which is funded by the Agence Universitaire de la Francophonie. She is the co-editor of *Adolescences et cultures: Pratiques, usages et réception à l'épreuve des genres* (2009) and a *Dossier Africultures* on the Cinemas of the Maghreb (2012).

EMMANUEL GODIN is principal lecturer in French and European studies in the School of Languages and Area Studies at the University of Portsmouth. His research is on French history, politics and society. He co-authored *France 1815–2003* with Martin Evans and has recently co-edited *The End of the French Exception?* with Tony Chafer and *Mapping the Extreme-Right in Contemporary Europe: From Local to Transnational* with Andrea Mammone and Brian Jenkins. He is currently working on a monograph on France since the 1970s.

JEAN-ROBERT HENRY is directeur de recherche at the CNRS, Institut d'Etudes et de Recherches sur le monde Arabe et Musulman (IREMAM), Aix-en-Provence, and has published widely on France, Algeria and the Mediterranean. His publications include 'L'invention des frontières méditerranéennes de l'Europe: une dérive culturaliste', *Après-demain* (avril 2004) and 'L'Europe et son Sud' *Ceras – revue Projet* (janvier 2008).

Notes on Contributors

BRUNO LEVASSEUR teaches in the French Department of the University of Birmingham. His research focuses on cultural and political representations of the *grands ensembles* in contemporary France.

JEAN-YVES MOISSERON is a socio-economist at the French Institute for Developing Countries (IRD). He is former director of the IRD office in Cairo and is currently professor at the University Paris 1 – Pathéon-Sorbonne and at the EHESS (Ecole des Hautes Etudes en Sciences Sociales – Paris). He has published several books and articles on development issues using pluri-disciplinary approaches. For several years he was involved in a research programme entitled 'Governance in the Euro-Mediterranean'. His work on the Mediterranean has been rewarded by the French Society of Geography.

ED NAYLOR received his PhD in 2011 from Queen Mary, University of London. His thesis, entitled 'The Politics of a Presence: Algerians in Marseille from independence to "immigration sauvage" (1962–1974)', was funded by the Arts and Humanities Research Council. His research interests lie in the fields of colonial legacies, migration, and the politics and society of post-war France. He currently teaches at the University of Portsmouth.

DAVID STYAN teaches in the Department of Politics, Birkbeck College, London. He has written widely on French foreign policy, notably towards the Middle East and Africa. He is the author of *France and Iraq: Oil, Arms and French Policy Making in the Middle East* (2006).

NATALYA VINCE received her PhD in 2008 from the University of London (Queen Mary) for her thesis 'To be a *moudjahida* in Independent Algeria: Itineraries and Memories of Women Veterans of the Algerian War'. Her research interests lie in twentieth-century French and North African history, politics and society. She is currently a lecturer in the School of Languages and Area Studies at the University of Portsmouth.

Index

Abbas, Mohamed Chérif 328
Accoyer, Bernard 116
African Union 8, 15
Albou, Karin 170
Alexandria 29, 133, 136, 143, 145, 146
Algeria 3, 14, 33, 39–40, 44, 48, 58, 72,
 79, 80, 84, 109, 156, 163–5, 170,
 172–3, 250, 257, 259, 261, 271,
 310–18, 326–9, 334–8, 339
Algerian War 12, 47, 138, 220, 225, 228,
 231, 239, 270, 275–99, 305, 307–8,
 310, 318–21, 337
Alleg, Henri 319
Alliance française 41, 140
Alliot-Marie, Michèle 2, 7, 105, 119–20,
 129
Allouache, Merzak 177
Amicale des Algériens 271, 314
Amari, Raja 166, 173, 177, 184
Amrani, Youssef 5
Andréani, Jacques 142
Anna Lindh Foundation 62, 146–7, 151,
 152, 155, 157
Arab-Israeli conflict 9, 46, 47, 53, 86, 92,
 136, 153, 156, 317
Arab League 47, 48, 53, 87, 146, 149,
 155
'Arab policy' *see politique arabe*
Arab Spring 1, 6, 15, 69, 98, 106
arms sales 41, 107, 110, 113–17, 118,
 119–20, 122, 124, 128
Assad (al-), Bashar 2, 52
Ateliers culturels méditerranéens 13, 137,
 142, 143–6, 150, 153

Audin, Josette 319
Audisio, Gabriel 42, 58
Aussaresses, Paul 318, 321
Azoulay, André 157

Baccar, Selma 171, 175, 176
Bachir Chouikh, Yamina 166, 170, 173,
 184
Bajolet, Bernard 327
banlieue 223–4, 233, 234–5, 236, 308, 325,
 326
Barcelona Process *see* Euro-
 Mediterranean Partnership
Barrot, Jacques 123
Bashir, Omar (al-) 125
Bedjaoui, Amal 170
Belhadj, Ali 73
Belkhadem, Abdelaziz 306
Ben Ali, Zine El Abidine 3, 5, 81, 105
Ben Bella, Ahmed 311
Bendjedid, Chadli 83, 313
Benghazi 105, 112–13, 118, 129
Benguigui, Yamina 170, 173, 185
Benhadj, Rachid 177
Benlyazid, Farida 169, 173, 175, 176, 184,
 186
Ben N'hia, Kaouther 185
Berlusconi, Silvio 6, 52, 91, 118
Berque, Jacques 42
Besson, Eric 328
Bigeard, Marcel 319
Blair, Tony 113, 118
Bockel, Jean-Marie 114
Borghezio, Mario 5

Bornaz, Kalthoum 171, 173, 177, 184
Bouchareb, Rachid 42, 305–6, 307, 333
Boughedir, Ferid 166
Bouhati (el-)Souad 185
Boumediene, Houari 258, 259, 261, 311, 312–14, 336
Bourdieu, Pierre 242
Bourquia, Farida 168, 169
Bouteflika, Abdelaziz 311, 315–16, 317, 318, 327, 318, 336
Boutros-Ghali, Boutros 47, 87
Braudel, Fernand 10, 42, 50, 147
Bugat, Alain 117
Bush, George W. 49, 54, 86, 110, 111

Cairo 58, 87, 137, 138
Camus, Albert 13, 50
Centre cinématographique marocain 168–9
Chad 8, 109, 123, 124–6
Chala, Samia 185
Chérabi, Nadia 173, 175
Chevalier, Michel 24–5
Cheysson, Claude 44
China 8, 120, 312
Chirac, Jacques 9, 46, 48, 50, 81–2, 84, 86, 95, 96, 119, 133, 134, 136, 137–8, 143, 146, 156, 158, 239, 249, 292, 305, 321, 325, 327
Chouikh, Mohamed 177
Chraïbi, Driss 283
civil society 15, 41, 56, 59–62, 98, 138–42, 143, 147, 149, 158, 250, 339
'civilizing mission' 26–7, 309, 331
'clash of civilizations' 13, 32, 34, 55, 77, 143, 156, 158, 295
Colin de Verdière, Hubert 327
colonial fracture 11–12, 308, 330–8
coopération 40, 81, 114, 139, 140, 142
Council of Europe 64, 87, 137, 146, 154–5, 157

Daeninckx, Didier 221, 225, 227, 228–32
Darfur 109, 125–6
Debray, Régis 158
Déby, Idriss 125
Defferre, Gaston 11, 251, 255–71
De Gaulle, Charles 40, 46, 47, 55, 76, 110, 158, 270, 281, 312
De Rohan, Josselin 142
Derrida, Jacques 221, 240
De Villepin, Dominique 32, 33, 327
Djebar, Assia 170
Domenech, Gabriel 252, 258

Egypt 2, 8, 23, 29–30, 39, 45, 69, 73, 82, 86–9, 96, 98, 105, 110, 129, 138
El Fani, Nadia 171, 173
Euro-Mediterranean Partnership 19, 28, 29, 30–2, 34–5, 38, 45, 46, 48, 50, 52, 57, 59–60, 63, 69, 71, 74–6, 84–5, 86, 88–9, 91–3, 96, 98–9, 124, 133, 140, 148, 178
European Commission 6, 84, 92, 94, 153, 154, 178
European Council 2, 3, 6, 51, 52
European Neighbourhood Policy 19, 32, 84, 93, 312
European Operational Rapid Force 125, 126
Europeanization 65, 70–1, 76, 82, 88, 91–4, 98–9

Falco, Hubert 306, 328
Fanon, Frantz 225, 275, 288
Feraoun, Mouloud 283
Ferrero-Waldner, Benita 112, 114
Fillon, François 106, 116, 121
Five + five (5+5) dialogue 44–6, 50, 88, 124, 127
Fondation du 8 mai 1945 324, 329, 334–5

Index 351

Fonds de la diversité 184
Fonds sud 180, 183–4
Francophonie 47, 78, 182, 183, 311
Front de libération nationale (FLN) 265, 293, 306, 314, 318–20, 328, 331, 334–7
Front national (FN) 5, 252, 256, 264, 267, 330–1

Gaddafi, Muammar 4, 5, 8, 95, 105–6, 108, 110–18, 121, 122, 124, 126, 128–9
Gamma't al-Islamiyya (al-) 73
gas (natural) 80, 311, 312
Gaudin, Jean-Claude 201
Gaullism 46, 48, 50, 70, 76, 81, 86, 92, 94, 96, 133, 138
Gaza 9, 53, 55, 57, 153, 156
Germany 4, 8, 38, 40, 49, 51–2, 57, 59, 72, 78, 81, 82, 84, 85, 87, 91, 92–3, 95, 113, 120, 146, 173, 313
Ghanem, Choukri 119
Giscard d'Estaing, Valéry 262
Global Mediterranean Policy 71–2, 77
González, Felipe 83, 85
Grande mosquée de Paris 314
Greece 6, 21, 26, 32, 45, 46, 124, 144, 278
Guaino, Henri 9, 50, 92, 149
Guéant, Claude 3, 96, 112, 114, 116, 123
grandeur 10, 46, 70, 76, 85, 100

Hadj, Messali 305, 325
Halimi, Gisèle 319
Hariri, Rafic 81
harkis 291–3, 323–4, 335
Hassan II 81
Hollande, François 115
Hortefeux, Brice 121
hybridity 12, 13–14, 197, 240

Ighilahriz, Louisette 318–20, 322
immigration 2, 5–6, 8, 11, 61, 79, 108, 109, 124, 189, 222–3, 228, 232, 233–43, 249–71, 313–14, 315, 326, 328, 330–4
International Monetary Fund 6, 313
Iraq 86, 109
Iran 11, 55, 73, 95, 331
Islam *see* religion
Islamism 7, 72–3, 75, 80, 84, 126, 165, 316–17, 322
Israel 8, 9, 46–7, 53, 55, 56, 57, 72, 95, 152, 156
Istanbul 154
Italy 2, 6, 28, 44, 51, 52, 60, 72, 81, 83, 89, 91, 105, 106, 109, 118–19, 120, 122–3, 128, 313

Jospin, Lionel 321
Juppé, Alain 3, 129

Kassari, Yasmine 184
Kilani, Leila 172, 184
Kohl, Helmut 85
Kouchner, Bernard 95, 108, 114, 116, 125, 328
Koussa, Moussa 119

Lakhdar-Hamina, Mohamed 177
Lampedusa 5–6, 122
Laskri, Amar 177
Lebanon 79
Legendre, Jacques 142
Leon, Bernardino 4–5
Le Pen, Jean-Marie 331, 264
Le Pen, Marine 5
Lévy, Bernard-Henri 3, 129
Libya 2, 3, 4, 8, 28, 40, 44, 69, 95, 105–29, 316
Loo, Charles-Emile 255, 268
Loos, François 119

Luca, Lionnel 306
Lureau, François 117

Madani, Abassi 73
Madrid 92
Maghreb 8, 14, 40, 41, 42–3, 45, 59, 79–81, 83–4, 86, 87–8, 107, 156, 163–8, 171, 173–4, 175–6, 178, 180–90
Malraux, André 158
Malte-Brun, Conrad 22–3
Mammeri, Mouloud 283
Marcellin, Raymond 249, 257–8
Marrakech 60
Marseille 15, 48, 52, 55, 57, 59, 61, 69, 89, 148–59, 193–214, 249–71
Massu, Jacques 318
Meddour, Mounira Anna 185
MEDIA programme 177–9
Mediterranean, conceptualizations of 7–9, 19–36, 42, 48, 188, 307, 339–40
Mediterranean Cultural Forum (Etats généraux culturels méditerranéens) 133, 134, 136, 141, 148–50, 151–9
Mediterranean Union *see* Union for the Mediterranean
memory 11–12, 40, 90, 137, 145, 226, 228, 231–2, 277, 279, 284–6, 287, 290–9, 307, 318–30, 338–40
Merkel, Angela 57, 91, 92
Messmer, Pierre 257
migration 10–11, 34, 60, 61, 79, 80, 93, 94, 107, 109, 115, 118, 121, 122, 123, 165, 188, 250, 257, 260, 261, 271, 313–14, 316
Ministry for Immigration, Integration, National Identity and Co-development 3, 58, 62, 94, 121, 328

Mitterrand, François 44, 80, 83, 85, 195, 265
Moknèche, Nadir 166
Moratinos, Angelo 63, 123
Morin, Edgar 42, 50
Morin, Hervé 120
Morocco 14, 32, 39, 44, 65, 79, 81, 83, 84, 97, 109, 122, 139, 149, 156, 157, 163–4, 168–9, 172–4, 176, 177, 181, 182, 184, 186, 190, 316
Moscovici, Pierre 113, 116
Mouvement des Indigènes de la République 42, 308, 326
Mouvement des Travailleurs Arabes 253, 261
Mubarak, Hosni 2, 87, 96, 105, 138

Napoleon Bonaparte 23, 39
Napoleon III 39
National Transitional Council (Libya) 3, 106, 128–9
Nejjar, Narjiss 169, 172, 177, 184
NGOs (non-governmental organizations) 30, 33, 34, 61–3, 125, 138–41, 147
North Atlantic Treaty Organization (NATO) 4, 6, 86, 95, 106, 113
nuclear power 41, 55, 107, 108, 109, 112, 114–17, 120

Obama, Barack 4, 58
'Odyssey' (Homer) 34, 278, 299
Ogres de Barback, Les 221, 225, 227, 237–41
oil 40, 47, 72, 108–9. 118–19, 122, 124, 126, 249, 312–13, 318
Organisation armée secrete (OAS) 323, 331
Organisation internationale de la Francophonie (OIF) 182–3, 311

Organization for Economic Cooperation and Development (OECD) 109, 117–20, 128
Ottoman Empire 29, 38–9, 41, 144

Papon, Maurice 287
pan-Africanism 8, 14, 124
Parti Socialiste (PS) 115, 125, 251, 262, 263, 328
petrol *see* oil
pieds noirs 265, 289, 291–2, 306, 320, 323–4, 328
Pierini, Marc 113
Pisani, Edgar 44
politique arabe 8, 44, 46–7, 311
Pompidou, Georges 46, 47, 72, 110, 258, 270
Poniatowski, Axel 116
Portugal 44, 81, 250
postcolonial theory 12, 224, 239, 307–8, 240, 307–10, 324–5, 339
Pouillot, Henri 320
Prodi, Romano 113

Raoult, Eric 234, 235
rayonnement 10, 135
Reclus, Elisée 25–6
religion 12, 14, 23, 29, 30, 33, 34, 41, 43, 57, 58, 62, 150, 153, 158, 144, 195, 206, 309, 315, 322, 330
republicanism 221–43, 314–15, 332–4
Roman empire 20–1, 26, 27, 28, 30, 144, 201
Russia 4, 8, 95, 120, 155, 312

Sahraoui, Djamila 170, 173, 177, 184
Saint Simonianism 24, 25, 26, 32, 39, 40
Sarkozy, Nicolas 1–4, 6, 9, 12, 14, 37, 40, 41, 42, 46, 47, 49–50, 54, 55, 57, 69, 71, 89, 90–100, 105–9, 111–14, 117, 121–3, 124–9, 134, 136, 148–9, 151, 156, 158, 285–6, 298, 312, 327, 339
Sartre, Jean-Paul 225, 228, 288
Schengen agreement 2, 6, 59, 64
Schmitt, Maurice 319, 320
security 7, 8, 46, 50, 54, 55, 61, 73, 75, 79, 80, 81, 83, 84, 87, 89, 90, 92, 93, 99, 106, 111, 115, 121, 126, 219, 229, 257, 313, 316
Seville 136, 143, 157, 195, 213
Sétif 305–6, 325, 327, 337
Spain 5, 8, 28–9, 37, 39, 44–5, 51–2, 62, 65, 81, 82–5, 88, 91, 122, 144, 156–7, 195, 213–14
Stora, Benjamin 11, 231, 307, 315, 326, 329–30
Sudan 8, 109, 125
Suez Canal 25, 39, 47, 138
Syria 2, 30, 52, 69, 95, 97

Tavernier, Bertrand 221, 233–7
Tillion, Germaine 319
Tlatli, Moufida 14, 166, 175, 177, 184, 186
Toulon 49, 50, 252, 306
Treaty of Friendship (Franco-Algerian) 40, 327
Tunisia 3, 7, 14, 44, 46, 69, 81, 84, 85, 105, 109, 129, 156, 163–5, 170–1, 173–7, 182–4, 187, 190, 276, 316, 336
Turkey 29, 45, 46, 50, 53, 56, 57, 58, 91, 312

Union for the Mediterranean (UfM) 2, 5, 9, 19, 25, 31, 32, 37, 42, 46–55, 61, 63, 69, 70–1, 89–100, 105, 107, 123–4, 126, 127, 133, 134, 143, 148–9, 151–3, 156, 159

United Nations (UN) 4, 134, 153
United Nations Educational, Scientific
    and Cultural Organization
    (UNESCO) 134, 146
Union pour un Mouvement Populaire
    (UMP) 3, 129, 323
United States 4, 8, 312

Valéry, Paul 26, 50
Védrine, Hubert 54, 144

Vidal-Naquet, Pierre 319

war veterans 13, 275–304, 334–8
Weltheim, Risto 63
Western Sahara 317

Yade, Rama 108, 114
Yacine, Kateb 283

Zamoum, Fatma Zohra 173

Modern French Identities

Edited by Peter Collier

This series aims to publish monographs, editions or collections of papers based on recent research into modern French Literature. It welcomes contributions from academics, researchers and writers in British and Irish universities in particular.

Modern French Identities focuses on the French and Francophone writing of the twentieth century, whose formal experiments and revisions of genre have combined to create an entirely new set of literary forms, from the thematic autobiographies of Michel Leiris and Bernard Noël to the magic realism of French Caribbean writers.

The idea that identities are constructed rather than found, and that the self is an area to explore rather than a given pretext, runs through much of modern French literature, from Proust, Gide and Apollinaire to Kristeva, Barthes, Duras, Germain and Roubaud.

This series reflects a concern to explore the turn-of-the-century turmoil in ideas and values that is expressed in the works of theorists like Lacan, Irigaray and Bourdieu and to follow through the impact of current ideologies such as feminism and postmodernism on the literary and cultural interpretation and presentation of the self, whether in terms of psychoanalytic theory, gender, autobiography, cinema, fiction and poetry, or in newer forms like performance art.

The series publishes studies of individual authors and artists, comparative studies, and interdisciplinary projects, including those where art and cinema intersect with literature.

Volume 1   Victoria Best & Peter Collier (eds): Powerful Bodies.
            Performance in French Cultural Studies.
            220 pages. 1999. ISBN 3-906762-56-4 / US-ISBN 0-8204-4239-9

Volume 2   Julia Waters: Intersexual Rivalry.
            A 'Reading in Pairs' of Marguerite Duras and Alain Robbe-Grillet.
            228 pages. 2000. ISBN 3-906763-74-9 / US-ISBN 0-8204-4626-2

Volume 3	Sarah Cooper: Relating to Queer Theory. Rereading Sexual Self-Definition with Irigaray, Kristeva, Wittig and Cixous. 231 pages. 2000. ISBN 3-906764-46-X / US-ISBN 0-8204-4636-X
Volume 4	Julia Prest & Hannah Thompson (eds): Corporeal Practices. (Re)figuring the Body in French Studies. 166 pages. 2000. ISBN 3-906764-53-2 / US-ISBN 0-8204-4639-4
Volume 5	Victoria Best: Critical Subjectivities. Identity and Narrative in the Work of Colette and Marguerite Duras. 243 pages. 2000. ISBN 3-906763-89-7 / US-ISBN 0-8204-4631-9
Volume 6	David Houston Jones: The Body Abject: Self and Text in Jean Genet and Samuel Beckett. 213 pages. 2000. ISBN 3-906765-07-5 / US-ISBN 0-8204-5058-8
Volume 7	Robin MacKenzie: The Unconscious in Proust's *A la recherche du temps perdu.* 270 pages. 2000. ISBN 3-906758-38-9 / US-ISBN 0-8204-5070-7
Volume 8	Rosemary Chapman: Siting the Quebec Novel. The Representation of Space in Francophone Writing in Quebec. 282 pages. 2000. ISBN 3-906758-85-0 / US-ISBN 0-8204-5090-1
Volume 9	Gill Rye: Reading for Change. Interactions between Text Identity in Contemporary French Women's Writing (Baroche, Cixous, Constant). 223 pages. 2001. ISBN 3-906765-97-0 / US-ISBN 0-8204-5315-3
Volume 10	Jonathan Paul Murphy: Proust's Art. Painting, Sculpture and Writing in *A la recherche du temps perdu.* 248 pages. 2001. ISBN 3-906766-17-9 / US-ISBN 0-8204-5319-6
Volume 11	Julia Dobson: Hélène Cixous and the Theatre. The Scene of Writing. 166 pages. 2002. ISBN 3-906766-20-9 / US-ISBN 0-8204-5322-6
Volume 12	Emily Butterworth & Kathryn Robson (eds): Shifting Borders. Theory and Identity in French Literature. VIII + 208 pages. 2001. ISBN 3-906766-86-1 / US-ISBN 0-8204-5602-0
Volume 13	Victoria Korzeniowska: The Heroine as Social Redeemer in the Plays of Jean Giraudoux. 144 pages. 2001. ISBN 3-906766-92-6 / US-ISBN 0-8204-5608-X

Volume 14	Kay Chadwick: Alphonse de Châteaubriant: Catholic Collaborator. 327 pages. 2002. ISBN 3-906766-94-2 / US-ISBN 0-8204-5610-1
Volume 15	Nina Bastin: Queneau's Fictional Worlds. 291 pages. 2002. ISBN 3-906768-32-5 / US-ISBN 0-8204-5620-9
Volume 16	Sarah Fishwick: The Body in the Work of Simone de Beauvoir. 284 pages. 2002. ISBN 3-906768-33-3 / US-ISBN 0-8204-5621-7
Volume 17	Simon Kemp & Libby Saxton (eds): Seeing Things. Vision, Perception and Interpretation in French Studies. 287 pages. 2002. ISBN 3-906768-46-5 / US-ISBN 0-8204-5858-9
Volume 18	Kamal Salhi (ed.): French in and out of France. Language Policies, Intercultural Antagonisms and Dialogue. 487 pages. 2002. ISBN 3-906768-47-3 / US-ISBN 0-8204-5859-7
Volume 19	Genevieve Shepherd: Simone de Beauvoir's Fiction. A Psychoanalytic Rereading. 262 pages. 2003. ISBN 3-906768-55-4 / US-ISBN 0-8204-5867-8
Volume 20	Lucille Cairns (ed.): Gay and Lesbian Cultures in France. 290 pages. 2002. ISBN 3-906769-66-6 / US-ISBN 0-8204-5903-8
Volume 21	Wendy Goolcharan-Kumeta: My Mother, My Country. Reconstructing the Female Self in Guadeloupean Women's Writing. 236 pages. 2003. ISBN 3-906769-76-3 / US-ISBN 0-8204-5913-5
Volume 22	Patricia O'Flaherty: Henry de Montherlant (1895–1972). A Philosophy of Failure. 256 pages. 2003. ISBN 3-03910-013-0 / US-ISBN 0-8204-6282-9
Volume 23	Katherine Ashley (ed.): Prix Goncourt, 1903–2003: essais critiques. 205 pages. 2004. ISBN 3-03910-018-1 / US-ISBN 0-8204-6287-X
Volume 24	Julia Horn & Lynsey Russell-Watts (eds): Possessions. Essays in French Literature, Cinema and Theory. 223 pages. 2003. ISBN 3-03910-005-X / US-ISBN 0-8204-5924-0
Volume 25	Steve Wharton: Screening Reality. French Documentary Film during the German Occupation. 252 pages. 2006. ISBN 3-03910-066-1 / US-ISBN 0-8204-6882-7
Volume 26	Frédéric Royall (ed.): Contemporary French Cultures and Societies. 421 pages. 2004. ISBN 3-03910-074-2 / US-ISBN 0-8204-6890-8
Volume 27	Tom Genrich: Authentic Fictions. Cosmopolitan Writing of the Troisième République, 1908–1940. 288 pages. 2004. ISBN 3-03910-285-0 / US-ISBN 0-8204-7212-3

Volume 28 Maeve Conrick & Vera Regan: French in Canada.
          Language Issues.
          186 pages. 2007. ISBN 978-3-03-910142-9

Volume 29 Kathryn Banks & Joseph Harris (eds): Exposure.
          Revealing Bodies, Unveiling Representations.
          194 pages. 2004. ISBN 3-03910-163-3 / US-ISBN 0-8204-6973-4

Volume 30 Emma Gilby & Katja Haustein (eds): Space.
          New Dimensions in French Studies.
          169 pages. 2005. ISBN 3-03910-178-1 / US-ISBN 0-8204-6988-2

Volume 31 Rachel Killick (ed.): Uncertain Relations.
          Some Configurations of the 'Third Space' in Francophone Writings
          of the Americas and of Europe.
          258 pages. 2005. ISBN 3-03910-189-7 / US-ISBN 0-8204-6999-8

Volume 32 Sarah F. Donachie & Kim Harrison (eds): Love and Sexuality.
          New Approaches in French Studies.
          194 pages. 2005. ISBN 3-03910-249-4 / US-ISBN 0-8204-7178-X

Volume 33 Michaël Abecassis: The Representation of Parisian Speech in
          the Cinema of the 1930s.
          409 pages. 2005. ISBN 3-03910-260-5 / US-ISBN 0-8204-7189-5

Volume 34 Benedict O'Donohoe: Sartre's Theatre: Acts for Life.
          301 pages. 2005. ISBN 3-03910-250-X / US-ISBN 0-8204-7207-7

Volume 35 Moya Longstaffe: The Fiction of Albert Camus. A Complex Simplicity.
          300 pages. 2007. ISBN 3-03910-304-0 / US-ISBN 0-8204-7229-8

Volume 36 Arnaud Beaujeu: Matière et lumière dans le théâtre de Samuel Beckett:
          Autour des notions de trivialité, de spiritualité et d'« autre-là ».
          377 pages. 2010. ISBN 978-3-0343-0206-8

Volume 37 Shirley Ann Jordan: Contemporary French Women's Writing:
          Women's Visions, Women's Voices, Women's Lives.
          308 pages. 2005. ISBN 3-03910-315-6 / US-ISBN 0-8204-7240-9

Volume 38 Neil Foxlee: Albert Camus's 'The New Mediterranean Culture':
          A Text and its Contexts.
          349 pages. 2010. ISBN 978-3-0343-0207-4

Volume 39 Michael O'Dwyer & Michèle Raclot: Le Journal de Julien Green:
          Miroir d'une âme, miroir d'un siècle.
          289 pages. 2005. ISBN 3-03910-319-9

Volume 40 Thomas Baldwin: The Material Object in the Work of Marcel Proust.
          188 pages. 2005. ISBN 3-03910-323-7 / US-ISBN 0-8204-7247-6

Volume 41	Charles Forsdick & Andrew Stafford (eds): The Modern Essay in French: Genre, Sociology, Performance. 296 pages. 2005. ISBN 3-03910-514-0 / US-ISBN 0-8204-7520-3
Volume 42	Peter Dunwoodie: Francophone Writing in Transition. Algeria 1900–1945. 339 pages. 2005. ISBN 3-03910-294-X / US-ISBN 0-8204-7220-4
Volume 43	Emma Webb (ed.): Marie Cardinal: New Perspectives. 260 pages. 2006. ISBN 3-03910-544-2 / US-ISBN 0-8204-7547-5
Volume 44	Jérôme Game (ed.): Porous Boundaries : Texts and Images in Twentieth-Century French Culture. 164 pages. 2007. ISBN 978-3-03910-568-7
Volume 45	David Gascoigne: The Games of Fiction: Georges Perec and Modern French Ludic Narrative. 327 pages. 2006. ISBN 3-03910-697-X / US-ISBN 0-8204-7962-4
Volume 46	Derek O'Regan: Postcolonial Echoes and Evocations: The Intertextual Appeal of Maryse Condé. 329 pages. 2006. ISBN 3-03910-578-7
Volume 47	Jennifer Hatte: La langue secrète de Jean Cocteau: la *mythologie personnelle* du poète et l'histoire cachée des *Enfants terribles*. 332 pages. 2007. ISBN 978-3-03910-707-0
Volume 48	Loraine Day: Writing Shame and Desire: The Work of Annie Ernaux. 315 pages. 2007. ISBN 978-3-03910-275-4
Volume 49	John Flower (éd.): François Mauriac, journaliste: les vingt premières années, 1905–1925. 352 pages. 2011. ISBN 978-3-0343-0265-4
Volume 50	Miriam Heywood: Modernist Visions: Marcel Proust's *A la recherche du temps perdu* and Jean-Luc Godard's *Histoire(s) du cinéma*. 277 pages. 2012. ISBN 978-3-0343-0296-8
Volume 51	Isabelle McNeill & Bradley Stephens (eds): Transmissions: Essays in French Literature, Thought and Cinema. 221 pages. 2007. ISBN 978-3-03910-734-6
Volume 52	Marie-Christine Lala: Georges Bataille, Poète du réel. 178 pages. 2010. ISBN 978-3-03910-738-4
Volume 53	Patrick Crowley: Pierre Michon: The Afterlife of Names. 242 pages. 2007. ISBN 978-3-03910-744-5
Volume 54	Nicole Thatcher & Ethel Tolansky (eds): Six Authors in Captivity. Literary Responses to the Occupation of France during World War II. 205 pages. 2006. ISBN 3-03910-520-5 / US-ISBN 0-8204-7526-2

Volume 55	Catherine Dousteyssier-Khoze & Floriane Place-Verghnes (eds): Poétiques de la parodie et du pastiche de 1850 à nos jours. 361 pages. 2006. ISBN 3-03910-743-7
Volume 56	Forthcoming.
Volume 57	Helen Vassallo: Jeanne Hyvrard, Wounded Witness: The Body Politic and the Illness Narrative. 243 pages. 2007. ISBN 978-3-03911-017-9
Volume 58	Marie-Claire Barnet, Eric Robertson and Nigel Saint (eds): Robert Desnos. Surrealism in the Twenty-First Century. 390 pages. 2006. ISBN 3-03911-019-5
Volume 59	Michael O'Dwyer (ed.): Julien Green, Diariste et Essayiste. 259 pages. 2007. ISBN 978-3-03911-016-2
Volume 60	Kate Marsh: Fictions of 1947: Representations of Indian Decolonization 1919–1962. 238 pages. 2007. ISBN 978-3-03911-033-9
Volume 61	Lucy Bolton, Gerri Kimber, Ann Lewis and Michael Seabrook (eds): Framed! : Essays in French Studies. 235 pages. 2007. ISBN 978-3-03911-043-8
Volume 62	Lorna Milne and Mary Orr (eds): Narratives of French Modernity: Themes, Forms and Metamorphoses. Essays in Honour of David Gascoigne. 365 pages. 2011. ISBN 978-3-03911-051-3
Volume 63	Ann Kennedy Smith: Painted Poetry: Colour in Baudelaire's Art Criticism. 253 pages. 2011. ISBN 978-3-03911-094-0
Volume 64	Sam Coombes: The Early Sartre and Marxism. 330 pages. 2008. ISBN 978-3-03911-115-2
Volume 65	Claire Lozier: De l'abject et du sublime: Georges Bataille, Jean Genet, Samuel Beckett. 327 pages. 2012. ISBN 978-3-0343-0724-6
Volume 66	Forthcoming.
Volume 67	Alison S. Fell (ed.): French and francophone women facing war / Les femmes face à la guerre. 301 pages. 2009. ISBN 978-3-03911-332-3
Volume 68	Elizabeth Lindley and Laura McMahon (eds): Rhythms: Essays in French Literature, Thought and Culture. 238 pages. 2008. ISBN 978-3-03911-349-1

Volume 69	Georgina Evans and Adam Kay (eds): Threat: Essays in French Literature, Thought and Visual Culture. 248 pages. 2010. ISBN 978-3-03911-357-6
Volume 70	John McCann: Michel Houellebecq: Author of our Times. 229 pages. 2010. ISBN 978-3-03911-373-6
Volume 71	Jenny Murray: Remembering the (Post)Colonial Self: Memory and Identity in the Novels of Assia Djebar. 258 pages. 2008. ISBN 978-3-03911-367-5
Volume 72	Susan Bainbrigge: Culture and Identity in Belgian Francophone Writing: Dialogue, Diversity and Displacement. 230 pages. 2009. ISBN 978-3-03911-382-8
Volume 73	Maggie Allison and Angela Kershaw (eds): *Parcours de femmes*: Twenty Years of Women in French. 313 pages. 2011. ISBN 978-3-0343-0208-1
Volume 74	Jérôme Game: Poetic Becomings: Studies in Contemporary French Literature. 263 pages. 2011. ISBN 978-3-03911-401-6
Volume 75	Elodie Laügt: L'Orient du signe: Rêves et dérives chez Victor Segalen, Henri Michaux et Emile Cioran. 242 pages. 2008. ISBN 978-3-03911-402-3
Volume 76	Suzanne Dow: Madness in Twentieth-Century French Women's Writing: Leduc, Duras, Beauvoir, Cardinal, Hyvrard. 217 pages. 2009. ISBN 978-3-03911-540-2
Volume 77	Myriem El Maïzi: Marguerite Duras ou l'écriture du devenir. 228 pages. 2009. ISBN 978-3-03911-561-7
Volume 78	Forthcoming.
Volume 79	Jenny Chamarette and Jennifer Higgins (eds): Guilt and Shame: Essays in French Literature, Thought and Visual Culture. 231 pages. 2010. ISBN 978-3-03911-563-1
Volume 80	Vera Regan and Caitríona Ní Chasaide (eds): Language Practices and Identity Construction by Multilingual Speakers of French L2: The Acquisition of Sociostylistic Variation. 189 pages. 2010. ISBN 978-3-03911-569-3
Volume 81	Margaret-Anne Hutton (ed.): Redefining the Real: The Fantastic in Contemporary French and Francophone Women's Writing. 294 pages. 2009. ISBN 978-3-03911-567-9

Volume 82   Elise Hugueny-Léger: Annie Ernaux, une poétique de la transgression.
269 pages. 2009. ISBN 978-3-03911-833-5

Volume 83   Peter Collier, Anna Magdalena Elsner and Olga Smith (eds): Anamnesia: Private and Public Memory in Modern French Culture.
359 pages. 2009. ISBN 978-3-03911-846-5

Volume 84   Adam Watt (ed./éd.): Le Temps retrouvé Eighty Years After/80 ans après: Critical Essays/Essais critiques.
349 pages. 2009. ISBN 978-3-03911-843-4

Volume 85   Louise Hardwick (ed.): New Approaches to Crime in French Literature, Culture and Film.
237 pages. 2009. ISBN 978-3-03911-850-2

Volume 86   Emmanuel Godin and Natalya Vince (eds): France and the Mediterranean: International Relations, Culture and Politics.
372 pages. 2012. ISBN 978-3-0343-0228-9

Volume 87   Amaleena Damlé and Aurélie L'Hostis (eds): The Beautiful and the Monstrous: Essays in French Literature, Thought and Culture.
237 pages. 2010. ISBN 978-3-03911-900-4

Volume 88   Alistair Rolls (ed.): Mostly French: French (in) Detective Fiction.
212 pages. 2009. ISBN 978-3-03911-957-8

Volume 89   Bérénice Bonhomme: Claude Simon : une écriture en cinéma.
359 pages. 2010. ISBN 978-3-03911-983-7

Volume 90   Barbara Lebrun and Jill Lovecy (eds): *Une et divisible?* Plural Identities in Modern France.
258 pages. 2010. ISBN 978-3-0343-0123-7

Volume 91   Pierre-Alexis Mével & Helen Tattam (eds): Language and its Contexts/ Le Langage et ses contextes: Transposition and Transformation of Meaning?/Transposition et transformation du sens ?
272 pages. 2010. ISBN 978-3-0343-0128-2

Volume 92   Alistair Rolls and Marie-Laure Vuaille-Barcan (eds): Masking Strategies: Unwrapping the French Paratext.
202 pages. 2011. ISBN 978-3-0343-0746-8

Volume 93   Michaël Abecassis et Gudrun Ledegen (éds): Les Voix des Français Volume 1: à travers l'histoire, l'école et la presse.
372 pages. 2010. ISBN 978-3-0343-0170-1

Volume 94   Michaël Abecassis et Gudrun Ledegen (éds): Les Voix des Français Volume 2: en parlant, en écrivant.
481 pages. 2010. ISBN 978-3-0343-0171-8

Volume 95   Manon Mathias, Maria O'Sullivan and Ruth Vorstman (eds): Display and Disguise.
237 pages. 2011. ISBN 978-3-0343-0177-0

Volume 96   Charlotte Baker: Enduring Negativity: Representations of Albinism in the Novels of Didier Destremau, Patrick Grainville and Williams Sassine.
226 pages. 2011. ISBN ISBN 978-3-0343-0179-4

Volume 97   Florian Grandena and Cristina Johnston (eds): New Queer Images: Representations of Homosexualities in Contemporary Francophone Visual Cultures.
246 pages. 2011. ISBN 978-3-0343-0182-4

Volume 98   Florian Grandena and Cristina Johnston (eds): Cinematic Queerness: Gay and Lesbian Hypervisibility in Contemporary Francophone Feature Films.
354 pages. 2011. ISBN 978-3-0343-0183-1

Volume 99   Neil Archer and Andreea Weisl-Shaw (eds): Adaptation: Studies in French and Francophone Culture.
234 pages. 2012. ISBN 978-3-0343-0222-7